4th International Conference on Advanced Technologies in Design, Mechanical and Aeronautical Engineering (ATDMAE 2020)

IOP Conference Series: Materials Science and Engineering
Volume 926

Kuala Lumpur, Malaysia
9 - 11 July 2020

ISBN: 978-1-7138-3194-5
ISSN: 1757-8981

Printed from e-media with permission by:

Curran Associates, Inc.
57 Morehouse Lane
Red Hook, NY 12571

Some format issues inherent in the e-media version may also appear in this print version.

Institute of Physics
Dirac House, Temple Back
Bristol BS1 6BE UK

Phone: 44 1 17 929 7481
Fax: 44 1 17 920 0979

techtracking@iop.org

Additional copies of this publication are available from:

Curran Associates, Inc.
57 Morehouse Lane
Red Hook, NY 12571 USA
Phone: 845-758-0400
Fax: 845-758-2633
Email: curran@proceedings.com
Web: www.proceedings.com

TABLE OF CONTENTS

MECHANICAL ENGINEERING

AEROSPACE ENGINEERING

DESIGN

IOP Conf. Series: Materials Science and Engineering **926** (2020) 011001 doi:10.1088/1757-899X/926/1/011001

PREFACE

Currently the entire world is struggling against the virulent pandemic COVID-19. Unfortunately, each of us is affected, either directly or indirectly. Our conference, 4th International Conference on Advanced Technologies in Design, Mechanical and Aeronautical Engineering (ATDMAE 2020) was not an exception.

In July, mass gatherings are not permitted by the government in China to protect people. It is uncertain when the COVID-19 would end, so it remains unclear for postponement time, while many scholars and researchers wanted to attend this long-waited conference and have academic exchanges with their peers. Therefore, in order to actively respond the call of the government, and meet author's request, the ATDMAE 2020, which was planned to be held in Kuala Lumpur, Malaysia from July 9 to 11, 2020, was changed to be held on July 10, 2020 online through Tencent VooV software. This approach not only reduces people gathering, but also meets their communication needs.

Each keynote speech lasted 40 minutes, invited speech 30 minutes and author presentation 15 minutes. Each presentation was packed with questions and answers. There were lively discussion at the meeting, which promotes the academic exchange. The success and prosperity of the conference is reflected high level of the papers received. ATDMAE 2020 became an effective communication platform for all the participants over the world.

ATDMAE 2020 was organized by Hong Kong Society of Mechanical Engineers, supported by Nazarbayev University. This conference aims to provide a platform for researchers and engineers to share their ideas, recent developments and successful practices in Advanced Technologies in Design, Mechanical and Aeronautical Engineering. More than 60 participants attended the meeting, they were from USA, Kazakhstan, UK, Malaysia, Korea, India, Australia, Algeria, Netherlands, China and more. Three renowned speakers given speeches about their latest research and reports. They are: Prof. Mohd Hamdi Bin Abd Shukor, University of Malaya, Malaysia; Prof. Weimin Huang, Nanyang Technological University, Singapore; Prof. Parvathy Rajendran, Universiti Sains Malaysia, Malaysia. The conference also had 3 technical sessions and 1 poster sessions.

The conference proceeding is a compilation of the accepted papers and represent an interesting outcome of the conference. This book covers 3 chapters: 1. Mechanical Engineering; 2. Aerospace engineering; 3. Design.

IOP Conf. Series: Materials Science and Engineering **926** (2020) 011001 doi:10.1088/1757-899X/926/1/011001

We would like to acknowledge all of those who supported ATDMAE 2020. Each individual and institutional help were very important for the success of this conference. Especially we would like to thank the committee chairs, committee members and reviewers, for their tremendous contribution in conference organizing and peer review of the papers.

We sincerely hope that ATDMAE 2020 will be a forum for excellent discussions that will put forward new ideas and promote collaborative researches. We are sure that the proceedings will serve as an important research source of references and the knowledge, which will lead to not only scientific and engineering progress but also other new products and processes. Finally, we would like to thank the organization and committee for all their hard work and support and hope to meet you all in person at our next conference.

Prof. Yong Zhao

Conference Chairman

IOP Conf. Series: Materials Science and Engineering **926** (2020) 011002 doi:10.1088/1757-899X/926/1/011002

COMMITTEES

Conference Chairman

Yong Zhao, Nazarbayev University, Kazakhstan

Conference Chair

Xiaohui Su, Dalian University of Technology, China

Cees de Bont, Loughborough University, UK

Conference Program Co-Chairs

Zijian Wang, University of Kansas, USA

Bin Wang, Brunel University London, UK

Tongming Zhou, The University of Western Australia, Australia

Wenming Yang, National University of Singapore, Singapore

Henri Christiaans, Ulsan National Institute of Science & Technology, Korea

Organizing Committee

Gopalan Jagadeesh, Indian Institute of Science, India

Parvathy Rajendran, Universiti Sains Malaysia, Malaysia

Local Committee

Wong Shaw Voon, Universiti Putra Malaysia, Malaysia

International Technical Commmittee

Weihua Liu, Nanjing University of Aeronautics and Astronautics, China (Chair)

Ab Aziz bin Mohd Yusof, MARA University of Technology (UiTM), Malaysia

B. T. N. Sridhar, Anna University, India

B.T Hang Tuah b. Baharudin, Universiti Putra Malaysia, Malaysia

IOP Conf. Series: Materials Science and Engineering **926** (2020) 011002 doi:10.1088/1757-899X/926/1/011002

Javier Munguia Valenzuela, Newcastle University, UK

Jos Gunsing, Avans University of Applied Science, Netherlands

José Rui Marcelino, Lisbon University, Portugal

Shujin Lin, SUN Yat-sen University, China

Petra Badke-Schaub, Delft University of Technology, Netherlands

Tung-Jung Sung, National Taiwan University of Science and Technology, Taiwan, China

IOP Conf. Series: Materials Science and Engineering **926** (2020) 011003 doi:10.1088/1757-899X/926/1/011003

Peer review declaration

All papers published in this volume of IOP Conference Series: Materials Science and Engineering have been peer reviewed through processes administered by the Editors. Reviews were conducted by expert referees to the professional and scientific standards expected of a proceedings journal published by IOP Publishing.

- Type of peer review: Double-blind
- Conference submission management system: CMT
- Number of submissions received: 71
- Number of submissions sent for review: 64
- Number of submissions accepted: 38
- Acceptance Rate (Number of Submissions Accepted / Number of Submissions Received X 100): 60%
- Average number of reviews per paper: 2
- Total number of reviewers involved: 24

Contact person for queries: Lyn Lee lynlee@smehk.org

IOP Conf. Series: Materials Science and Engineering **926** (2020) 012001 doi:10.1088/1757-899X/926/1/012001

Experimental Study of Scissor-like-structure Vibration Isolator

Liu Yang[1], Zhang Wei[1, 2, *], Li Shuangbao[3] and Zhu Hengjia[1]

[1] College of Aeronautical Engineering,Civil Aviation University of China, Jinbei Road 2898, Dongli District, Tianjin 300300, China
[2] CAAC Aviation Special Ground Equipment Research Base, Jinbei Road 2898, Dongli District, Tianjin 300300, China
[3]College of Science, Civil Aviation University of China, Jinbei Road 2898, Dongli District, Tianjin 300300, China

[*] E-mail: weizhang@cauc.edu.cn

Abstract. Considering the problem of low-frequency vibration isolation and disadvantages of existing models, an optimized model is established. Based on optimized model, the expression of displacement transmissibility is derived. Then an experimental model of scissor-like-structure vibration isolator (SLS-VI) is established. The comparison of experimental curve and theoretical curve proves the feasibility of model optimization. The effective vibration isolation bandwidth is 10Hz-16Hz which proves the SLS-VI have excellent vibration isolation performance at low frequency. Experimental results are basically consistent with theoretical analysis in displacement transmissibility of the vibration isolator. It proves the correctness of the theoretical analysis of scissor-like-structure at low frequency.

1.Introduction

Vibration is a common problem in nature, engineering, daily life and social life. Such as sound waves and ultrasound, mechanical vibration. From the smallest particle to huge celestial bodies, vibration problems exist in everywhere [1]. Although there are many related studies in the field of vibration isolation, most of them are based on traditional linear vibration isolation theory. Traditional linear vibration isolators have significant effects on high-frequency isolation, but still cannot solve low-frequency isolation problems such as the influence of vehicle seats on drivers [2].

For solving the problem of low-frequency vibration isolation, Carrella, Kovacic et al. [3-5] proposed a quasi-zero stiffness vibration isolation system with two or more oblique springs and a vertical spring. Through the interaction between springs, to make the overall stiffness of the system approximately equal zero. However, the traditional quasi-zero stiffness isolators are not very stable and can not to adjust structural parameters [6]. Then, Sun et al. proposed a scissor-like structure vibration isolator (SLS-VI) which has geometric nonlinear characteristics [7-8]. Studies have shown that the SLS-VI can obtained the ideal nonlinear vibration isolation performance by adjusted structural parameters. Zhang et al. [9] introduced unified expression of springs full-type installation into the structure of SLS-VI. Studies have shown that it is more convenient to obtain the ideal nonlinear vibration isolation performance, only through changing springs installation parameters [10].

Besides, there have many studies on principle, dynamic characteristics analysis and engineering application of zero-stiffness isolators. Such as, Wang et al. tested the performance of bi-annular-shaped permanent magnets type isolator and the tri-magnets type isolator. Results of tests were compared with the performance of equivalent linear vibration isolation system [11]. Hang et al. carried out an experimental study on the dynamic characteristics of aero-engine vibration isolators in two special materials [12]. However, there were few experiments to analyze the dynamic characteristics of the SLS-VI.

By referring to the relevant experimental methods of above studies, the SLS-VI theoretical formula was simplified. And simplified model can reduce theoretical calculation time. Scissor-like-structure experimental model and test platform were set up. Its purpose was to verify the correctness of theoretical results and the vibration isolation performance of SLS-VI at low frequency.

2.Theoretical analysis

For simplifying the existed mathematical model and reducing theoretical calculation time, the coulomb friction and the mass of scissor arms are not considered in this study. Besides the spring installation type 5 is selected as the research object. The mathematical model is shown in Figure.1.

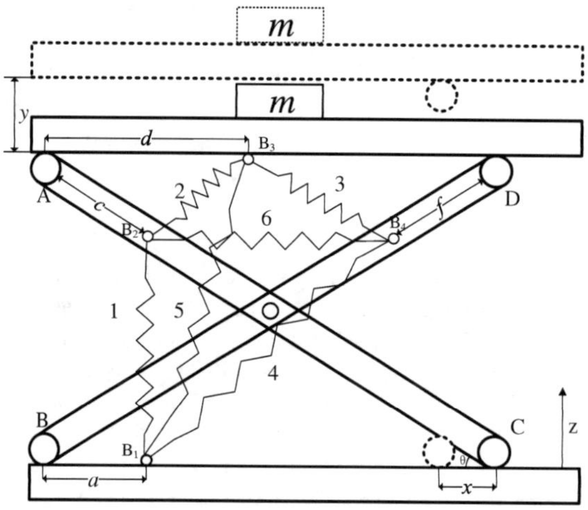

Figure 1. Simplified model of full-type configuration SLS-VI

The differential equation of motion of the vibration isolation system can be expressed as:

$$m\ddot{\hat{y}} + F_k + F_{C1} + F_{C2} = -m\ddot{z} \tag{1}$$

where $\dot{\delta} = (\frac{d\delta}{d\hat{y}})\cdot(\frac{d\hat{y}}{dt})$, $\dot{x} = (\frac{dx}{d\hat{y}})\cdot(\frac{d\hat{y}}{dt})$, δ is the rotation angle of each arm.

$$F_{C1} = 6c_1\dot{\delta}\frac{d\delta}{dy} = 6c_1(\frac{d\delta}{d\hat{y}})^2\dot{\hat{y}} = f_1\dot{\hat{y}} \tag{2}$$

$$F_{C2} = 2c_2\dot{x}\frac{dx}{dy} = 2c_2(\frac{dx}{d\hat{y}})^2\dot{\hat{y}} = f_2\dot{\hat{y}} \tag{3}$$

$$F_K = k(\Delta I - \lambda_0)\frac{d\Delta I}{dy} \tag{4}$$

The formula F_K , f_1 , f_2 are replaced by their Taylor expansions at $\hat{y}=0$, $\lambda_0 = 0$. And the coefficients are determined by structural parameters and spring installation parameters.

After the Taylor expansion was obtained, the perturbation method (PM) and the average method (AM) are combined to solve the dynamic equation. The displacement transmissibility equation is obtained:

$$T_d = \frac{\sqrt{a_0^2 + 2a_0 z_0 \cos\phi + z_0^2}}{z_0} = \sqrt{\frac{a_0^2}{z_0^2} + \frac{a_0(4y_0\gamma_2 - 2\sigma)}{z_0^2\Omega^2} + 1} \tag{5}$$

where γ_1, γ_2 are related to the spring installation parameters.

3. Experimental model

3.1 The main components

The experimental model of the SLS-VI is shown in Figure 2. It is made up of two scissor arms, two compression springs, top and bottom platforms. The left end of two arms can horizontally move along the slideway. Two scissor arm brackets are hinged together through a central connecting shaft to form an "X" shape. In this experimental model, the compression spring is hung vertically between top platform and bottom platform.

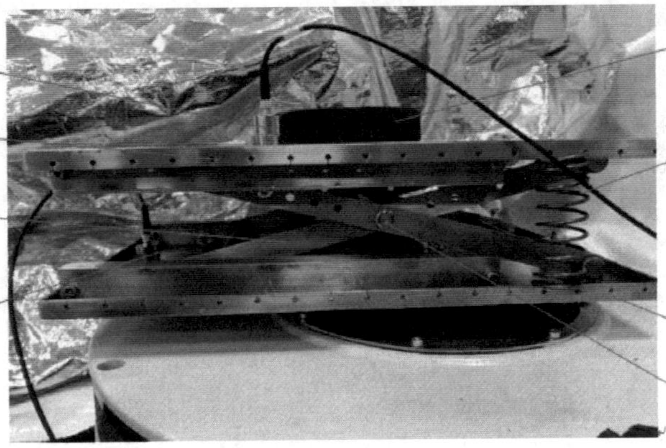

Figure 2. Experimental model of SLS-VI

3.2 Parameters of each parts

Silicon-manganese spring steel has good elastic limit and mechanical characteristics. So the material is 60Si2Mn spring steel, the shear modulus is G = 79000 MPa. Linear springs are simple and stable.

According to the spring design criteria and spring stiffness requirements in the mechanical design manual, the spring stiffness is:

$$k = \frac{Gd^4}{8nD^3} = 634N/m \tag{6}$$

Other important component parameters: top platform weight is 2.5kg, the scissor arm weight is 0.12kg, and the scissor arm length is 0.34m, spring installation parameters d=a=0.09m.

4.Experimental equipment

The experimental system consists of three parts: SLS vibration isolation system which is shown in Figure 2, excitation system which is shown in Figure 3, and signal acquisition system which is shown in Figure 4.

Excitation system includes computer (with vibration control software), signal controller, and power amplifier (in the control cabinet). Electric vibration table is shown in Figure 3. The computer is used to select the type of vibration and set the vibration parameters. Finally the built-in power amplifier was used to transmit the signal to start the vibration table.

Signal acquisition and processing system: It includes voltage sensor, signal acquisition instrument and analysis software(in the computer).Voltage sensors are used to collect the dynamic response signal of the system. The signal acquisition instrument can receive the electrical signals transmitted by sensors. Analysis software is installed on the computer. The data point information can be obtained, stored and processed in the computer.

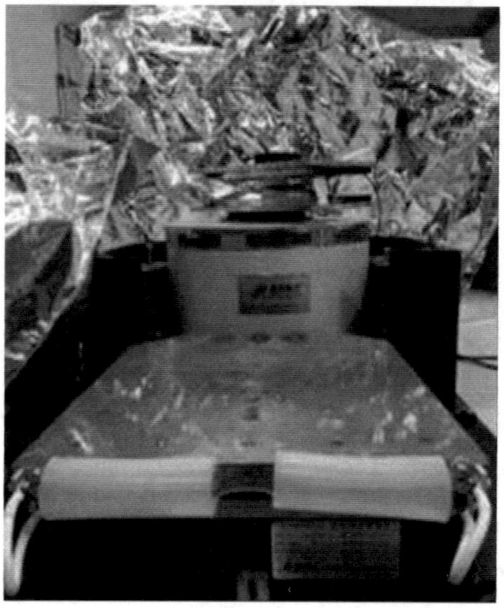

Figure 3. Vibration table

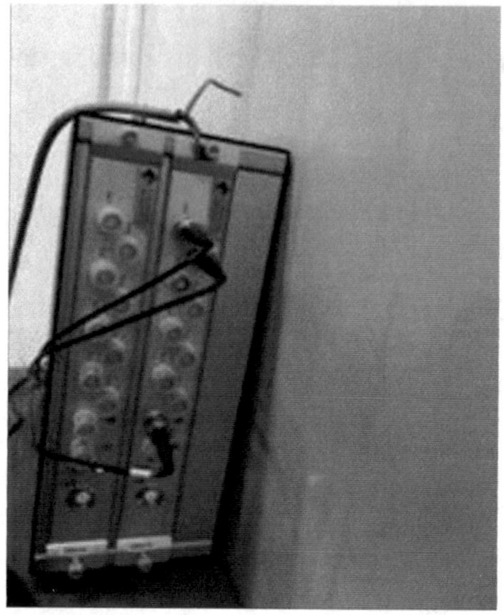

Figure 4. Signal acquisition instrument

Scissor-like-structure vibration isolation system is shown in Figure 2: it is made up of SLS vibration isolator, vibration-isolated object, and positioning bolts. The bottom platform is connected to the vibration table with positioning bolts. The vibration-isolated object is m,as shown in Figure 1. The bottom of mass is processed with internal threads and connected to the top platform with bolts. When the vibration excitation system starts to work, sensors receive displacement signals from bottom platform and top platform. Finally, the displacement transmissibility can be obtained

5. Experimental analysis of dynamic characteristics

5.1 sensors position

In experiment, two sensors are used to control the vibration table and collect dynamic characteristical signals of the SLS-VI. As shown in Figure 2, sensor a is placed on the bottom platform, it can be used to both control vibration table and collect input signals at the same time. The top platform and the mass are considered as a whole, with a total weight of 4.5kg.sensor b is placed on the right side of top platform, it is used to collect output signals of vibration isolation system.

5.2 Experimental conditions.

The vibration table simulates external excitation type which is applied to the SLS-VI. Then the dynamic characteristics of system is tested under different excitation amplitude and excitation frequency. The displacement transmissibility curve of experiment is compared with the curve of theoretical analysis. In addition, vibration isolation performance of the SLS-VI under low frequency excitation is analysed. After a series of experiments, a set of most practical experimental points is selected. Their frequency sweep range is 2-20 Hz and the peak-to-peak displacement limit is 0.6 mm. Experimental curve of displacement transmissibility is shown in Figure 5. Due to the influence of surrounding experimental environment, there is a certain amount of interference. So the most obvious part which is 3-10Hz was selected to observe. Comparison of displacement transmissibility is shown in Figure 6.

IOP Conf. Series: Materials Science and Engineering **926** (2020) 012001 doi:10.1088/1757-899X/926/1/012001

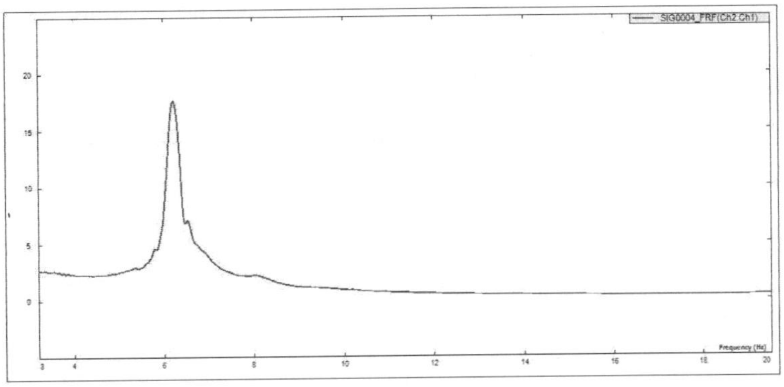

Figure 5. Experimental curve of displacement transmissibility

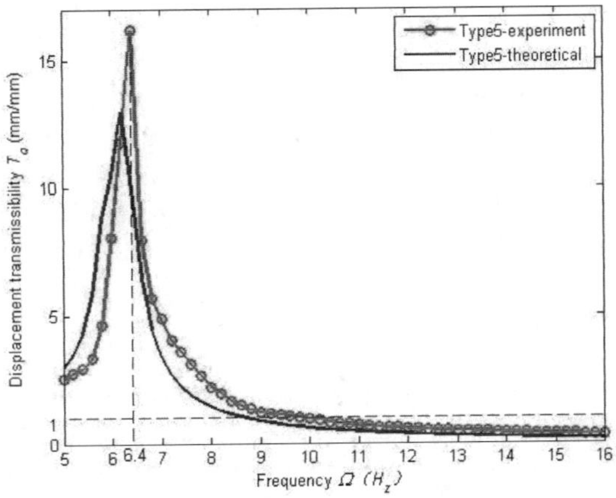

Figure 6. Comparison of displacement transmissibility

5.3 Analysis of experimental results

Through the comparison of output and input signals in Figure 7 and Figure 8, it can be seen that the scissor-like-structure vibration isolator has obviously vibration isolation effect on the excitation from bottom.

The dynamic characteristics of the SLS-VI are shown in the following figures:

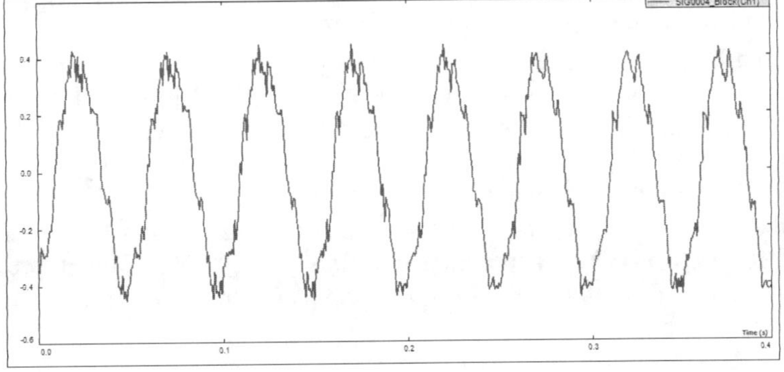

Figure 7. Input time domain signal

IOP Conf. Series: Materials Science and Engineering **926** (2020) 012001 doi:10.1088/1757-899X/926/1/012001

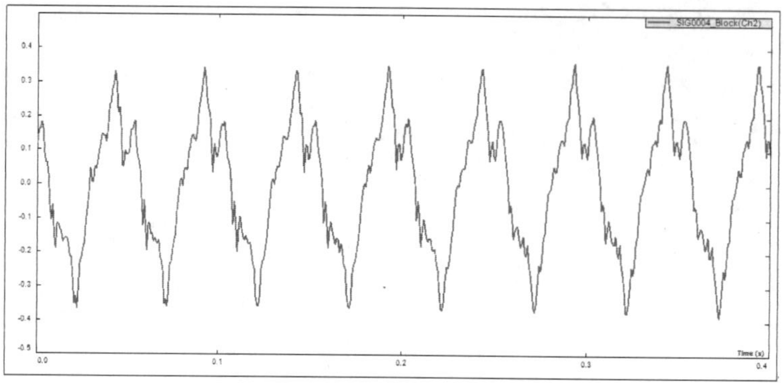

Figure 8. Output time domain signal

Considering the influence of environmental factors on sensor signal acquisition during experiment, the image is further processed. Some experimental data points are selected to compare with theoretical data points. It can be seen from Figure 6 that the experimental resonance frequency of 6.4 Hz is similar to theoretical resonance frequency of 6.2 Hz.

It is found that the theoretical analysis of vibration isolator is basically consistent with the experimental test results in displacement transmissibility. The reason for the deviation may be the disturbance of external environment during the signal acquisition. In addition, because the scissor-like-structure vibration isolator and the vibration table are connected by screws, there is a slight movement, which is also a reason for deviation.

It can be seen from Figure 6 that the experimental datas of SLS-VI agree well with the theoretical analysis, which proves the rationality of theoretical analysis. The displacement transmissibility of SLS-VI under spring installation type five increase first then decrease with the increase of excitation frequency. When the excitation frequency reaches 10Hz, the displacement transmissibility is less than 1,which can completely achieve low frequency vibration isolation effect.

6. Conclusion

In oder to solve the problem of low-frequency vibration isolation in engineering technology field, experimental system is established in combination with vibration table, signal acquisition equipment and experiment of SLS-VI.. The signal acquisition equipment is used to collect the vibration signals of bottom and top platform of the scissor-like-structure. The following results are got from experiment.

1. It can be seen from Figure 6 that the resonance frequency of isolator is 6.4 Hz, which is very close to the resonance frequency of 6.2 Hz in theoretical curve.And the peak value of displacement transmissibility is basically consistent.

2.The effective vibration isolation bandwidth is 10Hz-16Hz, which proves that the SLS-VI can achieves low frequency vibration isolation.

3.It is found that the resonance duration and peak value of SLS-VI under the excitation of reverse frequency sweep are higher than those under the forward frequency sweep.

4.Experimental results are basically consistent with theoretical analysis, which verify the correctness and feasibility of theoretical analysis.

In summary, the experimental system proposed in this study has reasonable structure and achieves the experimental requirements. The experimental results are obvious. The improved scissor-like-structure vibration isolator can effectively solve the problem of low frequency vibration isolation in the field of engineering technology.

References

[1] Liu Y Z and Chen L Q 2001 Nonlinear Vibrations (Beijing:Higher Education Press) pp1-2
[2] Yu S Z,et al.2000 *Automobile Theory* (Beijing:China Machine Press) pp56-59
[3] Carrella A, Brennan M J and Waters T P.2007 *J Sound and Vibration* 301(3-5) pp 678-689

IOP Conf. Series: Materials Science and Engineering **926** (2020) 012001 doi:10.1088/1757-899X/926/1/012001

[4] Kovacic I, Brennan M J and Waters T P.2008 *J Sound and Vibration* 315(3) pp700-711
[5] Carella A, Brennan M J, Kovacic I, et al.2012 *J Mechanical Sciences* 55 (1) pp22-29
[6] Sun X T, Jing X J, Xu J, et al.2014 *J Sound and Vibration* 333(9) pp 2404-2420
[7] Sun X T, Jing X J.2016 *Mechanical Systems and Signal Processing* 66-67723-742
[8] Xu J.2015 *Chinese Quarterly of Mechanics* 15(4) pp547-565
[9] Zhang W, Zhang X, Yan C, et al.2015 *J A M D S M* 9(3):JAMDSM0042-JAMDSM0042
[10] Zhang W, Zhao J B 2016 *Nonlinear Dynamics* 86(1) pp17-36
[11] Wang Y C, Chai K, Liu S Y and Yang Q C.2019 *Noise and Vibration Control* 39(5) pp223-230
[12] Hang C, Su E D, Y Q and Huang W C.2019 *J Experimental Mechanics* 34(01) pp160-168

IOP Conf. Series: Materials Science and Engineering **926** (2020) 012002 doi:10.1088/1757-899X/926/1/012002

Micro Dry Wire EDM: Kerf Investigation using Response Surface Methodology

Mohammad Yeakub Ali[1] *, Asfana Banu[2], Mohamed Abdul Rahman[2], Muataz Hazza Al Hazza[3] and Ahmed Ghalib Khan Chowdhury[2]

[1] Mechanical Engineering Programme Area, Faculty of Engineering, Universiti Teknologi Brunei, Tungku Highway, Gadong BE1410, Brunei Darussalam
[2] Department of Manufacturing and Materials Engineering, Faculty of Engineering, International Islamic University Malaysia, P.O. Box 10, 50728 Kuala Lumpur, Malaysia
[3] Mechanical and Industrial Engineering Department, School of Engineering, American University of Ras Al Khaimah, PO Box 10021, United Arab Emirates

*Contact E-mail: yeakub.ali@utb.edu.bn

Abstract. Dry electrical discharge machining (DEDM) is an environmentally friendly and green machining process where it uses gas as the dielectric fluid instead of liquid. It is an alternative to the conventional electrical discharge machining (EDM) process. However, kerf variation remains as a critical issue in micro dry wire EDM (µDWEDM) process. Therefore, the objective of this research is to investigate kerf in µDWEDM using response surface methodology (RSM). The experimental investigation was perform using an integrated multi process machine tool, DT 110 (Mikrotools Inc., Singapore). Stainless steel (SS304), tungsten wire, and compressed air were used as the workpiece, electrode, and dielectric fluid respectively. Central composite design (CCD), a type of RSM, was used to design the experiment using two controlled parameters which were capacitance and gap voltage. Analysis of variance (ANOVA) was used to analyse the results and to evaluate the adequacy of the developed model. The results were obtained by measuring the kerf using scanning electron microscope (SEM) (JEOL JSM-5600, Japan). An empirical model has been developed, and it was found that both parameters; capacitance and gap voltage have high influence on kerf. The optimum parameters for minimum kerf were found to be 0.1 nF capacitance and 91 V gap voltage. The developed model was found to be adequate since the percentage error was relatively small (≈ 2%).

1. Introduction

Dry electrical discharge machining (DEDM) is a type of green electrical discharge machining (EDM) process where gas is used as the dielectric fluid instead of liquid [1-3]. This process was introduced as an alternative to reduce fire hazard, health hazard to the operators, and environmental problems [3,4-5]. In DEDM, the gas is supplied through a pipe electrode where sparks are generated between the workpiece and the electrode [6]. This machining technique is also applicable in micro level such as micro dry EDM (µDEDM) and micro dry wire EDM (µDWEDM) [1-3, 6]. µDWEDM is able to produce high accuracy finishing cut since it has the ability to generate craters with small volumes. This situation happens due to the limited gap distance between the wire and the workpiece and negligible process reaction force during the machining operation [7].

Due to the non-contact in nature, µDWEDM is considered as a promising process for microfabrication of miniaturized products. However, one of the main concerns on this type of machining is the dimensional accuracy of the machined parts [8-9] which is related to kerf; width of the machined slots [10-11]. The corner errors and kerf variation are usually caused by the wire tool deflection and vibration in the discharge gap. These are the main factors that causes a lack of precision at the machined area which affects the dimensional accuracy. It frequently happens during rough cuts when high discharge energy is used [8-9, 12]. However, the wire behaviour during the machining operation is still unclear due to their unsteady and very high-speed phenomenon in narrow machined kerf [13]. Therefore, the main objective of this research is to investigate kerf in µDWEDM using response surface methodology.

2. Methodology

The workpiece and the electrode used were stainless steel (SS304) plate (30 mm × 20 mm × 0.5 mm (t)) and tungsten (W) wire (Ø 70 µm) respectively. Stainless steel is desirable to fabricate micro-fins for electronic components [1] while tungsten wire is desirable due to its ability in fabricating high tolerance small features with high accuracy [1]. Central composite design (CCD); a second-order design of response surface methodology (RSM); was utilized to investigate the kerf in µDWEDM using capacitance and gap voltage as the controlled parameters as shown in Table 1. The total number of experiments were 13 (Table 2) with four factorial points ($2^k = 2^2 = 4$), four axial points ($2k = 2(2) = 4$), and five centre points.

Table 1. µDWEDM experimental parameters using CCD

Controlled Parameters	Symbol	Coded Levels		
		-1	0	+1
Capacitance (nF)	c	0.1	1.0	10.0
Gap voltage (V)	v	80	95	110
Fixed Parameters				
Workpiece, thickness (µm)	Stainless steel (SS304), 500			
Wire electrode material, diameter (µm)	Tungsten, 70			
Machining length (µm)	300			
Dielectric fluid	Compressed air			
Dielectric fluid pressure (MPa)	0.0345			
Workpiece polarity	+ve			
Threshold voltage (%)	24			
Wire tension (N)	0.0809			
Wire feed rate (µm/sec)	0.2			
Wire speed (rpm)	0.6			

The workpiece was grounded manually using 320, 400, 600, and 800 grades of sand papers respectively for sample preparation. Then, the workpiece was cleaned using ultrasonic cleaning machine (BRANSON 2510, Virginia) for 5 minutes with ethanol as the cleaning media. After the completion of the sample preparation, µDWEDM process was conducted using an integrated multi process machine tool DT-110 (Mikrotools Inc., Singapore). Figure 1a shows the image of the machining setup during the machining process. The workpiece was cleaned once again in ethanol for 5 minutes using ultrasonic cleaning machine after the µDWEDM process was completed. The kerfs were measured at five different places along the microchannels (Figure 1b) using SEM (JEOL JSM-5600, Japan). The averages of the kerfs are tabulated in Table 2. The data analyses were carried out using ANOVA approach at 5% significance level [14].

Table 2. CCD for 13 runs of experiments

Expt. No.	Parameters		Responses
	c: Capacitance (nF)	*v*: Gap Voltage (V)	Average Kerf (µm)
1	1	95	83.20
2	1	110	96.74
3	0.1	80	86.04
4	0.1	95	82.34
5	1	95	85.32
6	1	95	86.64
7	10	110	103.60
8	10	80	92.46
9	10	95	91.86
10	1	95	88.38
11	0.1	110	87.12
12	1	95	87.34
13	1	80	87.54

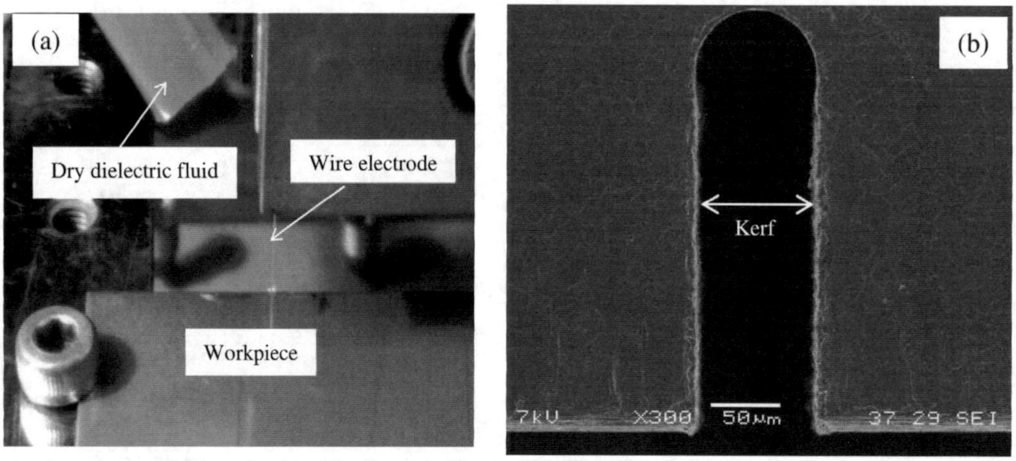

Figure 1. (a) Machining setup during µDWEDM and (b) kerf measurement

3. Results and Analysis

The adequacy of the developed statistical model was check using ANOVA. Eqn. 1 was developed based on the response surface quadratic model which was considered statistically significant; Prob > F equals to 0.0082 and F-value equals to 16.55. The adequacy and the validity of the model is determined as significant when the model terms with Prob > F is less than 0.05 [15]. Therefore, factors *c* (capacitance), *v* (gap voltage), c^2 (capacitance), and v^2 (gap voltage) are the significant model terms. In addition, the R^2 value of 0.922 indicates that the model is sufficient in explaining most of the variability in the experiment results. This is because the R^2 value is more than 0.75 [8, 15]. The Predicted R^2 is in reasonable agreement with the Adjusted R^2 since the difference is within 0.1851 which is less than 0.2 [15]. The adequate precision of 15.298 indicates that the signal-to-noise ratio is adequate since the value is greater than 4 [8, 10]. Hence, the model seems to be adequate to predict the behaviour of the kerf within the experimental constraints [8].

$$Kerf = 290.867 + 3.719c - 4.574v + 0.023cv - 0.477c^2 + 0.025v^2 \tag{1}$$

where, c = capacitance (nF) and v = gap voltage.

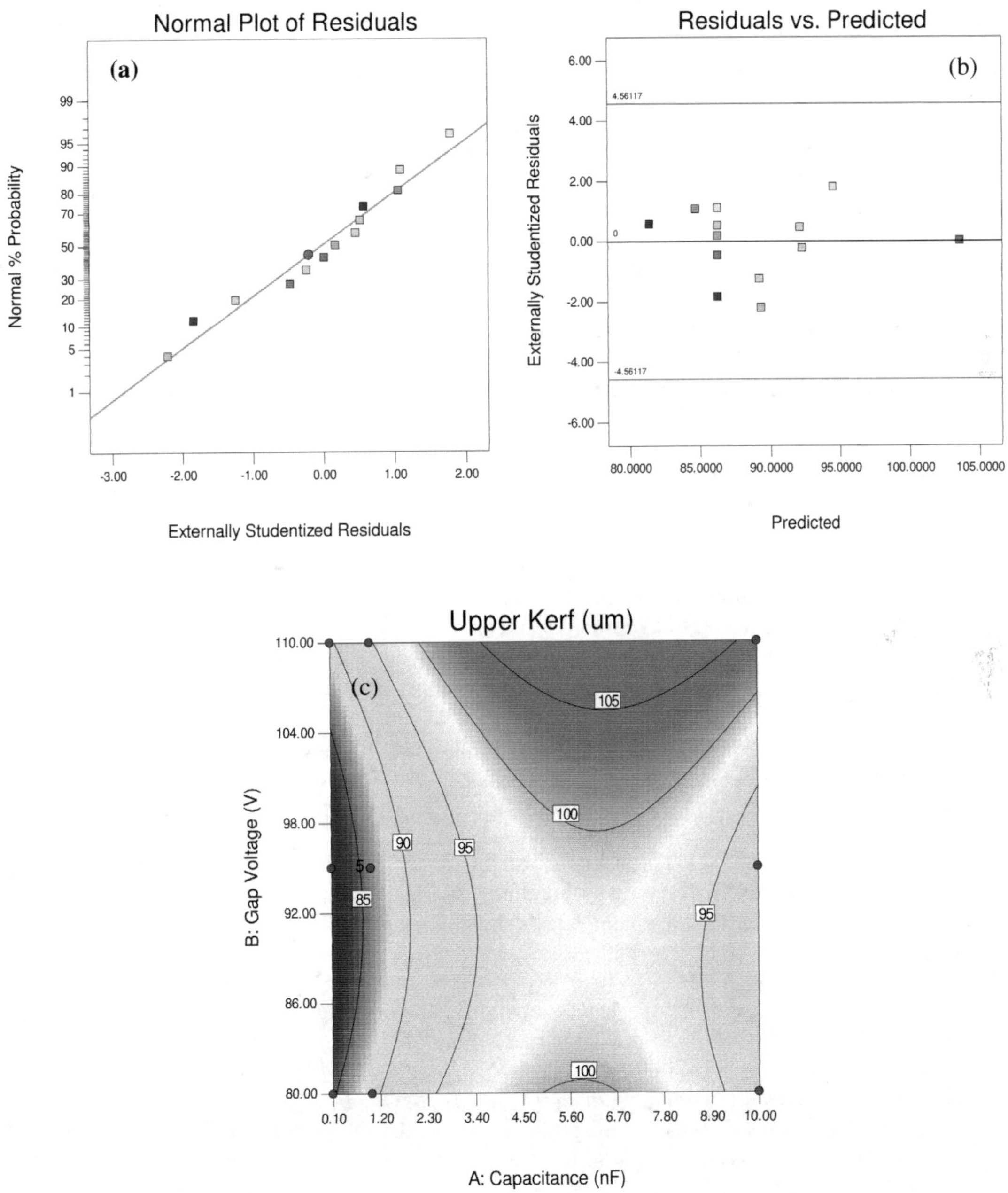

Figure 2. Diagnostic test for kerf based on (a) normal probability of residuals plot, (b) residuals vs. predicted plot; and (c) contour plot of kerf vs. gap voltage and capacitance.

Diagnostic test such as the analyses of the residuals were used to ascertain the adequacy of the developed model. Normally, adequate model shows the residual of the data is normally distributed and unstructured with constant variance [8]. Therefore, Figure 2a and Figure 2b shows the normal probability of residual plot and residual versus predicted plot respectively. In Figure 2a, the residual of the data is within the straight-line pattern where it shows the normality condition is satisfying. In addition, Figure 2b is considered adequate because the fitted values are dispersed randomly. Thus, it is anticipated that the variances of the original observations are constant for all the values [8].

Figure 2c shows the contour plot of kerf with gap voltage and capacitance. The figure shows, kerf is at the peak when the gap voltage is more than 104 V. It is due to the wire vibration [8, 10]. Usually, when higher discharge energy is employed, wire vibration will increase, and the kerf also increases causing poor accuracy [8-9]. Wire vibration occurs due to the forces acting during the machining process such as the hydrodynamic forces from the flushing system, the electrostatic forces that act on the wire, and the electromagnetic forces from the spark generation [1, 9]. Therefore, generally, in µDWEDM, the wire vibration is considered minimum compared to the conventional µWEDM [8].

The relationship between the controlled parameters; capacitance and gap voltage; together with the kerf were analysed for the optimal condition using Eqn. 1. The optimal condition suggested that minimum kerf, 81.20 µm is obtainable at 0.1 nF capacitance and 91 V gap voltage with the desirability of 97.8%. The model was verified by conducting the experiment based on the optimized parameters. From the experiment, the actual value of kerf (82.84 µm) is higher compared to the optimized kerf with maximum error of 2.02%. The percentage error is relatively small which indicates that the developed model is significant [8].

4. Conclusions

This research investigate kerf in µDWEDM using RSM. CCD, a type of RSM, was used to design the experiments using capacitance and gap voltage as the controlled parameters. An empirical model was developed. The parameters obtained from the optimization were also verified by another set of experiments. This investigation showed:

1. Capacitance and gap voltage have strong influence on kerf. Discharge energy is one of the reasons that causes the wire to vibrate and affect the accuracy of the kerf.
2. It was found that the optimum values of process parameters for minimum kerf are 0.1 nF capacitance and 91 V gap voltage.
3. The percentage error for the predicted value and the experimental value for kerf was found to be relatively small (≈ 2.02%). Therefore, the developed model is satisfactory.

Acknowledgments

This research was funded by MOSTI, under Research Grant SF15-016-0066. The authors are thankful to faculty and staff at Micromanufacturing laboratory and Metallographic laboratory at IIUM for their support.

References

[1] Banu A, Ali M Y, Rahman M A and Konneh M 2019 Investigation of process parameters for stable micro dry wire electrical discharge machining *Int. J. Adv. Manuf. Tech.* **103** 723-741

[2] Hoang K T and Yang S H 2015 A new approach for micro-WEDM control based on real-time estimation of material removal rate *Int. J. Precis. Eng. Man.* **16** 241-246

[3] Azhiri R B, Teimouri R, Baboly M G and Laseman Z 2014 Application of Taguchi, ANFIS and grey relational analysis for studying, modelling and optimization of wire EDM process while using gaseous media *Int. J. Adv. Manuf. Tech.* **71** 279-295

[4] Ali M Y, Banu A, Rahman M A, Hazza M and Adesta E Y T 2018 Precision control of kerf in metal cutting using dry micro WEDM: issues and challenges *Key Eng. Mater.* **775** 499-505

[5] Banu A and Ali M Y 2016 Electrical discharge machining (EDM): a review *Int. J. Eng. Mater. Manuf.* **1** 3-10

[6] Skrabalak G and Kozak J 2010 Study on dry electrical discharge machining *Wear* **5** 7
[7] Abbas M N, Solomon D G and Fuad Bahari M 2007 A review on current research trends in electrical discharge machining (EDM) *Int. J. Mach. Tool. Manu.* **47** 1214-1228
[8] Ali M Y, Banu A, Salehan M, Adesta E Y T, Hazza M and Shaffiq M 2018 Dimensional accuracy in dry micro wire electrical discharge machining *J. Mech. Eng. Sc.* **12** 3321-3329
[9] Conde A, Arriandiaga A, Sanchez J A, Portillo E, Plaza S and Canabes I 2018 High-accuracy wire electrical discharge machining using artificial networks and optimization techniques *Robot. Cim-Int. Manuf.* **49** 24-38
[10] Banu A, Bakar M A, Ali M Y and Adesta E Y 2017 Analysis of WEDM process parameters on surface roughness and kerf using Taguchi method *Int. J. Eng. Mater. Manuf.* **2** 103-109
[11] Hoang K T and Yang S H 2015 Kerf analysis and control in dry micro-wire electrical discharge machining *Int. J. Adv. Manuf. Tech.* **78** 1803-1812
[12] Di S, Chu X, Wei D, Wang Z, Chi G and Liu Y 2009 Analysis of kerf width in micro-WEDM *Int. J. Mach. Tool. Manu.* **49** 788-792
[13] Kawata A, Okada A, Okamoto Y and Kurihara H 2017 Influence of nozzle jet flushing on wire breakage in 1st-cut wire EDM from start hole *Key Eng. Mater.* **749** 130-135
[14] Ndaliman M B, Khan A A, Ali M Y and Wahid Z 2013 Determination of influential factors on EDMed surface properties using Placket-Burman design *World Appl. Sci. J.* **21** 88-93
[15] Adebisi A A and Ndaliman M B 2015 Mathematical modelling of stir casting process parameters for AlSiCp composite using central composite design *Proc. Int. Conf. of the Nigerian Institution for Mechanical Engineers*

IOP Conf. Series: Materials Science and Engineering **926** (2020) 012003 doi:10.1088/1757-899X/926/1/012003

Non-contact Detection of Delamination in Cu/Al Laminated Plate with Laser Ultrasonic

Baoping Ji[1,*], Qingdong Zhang[1], Liyuan Zhang[1], Jianshu Cao[2] and Boyang Zhang[1]

[1]School of Mechanical Engineering, University of Science and Technology Beijing, Beijing 100083, China
[2]College of Mechanical Engineering, Beijing Institute of Petrochemical Technology, Beijing 102617, China

[*]jibaoping08@163.com

Abstract. In this paper, a non-contact non-destructive inspection method based on laser ultrasonic was proposed for the detection of interface delamination in Cu/Al laminated plates. The finite element model of laser ultrasonic Cu/Al laminated plates were built to analyze the propagation of laser ultrasound in Cu/Al laminated plates and the effects of interface delamination defects on the ultrasonic field distribution. The simulation results were verified by the laser ultrasonic detection system. The C-scan detection results realizes the visual detection of delamination defects, and provides precise information about the composite quality of the Cu/Al laminated plates. Experimental results show that laser ultrasonic technology has great potential for online non-contact detection of Cu/Al laminated plates.

1. Introduction

Cu/Al laminated plates not only have the properties of low density, easy processing and corrosion resistance of aluminum, but also have the properties of high electrical conductivity and high thermal conductivity of copper, which have been applied to modern industries such as heat dissipation, power and electricity, communication shielding extensively[1-4]. Cu/Al laminated plates have very important economic benefits, which not only reduces the quality of the products, but also greatly saves copper resources and alleviates the current situation of copper resource shortages[1,5]. In the manufacturing process, serious defects such as slags, voids and delamination, occur at the interface of Cu/Al laminates on account of the improper production process, which seriously impacts the compound quality. In addition, some factors in the process of use, such as static load, dynamic load, mechanical damage, fatigue and creep, will also cause the emergence of interface separation in the laminated plates, resulting in the deterioration of material properties and endangering the safety and reliability of major equipment[6]. Therefore, effective non-destructive testing methods must be used for detecting interface delamination defects in laminated plates.

Ultrasonic method, eddy current method and infrared thermography are the common methods to detect the interface separation defects of laminated plates [7-8]. In these non-destructive detection techniques, the ultrasonic methods are widely used for evaluation of metal laminated plates. However, the traditional ultrasonic methods are difficult to evaluate the composite quality of thinner laminates due to the limitation of the blind zone of ultrasonic testing. As a novel testing method, laser ultrasonic technique has great potential for developing non-contact detection and quantitative evaluation for the delamination defects of metal laminated plates. Recently, researchers have done a lot of theoretical and experimental investigation on the application of laser ultrasonic techniques to detect defects in Multilayer structural material. Watanabe et al. monitored the interface separation of the surface coating by using laser ultrasonic technique, and visualized the delamination area by C-scan imaging [9]. The laser-ultrasonic was applied to a thin carbon fiber reinforced polymer test sample, where two different depth delamination have been created, and the delamination were detected by B-scan and C-scan[10]. Zhang et al. detected the disbands in steel-lead bonded structure based on laser ultrasonic with transmission mode and pulse-echo mode, and proposed quantitative methods to evaluate the disbands sizes[11-12].

However, there are few studies on the testing of laminated plates with laser ultrasound. In this paper, the characteristics of delamination defects in laminated plates had been investigated. The finite element model of laser ultrasonic was built to analyze the propagation of laser ultrasound in Cu/Al laminated plates and the effects of interface delamination defects on the ultrasonic field distribution. Finally, C-scans were performed to visualize the interface delamination.

2. Theoretical model

2.1. Generation of laser ultrasonic

When the pulse laser is induced on the laminated plates, part of the laser energy is absorbed into the thin surface of the laminated plates, and the temperature of the irradiated area increases, resulting in a large temperature gradient field, which causes thermal expansion and generates ultrasonic waves. The physical process of the thermoelastic mechanism can be explained by the classical thermoelastic coupling theory. The thermoelastic coupling equations for isotropic materials is:

$$[K]\{T\}+[C]\{\dot{T}\}=\{R_q\}+\{R_Q\} \tag{1}$$

$$[M]\{\ddot{U}\}+[K]\{U\}=\{R_{ext}\} \tag{2}$$

where $[K]$ represents the conductivity matrix, $\{T\}$ is the temperature vector, $\{\dot{T}\}$ is the temperature change rate vector, $\{R_q\}$ is the heat flux vector, $\{R_Q\}$ is the heat source vector, $[C]$ is the heat capacity matrix, $[M]$ is the mass matrix, $\{R_{ext}\}$ is the external force vector, $\{U\}$ is the displacement vector, $\{\ddot{U}\}$ is the acceleration vector[7].

In the finite element calculation process, considering the numerical solution precision and calculation time, the time step and element size must be selected for the appropriate values, the selection basis of the time step and element size are as follows[7]:

$$\Delta t \le \frac{1}{180 f_{\max}} \tag{3}$$

$$L_e \le \frac{\lambda_{\min}}{20} \tag{4}$$

where Δt is the time step, $f_{\max}$ is the maximum center frequency of laser ultrasound, L_e is the element size, $\lambda_{\min}$ is the shortest wavelength of laser ultrasound.

2.2. Finite element model

As shown in Figure 1, to reduce the calculation load and shorten the simulation time, the 2D axisymmetric models have been built to simulate the Cu/Al laminated plate with various delamination defects. The dimensions of the Cu layer and the Al 1050 layer were 20mm×0.5mm and 20mm×2.5mm. Some elements on the aluminum side of the interface are removed to simulate the interface delamination defects. The physical parameters of Al 1050 and Cu used in the finite element model are listed in Table 1. In the numerical simulation, the spot diameter of the pulsed laser is 1mm, the duration of the pulsed laser is 8ns, and the power density of the pulsed laser is 5MW/cm^2. The grid size of the laser irradiation region is 5μm, and the maximum grid size of other regions is 25μm with the gradual grid division technology. To ensure the accuracy of numerical simulation, the integration time step is 0.5ns.The simulation time was 10μs. The coupled thermo-mechanical analysis module of the ABAQUS software was used for simulation.

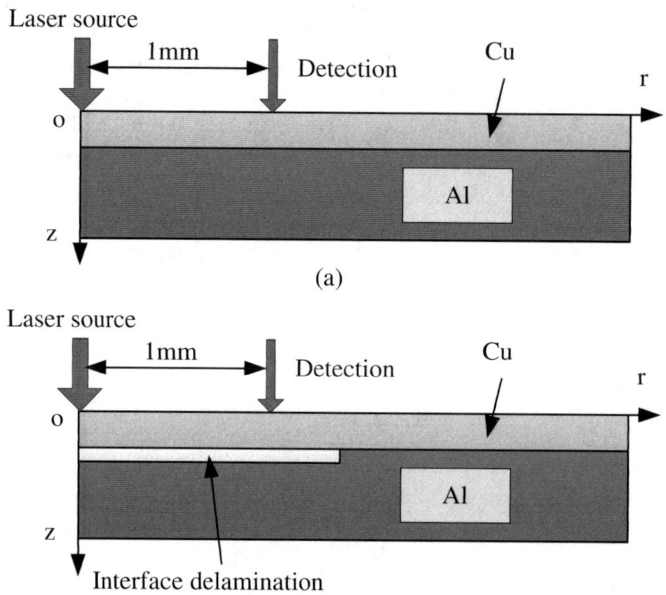

Figure 1. Laser ultrasonic in Cu/Al laminated plate: (a) a good model, and (b) a model with delamination.

Table 1. The material parameters of Cu/Al laminated plate used in numerical calculation.

	Cu	Al 1050
Thermal conductivity (W m^{-1} K^{-1})	386.4	209
Density(g cm^{-3})	8.96	2.71
Poisson's ratio	0.326	0.33
Thermal expansion coefficient (10^{-6}K^{-1})	17.2	23.6
Elastic Young's modulus (GPa)	119	68
Specific heat (J kg^{-1} K^{-1})	394	880

2.3. Simulation results and discussion

Figure 2 shows the distribution of the wave field of laser ultrasonic in a good model and the one with the delamination size of 2mm at 1μs. By comparing Figure 2(a) and (b), the interface delamination defects have a great influence on the distribution of wave field of laser ultrasonic, the displacement in the area with delamination is much higher than that in a good model, because the energy is reflected from the interface delamination. Figure 3 shows the out-of-plane displacement waveforms of the nodes located at 1mm from the laser source in a good area and the areas with delamination, with the delamination width of 2, 4, 6, 8 and 10 mm, respectively. The out-of-plane displacement amplitude gradually increases with increase of delamination sizes, and they can be used to characterize delamination defects.

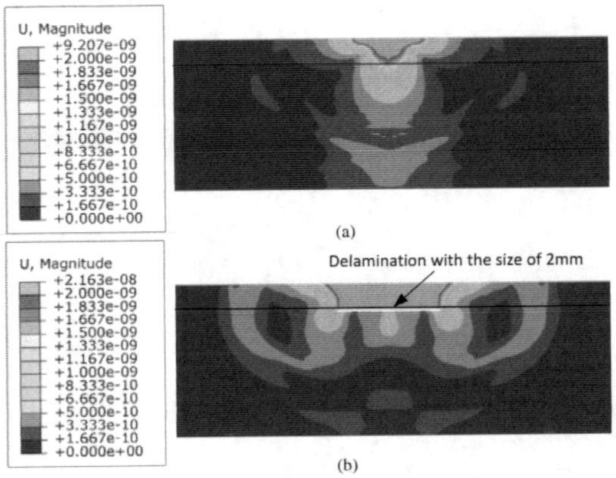

Figure 2. Distribution of the wave field of laser ultrasonic in Cu/Al laminated plates at 1μs: (a) a good model, and (b) a model with delamination size of 2mm.

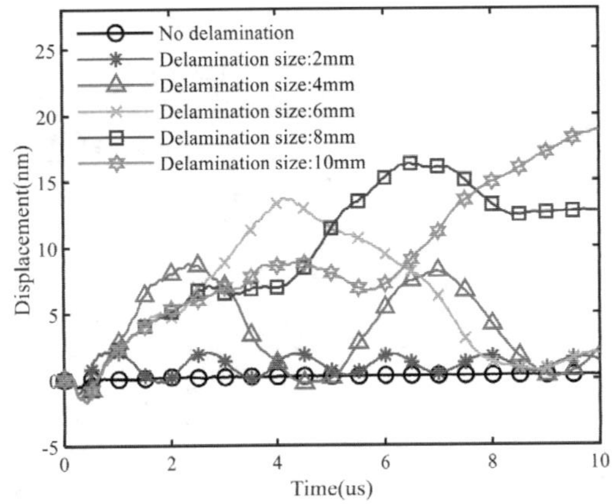

Figure 3. The relationship between delamination sizes and out-of-plane displacements measured by simulation.

3. Experimental analysis

3.1. Laminated plate sample

The Cu/Al laminated plate sample was manufactured by rolling cladding method, as shown in Figure 4. The dimensions of the Cu layer and the Al 1050 layer were 80mm×40mm×0.5mm and 80mm×40mm×2.5mm. The artificial T slots were finished by high-speed wire electrical discharge machining with the width of 0.2 mm and delamination width of 2, 4, 6, 8 and 10 mm, respectively.

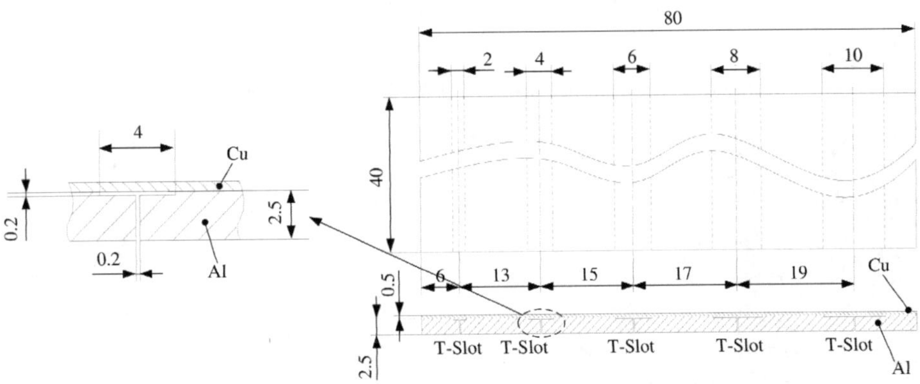

Figure 4. Cu/Al laminated plate sample.

3.2. Experimental setup

The laser ultrasonic experimental setup is shown in Figure 5. The Q switch Nd:YAG pulse laser produced by Quantel is focused onto the Cu/Al laminated plate to excite the ultrasonic. The wavelength of the pulsed laser is 1064nm, the duration of the pulsed laser is 8ns, the maximum repetition frequency is 20Hz, and the energy range of a single pulse laser is 0mJ-50mJ. The two-wave mixing laser interferometer manufactured by Intelligent Optical Systems Inc is used to detect ultrasonic signals. The wavelength of detection laser is 1550nm, the range of laser power is 0W-2W, and the spectrum range of interferometer is 50kHz -125MHz. The program controls the precise two-dimensional displacement platform to drive the sample to move and complete the C-scan detection and imaging.

IOP Conf. Series: Materials Science and Engineering **926** (2020) 012003 doi:10.1088/1757-899X/926/1/012003

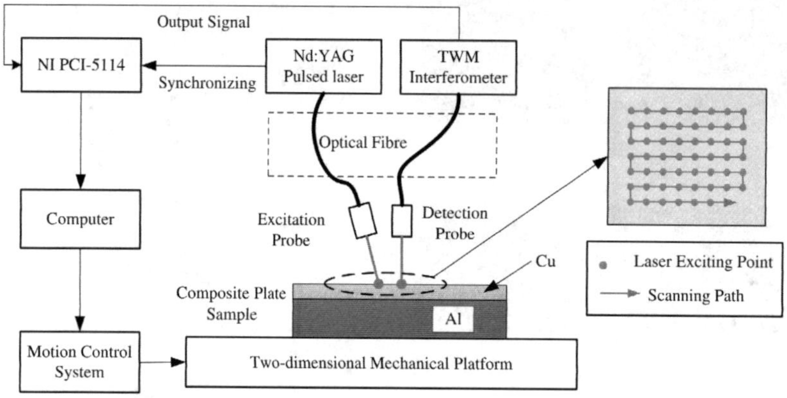

Figure 5. Schematic of the experimental setup.

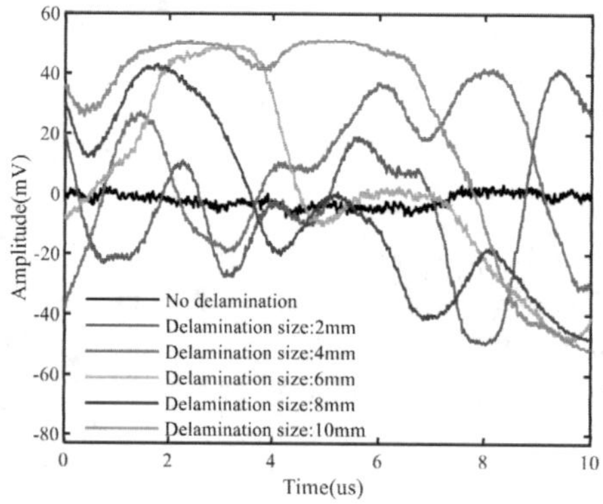

Figure 6. Experimental signals at different delamination areas.

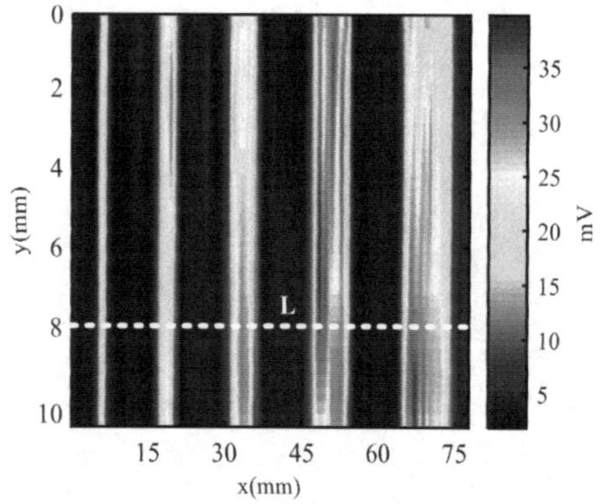

Figure 7. C-scan results of the Cu/Al laminated plate.

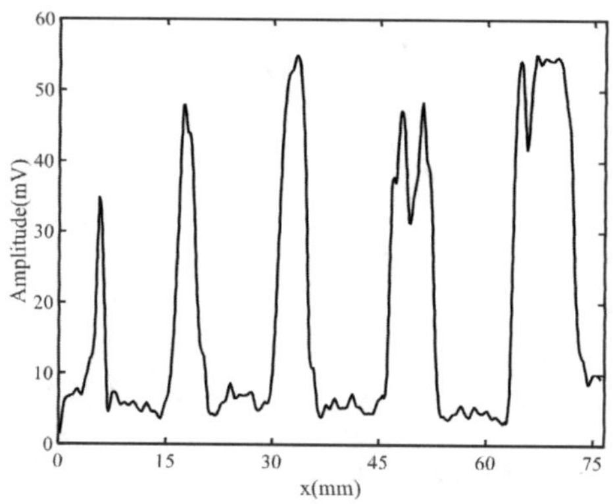

Figure 8. The amplitude curve at y=8mm in the C-scan result.

3.3. Experimental results

The ultrasonic wave in the Cu/Al laminated plate is generated exploiting a pulsed laser (λ=532nm, energy=10mJ, pulse duration=8ns, spot size=1mm). The detection laser power is 0.2W and the spot diameter is about 200μm. The laser ultrasonic signals were collected at a distance of 1mm from the irradiation point of the excitation laser. As shown in Figure 6, the amplitudes of laser ultrasonic signals in the areas with delamination are much higher than that in the areas with no delamination, which is consistent with the trend of numerical simulation results shown in Figure 3. The C-scan imaging result obtained from the 2D scan over the delamination defects is shown in Figure 7. Figure 8 shows the amplitude curve at y=8mm in the C-scan result. The delamination areas are easily distinguished from the normal areas, the detection results of the width and position of delamination defects are consistent with the simulated delamination defects in the sample as shown in Fiure. 4.

4. Conclusion

The out-of-plane displacement amplitude can effectively characterize delamination defects. The experimental results verify the correctness of the simulation results. The C-scan results of the delamination defects are consistent with the simulated delamination defects in the Cu/Al laminated plate sample. The laser ultrasonic technology can realize long-distance non-contact non-destructive detection of interface delamination defects in laminated plates, it has great potential for online non-contact inspection of Cu/Al laminated plates.

Acknowledgment

This work is supported by National Natural Science Foundation of China(No. 51575040), the Beijing Natural Science Foundation(No. 3182008), and the Fundamental Research Funds for the Central Universities Project(FRF-TP-19-039A2Z).

References

[1] Lee K S, Kwon Y N. Solid-state bonding between Al and Cu by vacuum hot pressing[J]. Transactions of Nonferrous Metals Society of China, 2013, 23(2):341-346.

[2] Xue P, Xiao B L, Wang D and Ma Z Y. Achieving high property friction stir welded aluminium/copper lap joint at low heat input[J]. Science and Technology of Welding and Joining, 2011, 16: 657–661.

[3] Furkan S. Weld ability Characteristics of Dissimilar Al/Cu Friction Stir Weld Joints[J]. Materials Testing, 2012, 54(2):85-91.

[4] Xia C Z, Li Y J, Wang J and Ma H J. Microstructure and phase constitution near interface of Cu/Al vacuum brazing[J]. Materials Science & Technology, 2007, 23(7):815-818.

[5] Kim I K, Hong S I. Effect of heat treatment on the bending behavior of tri-layered Cu/Al/Cu composite plates[J]. Materials & design, 2013, 47:590-598.

[6] Zheng G F, Xia W D, Zhou K F and Yan Y C. Study on the defects of bimetallic composite plate based on ultrasonic phased array technology[J]. Piezoelectrics & Acoustooptics. 2017, 39(03): 437-441.

[7] Sun G K, Zhao L M, Dong M L, Lou X P and Zhu L Q. Non-contact characterization of debonding in lead-alloy steel bonding structure with laser ultrasound[J]. Optik. 2018, 164: 734-744.

[8] LI L, Zhang X J, Liu H Y and Zhu Z C. Methods for evaluating interfacial bonding of laminated metal composites[J]. Journal of Wuhan University of Science and Technology, 2013, 36(3): 195-199.

[9] Watanabe Y, Fujisawa S, Yonezu A and Chen X. Quantitative evaluation of adhesion quality of surface coating by using pulse laser-induced ultrasonic waves[J]. Surface and Coatings Technology, 2016, 286, 231-238.

[10] Martarelli M, Chiariotti P, Pezzola M and Castellini P. Delamination detection in composites by laser ultrasonics[C]. AIP Conf. Proc. 2014, 1600:405-412.

[11] Zhang K S, Li S C, Zhou Z G. Quantitative analysis of voids in multi-layer bonded structures based on transmitted laser ultrasonic waves[J]. Archives of Civil and Mechanical Engineering. 2019, 19(1): 79-90.

[12] Zhang K S, Li S C, Zhou Z G. Detection of disbonds in multi-layer bonded structures using the laser ultrasonic pulse-echo mode[J]. Ultrasonics. 2019, 94: 411-418.

IOP Conf. Series: Materials Science and Engineering **926** (2020) 012004 doi:10.1088/1757-899X/926/1/012004

Kinematics of Planar Quabody Mechanism with More Kinematics Bifurcation Positions

C M Li[1,2], H Cao[1], Q Liu[1] and X Liu[1]

[1] Shengli College in China University of Petroleum, China University of Petroleum (East China), Dongying 257061, China

[2] School of Mechanical and Electronic Engineering, China University of Petroleum (East China), Qingdao 266580, China

lchming@126.com

Abstract. Based on multibody dynamics, some definitions of mechanism are defined. The kinematics bifurcation position is a cold researching spot for many many years. It is common in planar quabody mechanisms but has not been paid enough attention to. The kinematics bifurcation position refers to the mechanism position with a pressure angle of 90° of the follower, including the stuck position (dead point). In this paper, a planar quabody mechanism is studied. In a motion cycle, there are three kinematics bifurcation positions in the mechanism. The third derivative of angle to time, angular velocity, angular acceleration of the connecting rod in kinematics bifurcation positions is obtained by solving the kinematics parameters of kinematics bifurcation positions according to L'Hopital's rule. If the driving body swings in sinusoidal law, the motion of the mechanism is continuous. The kinematics characteristics of the mechanism are independent of the simulation convergence and response bifurcation of nonlinear dynamics. The length of the rocker is 1 m, the length of the connecting rod is 0.5 m, the distance between the track of the slider and the fixed point of the rocker is 0.5 m, and the rocker swings according to the sine law with a period of 9 seconds. The computer simulation shows that the speed and acceleration are continuous and so there is no impact in the kinematics bifurcation positions.

1. Introduction

Machinery is the main tool to replace or reduce human labor and increase labor productivity. It is a theoretical model or entity device composed of several parts with definite mutual motion and certain functions. From the motion of one object to get the motion of other objects, and there is a certain relationship between the motions of every two objects. This kind of motion is called mechanical motion. A machine is a movable device that performs mechanical motions. A mechanism is the theoretical model that performs mechanical motions.

Mechanical principle focuses on theory, so its concept should be consistent with theoretical mechanics and multibody dynamics. A body (component) is an independent object moving in a mechanism or machine. The connecting parts between the bodies are called motion pairs. In the mechanism, the body is rigid, which can be abstracted as the object formed by rigid connection of each motion pair element. A body has three freedoms (independent motions) in plane motion, two mutual vertical motions and one rotation. When one body rotates with respect to the other, two freedoms are constrained. This kind of motion pair is the rotating pair and the biconstraint pair (low

IOP Conf. Series: Materials Science and Engineering **926** (2020) 012004 doi:10.1088/1757-899X/926/1/012004

pair). When a body moves relative to another body, which is equivalent to rotating around infinity far, a motion freedom and a rotation freedom are constrained. This kind of motion pair is the translational pair and also the biconstraint pair. When one body rolls and slides with respect to the other, a motion freedom perpendicular to the rolling direction is constrained. The motion pair between the two bodies is the rolling-sliding pair and the monoconstraint pair (high pair). The body with two or more rotating pairs is called a rod (crank, rocker, connecting rod, multipair rod, etc.). The body with one rotating pair is called a block (slider, track rod, swing block, etc.). The body without rotating pair is called a board. Prefix multi, mono, bi and qua mean many, one, two and four respectively.

For the crank rocker mechanism, if the rocker can bypass the fixed body, and if the crank rotates at a constant speed, at the position where the connecting rod and the rocker are collinear, the rocker can rotate in the counterclockwise direction, or in the clockwise direction, which is called the kinematics bifurcation position. If the rocker is an active body, and if it swings according to the sinusoidal law, the position of the collineation between the crank and the connecting rod is also the kinematics bifurcation position. It can be seen that the pressure angle of the follower (the angle between the force direction and the motion direction) is 90 degrees in the kinematics bifurcation position. For the static kinematics bifurcation position, if no matter how large the driving force can not make the mechanism move, the position is also called the stuck position (dead point [1-2]).

The kinematics bifurcation position is common in planar mechanism. This motion characteristic is also called motion bifurcation. It is independent of simulation convergence and response bifurcation of nonlinear dynamics. However, in mechanism kinematics, the kinematics bifurcation position has not been paid enough attention to for many years[3].

The kinematic study[4] is the basis of the dynamic study[5-6]. In the searched single slider quabody mechanism, the connecting rod is not the longest rod, and it has three kinematics bifurcation positions. Assuming that the driven slider is affected by a certain flow rate, the motion law of the driven connecting rod and the driven slider is studied.

2. The rocker slider mechanism

In the rocker slider mechanism shown in figure 1, the rocker swings within a certain range. The rotating pair between rocker *AB* and fixed body is *A*, and the rotating pair of connecting rod *BC* and slider is *C*. The track path of the slider refers to a straight line or curve that passes through another motion pair and is parallel to the motion track. The distance between the track path and point *A* is half of the length of the rocker.

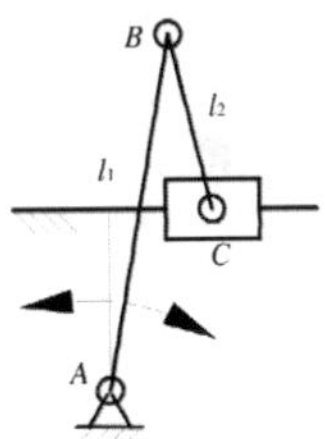

Figure 1. Rocker-slider mechanism with kinematic bifurcation positions

In this mechanism, the fixed body can be regarded as an infinitely long rod with *A* pointing to the infinitely far end which is vertical to the slide track, and the slide can be regarded as an infinitely long rod with a rotating pair *C* pointing to the infinitely far end perpendicular to the track. Point *A* is taken as the coordinate origin. The horizontal right direction is taken as the positive direction of x_1 axis. The vertical up direction is taken as the positive direction of x_2 axis. So a rectangular coordinate system is established. l_1 and l_2 are the length of rocker *AB* and connecting rod *BC* respectively, and the diameter vectors of the two bodies are $\boldsymbol{l}_1$ and $\boldsymbol{l}_2$ respectively. d is the vertical distance from *A* to the slider track. Both l_2 and d are half of l_1. The diameter vector of point *C* relative to the coordinate origin is $\boldsymbol{l}_3$.

According to the geometric relationship between the bodies and the geometric meaning of the vector equation (the vectors on both sides of the equal sign are connected with each other to form two vector chains. If their starting point is the same, their ending point is coincident). The vector equation of the mechanism is as the following[7-8].

$$\boldsymbol{l}_1 + \boldsymbol{l}_2 = \boldsymbol{l}_3 \tag{1}$$

If θ_1 and θ_2 are taken as the included angles of $\boldsymbol{l}_1$ and $\boldsymbol{l}_2$ with respect to the positive direction of x_1 axis. There is, $\theta_1 \in [0 \quad \pi]$.

3. Projection equations of mechanism vector equation

By projecting the vector equations (1) onto two coordinate axes respectively, a projection equations set can be obtained as the following.

$$l_1 \cos\theta_1 + l_2 \cos\theta_2 = x_{1,C},\ l_1 \sin\theta_1 + l_2 \sin\theta_2 = d \tag{2}$$

If the rocker is an active part, its motion law is known. The unknowns of equations (2) are θ_2 and $x_{1,C}$. θ_2 can be obtained from the second equation of equations (2) as the following.

$$\theta_2 = \arcsin\left(\frac{d - l_1 \sin\theta_1}{l_2}\right),\quad \theta_2 = \pi - \arcsin\left(\frac{d - l_1 \sin\theta_1}{l_2}\right) \tag{3}$$

According to θ_2 and $\dot{\theta}_2$ of the previous position, as well as $\dot{x}_{1,C}$, θ_2 can be obtained from formula (3). Then, $x_{1,C}$ can be obtained from the first equation of equations (2).

Calculating the first derivative of time, equations (s) changes to equations (4). It can be reorganized to matrix equation.

$$\begin{cases} -l_1\dot{\theta}_1 \sin\theta_1 - l_2\dot{\theta}_2 \sin\theta_2 = \dot{x}_{1,C} \\ l_1\dot{\theta}_1 \cos\theta_1 + l_2\dot{\theta}_2 \cos\theta_2 = 0 \end{cases},\quad \begin{bmatrix} -l_2 \sin\theta_2 & -1 \\ l_2 \cos\theta_2 & 0 \end{bmatrix}\begin{bmatrix} \dot{\theta}_2 \\ \dot{x}_{1,C} \end{bmatrix} = -\dot{\theta}_1 l_1 \begin{bmatrix} -\sin\theta_1 \\ \cos\theta_1 \end{bmatrix} \tag{4}$$

The second equation of equation group (4) has only one unknown variable, so it can be solved in turn without the simplified formula of the second-order matrix inversion to solve matrix equation (4). From the downer equation of equations (4), we can get:

$$\dot{\theta}_2 = \frac{-\dot{\theta}_1 l_1 \cos\theta_1}{l_2 \cos\theta_2} \tag{5}$$

$\dot{x}_{1,C}$ can be obtained from the upper equation of equations (4). Calculating the first derivative of time, equations (4) changes to equations (6). It can be reorganized to matrix equation (7).

$$\begin{cases} -l_1\ddot{\theta}_1 \sin\theta_1 - l_1\dot{\theta}_1\dot{\theta}_1 \cos\theta_1 - l_2\ddot{\theta}_2 \sin\theta_2 - l_2\dot{\theta}_2\dot{\theta}_2 \cos\theta_2 = \ddot{x}_{1,C} \\ l_1\ddot{\theta}_1 \cos\theta_1 - l_1\dot{\theta}_1\dot{\theta}_1 \sin\theta_1 + l_2\ddot{\theta}_2 \cos\theta_2 - l_2\dot{\theta}_2\dot{\theta}_2 \sin\theta_2 = 0 \end{cases} \tag{6}$$

$$\begin{bmatrix} -l_2 \sin\theta_2 & -1 \\ l_2 \cos\theta_2 & 0 \end{bmatrix}\begin{bmatrix} \ddot{\theta}_2 \\ \ddot{x}_{1,C} \end{bmatrix} = \begin{bmatrix} l_2 \cos\theta_2 & 0 \\ l_2 \sin\theta_2 & 0 \end{bmatrix}\begin{bmatrix} \dot{\theta}_2^2 \\ \dot{x}_{1,C} \end{bmatrix} - l_1\ddot{\theta}_1 \begin{bmatrix} -\sin\theta_1 \\ \cos\theta_1 \end{bmatrix} + l_1\dot{\theta}_1^2 \begin{bmatrix} \cos\theta_1 \\ \sin\theta_1 \end{bmatrix} \tag{7}$$

$\ddot{\theta}_2$ is obtained from the downer equation of equation set (6).

$$\ddot{\theta}_2 = \left(l_2\dot{\theta}_2^2 \sin\theta_2 - l_1\ddot{\theta}_1 \cos\theta_1 + l_1\dot{\theta}_1^2 \sin\theta_1\right)\left(l_2 \cos\theta_2\right)^{-1} \tag{8}$$

Then, $\ddot{x}_{1,C}$ can be obtained from the upper equation of equations (6).

By calculating the first derivative of equations (6) for time, the third derivative of vector equation (1) in x_2 axial projection can be obtained. Then $\dddot{\theta}_2$ can be arranged out.

$$l_1\dddot{\theta}_1\cos\theta_1 - 3l_1\dot{\theta}_1\ddot{\theta}_1\sin\theta_1 - l_1\dot{\theta}_1^3\cos\theta_1 + l_2\dddot{\theta}_2\cos\theta_2 - 3l_2\dot{\theta}_2\ddot{\theta}_2\sin\theta_2 - l_2\dot{\theta}_2^3\cos\theta_2 = 0 \tag{9}$$

$$\dddot{\theta}_2 = \left(-l_1\dddot{\theta}_1\cos\theta_1 + 3l_1\dot{\theta}_1\ddot{\theta}_1\sin\theta_1 + l_1\dot{\theta}_1^3\cos\theta_1 + 3l_2\dot{\theta}_2\ddot{\theta}_2\sin\theta_2 + l_2\dot{\theta}_2^3\cos\theta_2\right)\left(l_2\cos\theta_2\right)^{-1} \tag{10}$$

The third derivative of the vector equation projected on the x_1-axis is also obtained. At the same time, $\dddot{x}_{1,C}$ is obtained.

$$\dddot{x}_{1,C} = -l_1\dddot{\theta}_1\sin\theta_1 - 3l_1\dot{\theta}_1\ddot{\theta}_1\cos\theta_1 + l_1\dot{\theta}_1\dot{\theta}_1\dot{\theta}_1\sin\theta_1 - l_2\dddot{\theta}_2\sin\theta_2 - 3l_2\dot{\theta}_2\ddot{\theta}_2\cos\theta_2 + l_2\dot{\theta}_2\dot{\theta}_2\dot{\theta}_2\sin\theta_2 \tag{11}$$

4. Angular velocity and acceleration of the connecting rod in kinematics bifurcation positions

For the mechanism shown in figure 1, there are motion bifurcations in three positions $\theta_1 = 0$, $\theta_1 = \pi/2$ and $\theta_1 = \pi$, which are the kinematics bifurcation positions of the mechanism. In the position of $\theta_1 = \pi/2$, the connecting rod is pulled by the rocker without the feature of stuck position, which belongs to the general kinematics bifurcation position. In the other two positions, because the connecting rod is pressed by the rocker, if the sliding block remains stationary, the mechanism remains stationary no matter how much driving force, so both positions are the stuck positions of the mechanism.

If the rocker swings according to the sinusoidal law, T is the period, θ_1 and its derivatives from first to forth are as the following.

$$\begin{aligned}&\theta_1 = \frac{\pi-0}{2}\sin\left(\frac{2\pi}{T}t\right) + \frac{\pi+0}{2},\ \dot{\theta}_1 = \frac{\pi^2}{T}\cos\left(\frac{2\pi}{T}t\right)\\ &\ddot{\theta}_1 = -\frac{2\pi^3}{T^2}\sin\left(\frac{2\pi}{T}t\right),\ \dddot{\theta}_1 = -\frac{4\pi^4}{T^3}\cos\left(\frac{2\pi}{T}t\right),\ \ddddot{\theta}_1 = \frac{8\pi^5}{T^4}\sin\left(\frac{2\pi}{T}t\right)\end{aligned} \tag{12}$$

The connecting rod BC can rotate all the way. In the position of $\theta_1 = 0$, there are,

$$t = \frac{3}{4}T + nT,\ \frac{2\pi}{T}t = \frac{3}{2}\pi + 2n\pi,\ \cos\left(\frac{2\pi}{T}t\right) = 0,\ \dot{\theta}_1 = 0,\ \dddot{\theta}_1 = 0,\ \theta_2 = \frac{\pi}{2} \tag{13}$$

According to the previous research, $\dot{\theta}_2$ of formula (5), $\ddot{\theta}_2$ of formula (8) and $\dddot{\theta}_2$ of formula (10) are all infinitive of type 0/0. The value of $\dot{\theta}_2$ can be determined according to L'Hopital's rule as the following[9].

$$\dot{\theta}_2 = \left(-\ddot{\theta}_1 l_1\cos\theta_1 + \dot{\theta}_1^2 l_1\sin\theta_1\right)\left(-l_2\dot{\theta}_2\sin\theta_2\right)^{-1} = \left(\ddot{\theta}_1 l_1\right)\left(l_2\dot{\theta}_2\right)^{-1},\ \dot{\theta}_2 = \mp\sqrt{\ddot{\theta}_1 l_1 l_2^{-1}} \tag{14}$$

Where, "$\mp$" is the same as the $\dot{\theta}_2$ symbol of the previous position, but opposite to the $\dot{x}_{1,C}$ symbol of the previous position.

If $\ddot{\theta}_1 = 0$, the mechanism remains stationary. There are, $\dot{\theta}_2 = 0$, $\ddot{\theta}_2 = 0$. From (4), $\dot{x}_{1,C}$ is:

$$\dot{x}_{1,C} = -l_2\dot{\theta}_2 \tag{15}$$

According to L'Hopital's rule, $\ddot{\theta}_2$ can be obtained from formula (8).

$$\ddot{\theta}_2=\frac{2l_2\dot{\theta}_2\ddot{\theta}_2\sin\theta_2+l_2\dot{\theta}_2^3\cos\theta_2-l_1\ddot{\theta}_1\cos\theta_1+3l_1\dot{\theta}_1\ddot{\theta}_1\sin\theta_1+l_1\dot{\theta}_1^3\cos\theta_1}{-l_2\dot{\theta}_2\sin\theta_2}=\frac{2l_2\dot{\theta}_2\ddot{\theta}_2-l_1\ddot{\theta}_1+l_1\dot{\theta}_1^3}{-l_2\dot{\theta}_2} \quad (16)$$

$$\ddot{\theta}_2=\frac{l_1\ddot{\theta}_1-l_1\dot{\theta}_1^3}{3l_2\dot{\theta}_2}=0 \quad (17)$$

According to L'Hopital's rule, $\dddot{\theta}_2$ can be obtained from formula (10).

$$\dddot{\theta}_2=\frac{\begin{array}{c}-l_1\dddot{\theta}_1\cos\theta_1+l_1\dot{\theta}_1\ddot{\theta}_1\sin\theta_1+3l_1\left(\ddot{\theta}_1\ddot{\theta}_1+\dot{\theta}_1\dddot{\theta}_1\right)\sin\theta_1+3l_1\dot{\theta}_1^2\ddot{\theta}_1\cos\theta_1+3l_1\dot{\theta}_1^2\ddot{\theta}_1\cos\theta_1\\-l_1\dot{\theta}_1^4\sin\theta_1+3l_2\left(\ddot{\theta}_2\ddot{\theta}_2+\dot{\theta}_2\dddot{\theta}_2\right)\sin\theta_2+3l_2\dot{\theta}_2^2\ddot{\theta}_2\cos\theta_2+3l_2\dot{\theta}_2^2\ddot{\theta}_2\cos\theta_2-l_2\dot{\theta}_2^4\sin\theta_2\end{array}}{-l_2\dot{\theta}_2\sin\theta_2} \quad (18)$$

$$\dddot{\theta}_2=\left(-l_1\dddot{\theta}_1+6l_1\dot{\theta}_1^2\ddot{\theta}_1+3l_2\left(\ddot{\theta}_2\ddot{\theta}_2\right)-l_2\dot{\theta}_2^4\right)\left(-4l_2\dot{\theta}_2\right)^{-1} \quad (19)$$

In $\theta_1=\pi$, there are, $t=\frac{T}{4}+nT$, $\frac{2\pi}{T}t=\frac{\pi}{2}+2n\pi$, $\cos\left(\frac{2\pi}{T}t\right)=0$, $\dot{\theta}_1=0$, $\dddot{\theta}_1=0$, $\theta_2=\frac{\pi}{2}$

$\dot{\theta}_2$ and $\ddot{\theta}_2$ are also 0/0 infinitives type. According to L'Hopital's rule, from formula (5), we can get the value of $\dot{\theta}_2$ as the following.

$$\dot{\theta}_2=\pm\sqrt{-\ddot{\theta}_1 l_1 l_2^{-1}} \quad (20)$$

Where, "$\mp$" is the same as the $\dot{\theta}_2$ symbol of the previous position, but opposite to the $\dot{x}_{1,C}$ symbol of the previous position.

If $\ddot{\theta}_1=0$, the mechanism remains stationary, there are, $\dot{\theta}_2=0$, $\ddot{\theta}_2=0$. From equation set (4), formula (15) can be obtained. According to L'Hopital's rule, $\ddot{\theta}_2$ can be obtained from formula (8).

$$\ddot{\theta}_2=\left(-l_1\ddot{\theta}_1+l_1\dot{\theta}_1^3\right)\left(3l_2\dot{\theta}_2\right)^{-1}=0 \quad (21)$$

According to L'Hopital's rule, $\dddot{\theta}_2$ can be obtained from formula (10).

$$\dddot{\theta}_2=\left(l_1\dddot{\theta}_1-6l_1\dot{\theta}_1^2\ddot{\theta}_1+3l_2\left(\ddot{\theta}_2\ddot{\theta}_2\right)-l_2\dot{\theta}_2^4\right)\left(-4l_2\dot{\theta}_2\right)^{-1} \quad (22)$$

In the position of $\theta_1=0.5\pi$, there are,

$t=\frac{T}{2}+nT$ or $t=nT$, $\frac{2\pi}{T}t=\pi+2n\pi$ or $\frac{2\pi}{T}t=2n\pi$, $\sin\left(\frac{2\pi}{T}t\right)=0$, $\ddot{\theta}_1=0$, $\ddddot{\theta}_1=0$, $\theta_2=\frac{3\pi}{2}$

According to L'Hopital's rule, $\dot{\theta}_2$ can be obtained from formula (5) as the following.

$$\dot{\theta}_2=\left(-l_1\cos\theta_1\ddot{\theta}_1+l_1\sin\theta_1\dot{\theta}_1^2\right)\left(-l_2\sin\theta_2\dot{\theta}_2\right)^{-1}=\left(\dot{\theta}_1^2 l_1\right)\left(l_2\dot{\theta}_2\right)^{-1} \quad (23)$$

$$\dot{\theta}_2=\pm\sqrt{\dot{\theta}_1^2 l_1 l_2^{-1}} \quad (24)$$

Where, "$\mp$" is the same symbol as the $\dot{\theta}_2$ of previous position, otherwise a flexible impact occurs.

If $\dot{\theta}_1 = 0$, then $\dot{\theta}_2 = 0$. $\dot{x}_{1,C}$ can be obtained from equations (4).

$$\dot{x}_{1,C} = -l_1\dot{\theta}_1 + l_2\dot{\theta}_2 \tag{25}$$

According to L'Hopital's rule, $\ddot{\theta}_2$ can be obtained from formula (8).

$$\ddot{\theta}_2 = \frac{2l_2\dot{\theta}_2\ddot{\theta}_2\sin\theta_2 + l_2\dot{\theta}_2^3\cos\theta_2 - l_1\dddot{\theta}_1\cos\theta_1 + 3l_1\dot{\theta}_1\ddot{\theta}_1\sin\theta_1 + l_1\dot{\theta}_1^3\cos\theta_1}{-l_2\dot{\theta}_2\sin\theta_2} = \frac{-2l_2\dot{\theta}_2\ddot{\theta}_2\sin\theta_2 + 3l_1\dot{\theta}_1\ddot{\theta}_1\sin\theta_1}{l_2\dot{\theta}_2\sin\theta_2}$$

$$\ddot{\theta}_2 = \left(l_1\dot{\theta}_1\ddot{\theta}_1\right)\left(l_2\dot{\theta}_2\right)^{-1} = \pm\sqrt{l_1 l_2^{-1}}\,\ddot{\theta}_1 = 0 \tag{26}$$

Where, "$\mp$" is the same as that of formula (24).

If $\ddot{\theta}_1 = 0$, then $\ddot{\theta}_2 = 0$. When $\dot{\theta}_1 = 0$ and $\ddot{\theta}_1 = 0$ are held at the same time, the mechanism remains stationary. According to L'Hopital's rule, $\dddot{\theta}_2$ can be obtained from formula (10).

$$\dddot{\theta}_2 = \frac{+3l_1\left(\ddot{\theta}_1\ddot{\theta}_1 + \dot{\theta}_1\dddot{\theta}_1\right) + l_1\dot{\theta}_1\dddot{\theta}_1 - l_1\dot{\theta}_1^4 - 3l_2\left(\ddot{\theta}_2\ddot{\theta}_2 + \dot{\theta}_2\dddot{\theta}_2\right) + l_2\dot{\theta}_2^4}{l_2\dot{\theta}_2} \tag{27}$$

$$\dddot{\theta}_2 = \frac{+3l_1\left(\ddot{\theta}_1\ddot{\theta}_1\right) + 4l_1\dot{\theta}_1\dddot{\theta}_1 - l_1\dot{\theta}_1^4 - 3l_2\left(\ddot{\theta}_2\ddot{\theta}_2\right) + l_2\dot{\theta}_2^4}{4l_2\dot{\theta}_2} \tag{28}$$

5. Computer simulation results

5.1. The mechanism parameters

The lengths of the active rocker and the connecting rod are $l_1 = 1.0\text{m}$ and $l_2 = 0.5\text{m}$, respectively. The distance between the slide track and the rocker fixed point is $d = 0.5\text{m}$. The swing period of the rocker is T=9.0s.

5.2. Simulation result

The time history of kinematics parameters of each body is shown in figure 2 and figure 3. There is no sudden change of velocity and acceleration. So there is no impact in the kinematics bifurcation position[10-11]. This mechanism has motion continuity[12].

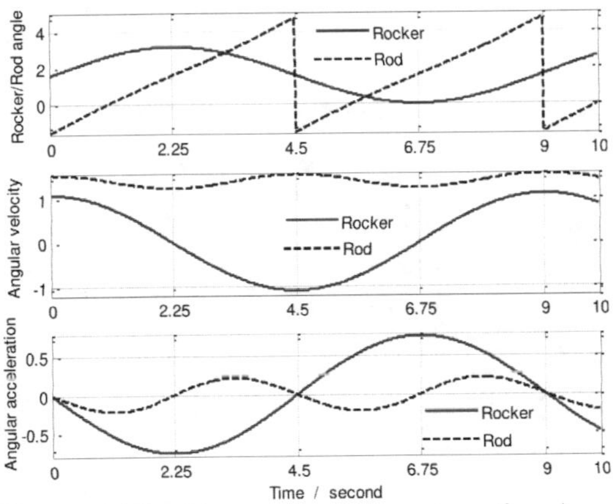

Figure 2. The kinematics parameters of rocker and connecting rod

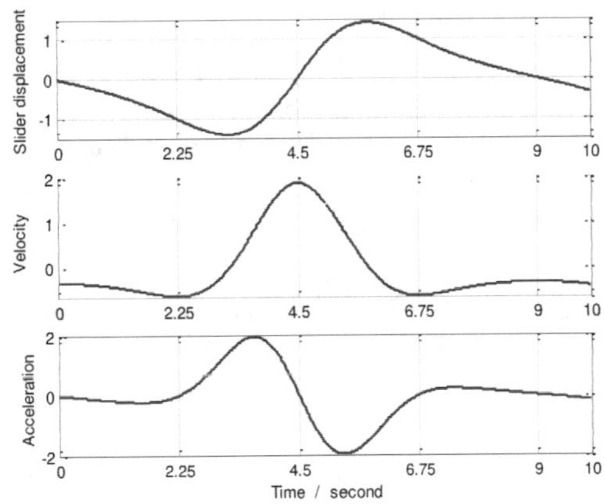

Figure 3. The kinematics parameters of slider

6. Conclusion

There are 3 points mainly as the following.

1)In the kinematics bifurcation position, the calculation formula of kinematics parameters is infinitive. If the corresponding calculation formula is not deduced in advance, it will be divided by zero and the calculation fails in the dynamic simulation of the mechanism.

2)For a planar quabody mechanism with three kinematics bifurcation positions in one motion cycle, this paper deduces the calculation formula of kinematics parameters of each body, which provides the research basis for the dynamic programmed computer simulation.

3)The given basic concepts of mechanical principle are in line with theoretical mechanics and multibody dynamics, such as mechanism, machine, monoconstraint pair, biconstraint pair, rod, block and board. Those concepts will make the subject more scientific.

Acknowledgments

Online course construction project of Shengli College in China University of Petroleum [XJKC201804].

References

[1] H. ALT. *Ueber die totlagen gelenk vicrecks zamm*, 1925: 5.

[2] X R Li, P F Wu 1988 The ALT means of designing oscillation-crank mechansis *Journal of Qiqihaer Institute of Light Industry* **4(3)** 47-52

[3] C M Li, J Chen 2011 The structure parameter optimization of double crank and groove wheel series mechanism *Manufacturing automation in China* **33(7)** 113-115

[4] H Cao, C M Li, Q Liu, et al. 2020 Dynamics of planar mechanism with kinematics bifurcation positions *Applied science and technology* **DOI**: 10.11991/yykj.202002005

[5] D F Song, Q R Yun, N N Yang, X H Zeng, X Q Wang 2019 Model predictive dynamic coordinated control of planetary hybrid electric bus *Journal of Harbin Institute of Technology* **51(01)** 150-156+161

[6] L H Yang, C Zhang, Y Y Qu, H Y Gao 2017 Kinetic analysis on the transmission of a controlled differential CVT *Applied science and technology* **44(05)**70-74+78

[7] C M Li, X L Yin, P L Yun, Y Guo, Y J Cui 2019 Crank slider mechanism dynamics and its related problem of a machine *Chinese Journal of Dezhou University* **35(6)** 40-46

[8] X L Yin, C M Li 2021 On the programmed kinematics computation of crank rocker mechanism based on harmonic composition and decomposition *Chinese Journal of Gansu Sciences* **33(1)** (*Preprint* gr-qc/20190326)

[9] X L Yin, C M Li 2020 On the programmed kinematics computation of crank rocker mechanism based on the stuck position analysis *Chinese Journal of Applied Mathematics and Mechanics* **41(4)** 367-375

[10] C M Li 2009 The quartic polynomial motion law of the cam mechanism follower *Beijing: China Scienpaper Online* **http**://www.paper.edu.cn

[11] C M Li 2010 On the dynamics of bull-nose plane six-rod mechanism considering some practical factor *Key Engineering Materials* **426-427** 65-69

[12] W W Liu, N L Lu, H X Liu 2020 Kinematic characteristics analysis and parametric solving of snow melting agent throwing mechanism with variable crank length *Journal of Harbin Institute of Technology (New series)* **http**://kns.cnki.net /kcms/detail/23.1378.T.20191217.1309.002.html

An Approach to Suppress Flow Separation by Plasma Vortex Generator as a Combined Active and Passive Control Mechanism

A N M M I Mukut[1, *], H M M Afroz[1], H Mizunuma[2] and O Hiromichi[2]

[1]Department of Mechanical Engineering, DUET, Gazipur-1707, Bangladesh
[2]Department of Mechanical Engineering, TMU, 1-1, Minami-Osawa, Tokyo, Japan

[*]Email: mukut@duet.ac.bd

Abstract. Flow control is an important issue in fluid mechanics and aerodynamics as separated flow introduce undesirable impact on overall system performance. Several active and passive techniques have been used to control the flow when necessary. But in passive control techniques, the actuator has to be fixed on the surface so that it can create vortices along streamline to enhance the mixing in the separated region to delay or suppress separation. Among passive flow control system vortex generators (VGs) are generally used but it control the flow with the cost of drag increase. It would be advantageous by combining both active and passive flow techniques together. In this regard plasma actuator (PA, active flow control device) and VGs are combined together with a view to not only suppress the separation but also minimizing the drag that could arise by using conventional passive control techniques. The combination flow control device is named as Plasma Vortex Generator (PVG) which proves its effectiveness to control separation as conventional VGs do but without any drag penalty.

1. Introduction

Due to the separation of flow from the wall, the control of separation is necessary to ensure the performances of various practical devices. As a result, flow control has great practical importance and becomes the rapidly growing field in fluid mechanics which has several engineering benefits but not limited to turbulence management, separation delay or suppression, drag reduction, lift enhancement, fuel economy, mixing improvement, noise control, environmental compliance and so on. Both active and passive flow control techniques are used to control the flow, although all of them have some limitation. Among various passive control devices, vortex generator (VG) is most effective which introduced by Taylor [1] in the late 1940s. These VG are simply a series of thin plates of various shapes (rectangular vanes [2], doublets & wishbone [3], trapezoid vanes & delta vanes [4-8], triangular vanes [9], cylinder [10]) that are placed normal to the surface with an angle along flow direction. Due to this angle of incidence incoming air creates vortices which adds momentum to the wall surface thus mixing is enhanced. This enhancement suppress or delay the separation. The conventional vane-type VG has a height of the order of the boundary layer thickness, and has a penalty of drag increase as they are with certain angle along flow direction. This penalty is critical for shorter flow separation control situation. In order to minimize the drag penalty, smaller VGs submerged in the boundary layer have been developed and investigated [11]. To overcome these limitation, a novel approach has been taken to combine both active and passive flow control techniques together. Among various active flow control methods plasma actuator (PA) are prominent due to its simplicity and quick response capability which is consist of a exposed electrode and an embedded electrode, separated by a dielectric material. A set of configurations of PAs have been investigated both experimentally and numerically to figure out the optimum configuration to make the so called PVG. Among tested three configurations as shown in Figure 1, the leading edge type of PA has been modified to construct the PVG as this one gives maximum induced flow which is 1.81 m/s [12].The device is named as plasma vortex generator (PVG) which combines both plasma actuator and conventional vortex generator together. These PVGs are placed normal to the surface but without any angle of incidence along the flow direction. The plasma induced flow from the vane tip to the hub produces a pair of stream-wise counter-rotating vortices downstream.

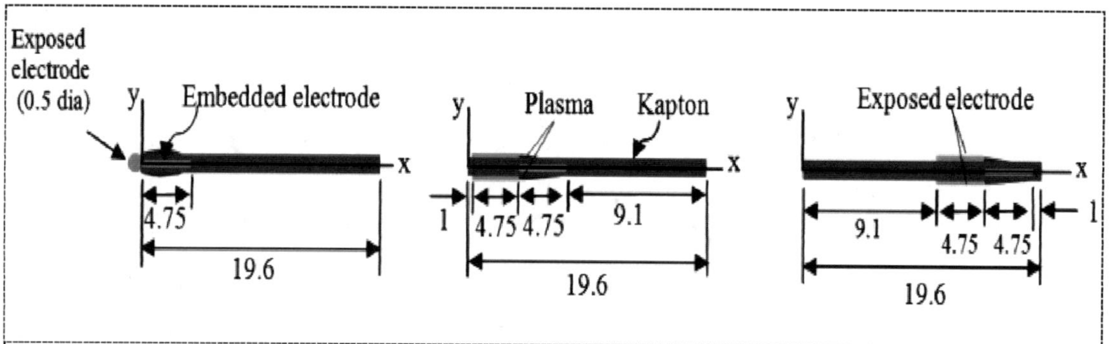

Figure 1. Three configurations of PA: L Type, M Type and T Type (from left to right) [12].

The angle of the PVG (β) is zero against a main flow, and thus the drag penalty is expected to be low if the plasma is switched off. The basic difference of vortices formation between conventional VGs and PVGs is shown in Figure 2.

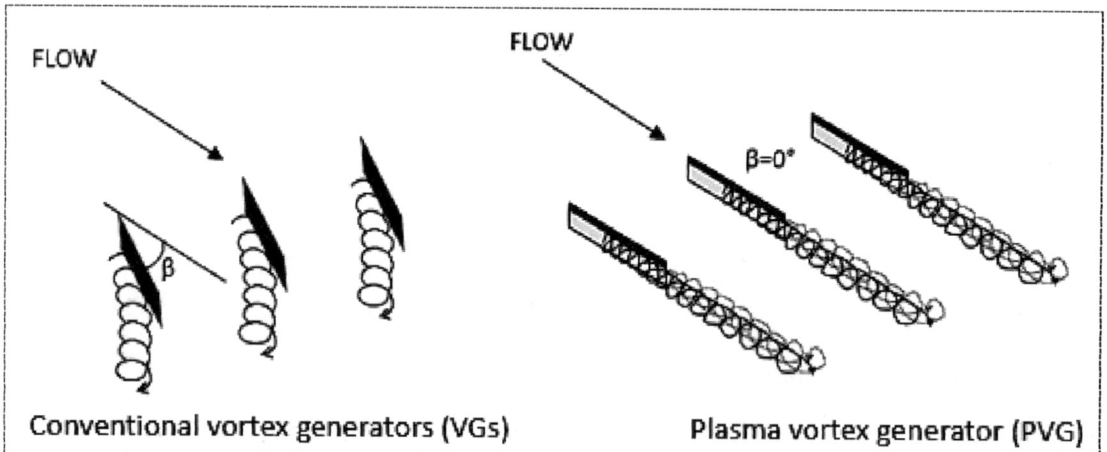

Figure 2. Vortices formation from VG and PVG.

2. Experimental setup and simulation

The experiment is carried out in an open type wind tunnel of Tokyo Metropolitan University (TMU), Tokyo, Japan with a free stream velocity of 4 m/s. Flow separation is created in a 20° slope of diffuser section of the wind tunnel. All measurements haven been conducted along B. Details of measurements and dimension (not to scale) shown in Figure 3.

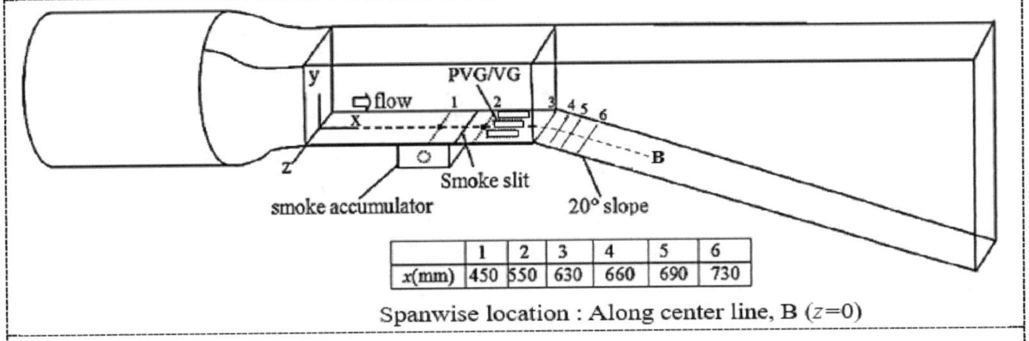

Figure 3. Schematic of wind tunnel (not to scale).

The streamwise lengths of straight section is 0.63 m and the diffuser section is 1.4m. The cross section of the entrance was 0.2 m by 0.2 m. A tripping wire of 2 mm diameter was fixed at x = 50 mm in the straight entrance section to develop a turbulent boundary layer. To confirm the development of turbulent boundary layer velocity profile measured in some stream-wise locations and compare with 1/7th law [13]. Dimensions of VGs' have been characterized from Lin [2]. As the comparison has been made between PVG and VG on separation, hence same configurations have been considered for both except the angle of incidence (Figure 2) which is shown in Table 1.

Table 1. Dimension of VG and PVG

Parameters	Dimensions (mm)
Height, h	7
Length, l	28
Spanwise spacing, d	64
Distance from separation, Δx_{VG}	70

Three sets of experiments, namely (1) no actuator (2) with PVG and (3) VG, have been conducted to evaluate the performance of PVG on flow control by comparing it with conventional VGs. Flow profiles have been visualized by using hot-wire anemometer with also smoke sheet, smoke wire visualization as the hot wire sensor used in this research has the ability to give only the magnitude of velocity not the direction. The velocity profiles obtained from hot-wire have been overlapped with the smoke visualization to ensure the near wall reverse flow. The separation phenomena for case (i) has been investigated numerically by using commercial CFD to confirm the experimental result. k-ε (RNG) with enhance wall treatment has been implement in simulation. The flow domain is shown in Figure 4 and corresponding boundary conditions has been shown in Table 2.

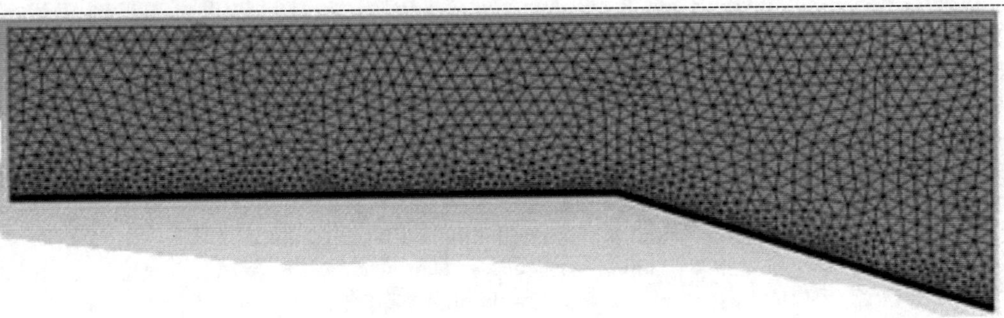

Figure 4. Flow domain for case (1).

Table 2. Boundary conditions and solver setup

Parameters	Case (1)
Model	k-ε (RNG) with enhance wall treatment
Inlet B.C.	Velocity inlet (4m/s)
Outlet B.C.	Pressure outlet
Bottom Wall B.C.	No slip wall
Solution Method	Pressure velocity coupling
Pressure	Second order
Momentum	Second order upwind
Turbulent KE	Second order upwind
Turbulent Dissipation Rate	Second order upwind

3. Results and discussions

Flow separation have been observed by passing a sheet of smoke in the wind tunnel and flow is separated at x=640 mm i.e. 10 mm downstream from the location B3. Flow separation has been shown in Figure 5 and 6 by smoke sheet and smoke wire visualization respectively.

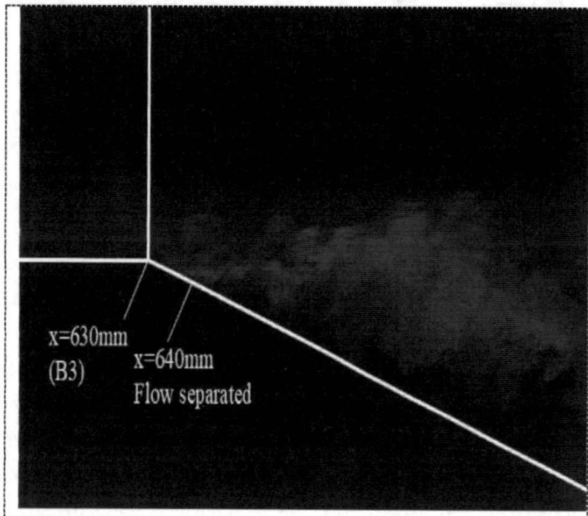

Figure 5. Smoke sheet visualization (case-1).

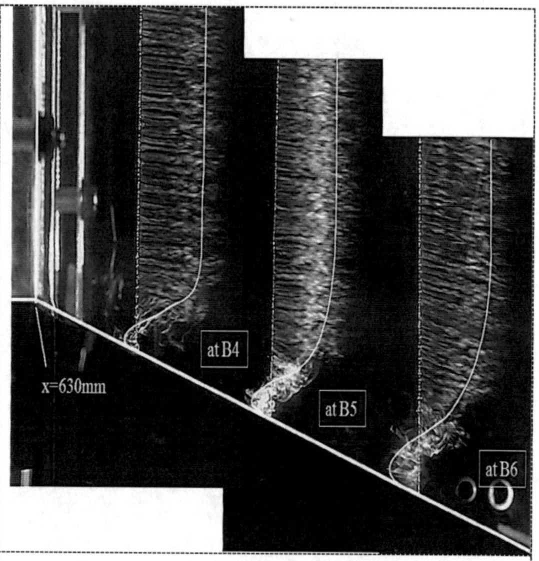

Figure 6. Smoke wire visualization (case-1).

As hot-wire probe only give the magnitude of local velocity, hence the profile obtained from time averaged velocity by the hot-wire probe is overlapped with smoke wire data and simulated results in Figure 7 in which experimental velocity profiles showed good agreement with the numerical results. Momentum addition by both PVG and VG suppress the separation further downstream as shown Figure 8 by smoke sheet visualizations. Velocity profiles obtained by hot-wire probe and smoke wire by using VG and PVG have been overlapped together to ensure the repeatability of the experiment and a good agreement have been found which is shown in Figure 9. As PVG is controllable by switching OFF and ON as per flow situation and it is found that at switching OFF condition of PVG provides the same velocity profile that is obtained without actuator case. That's mean, in switching OFF condition it doesn't affect the normal flow structure. On the other hand the VG is always attached to surface with an angle of incident along flow direction. In case of no controlled case (when it not necessary to control the flow), it appends a drag due to its angle of incident which is undesirable. The velocity profiles of PVG switching OFF and no actuator case are almost identical which proves that no drag penalty has been occurred which is shown in Figure 10 at location B4.

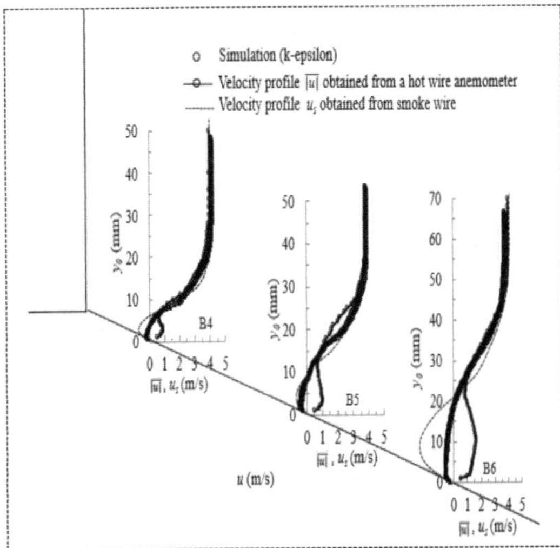

Figure 7. Velocity profiles obtained from smoke wire, hot-wire and numerical results for case-1.

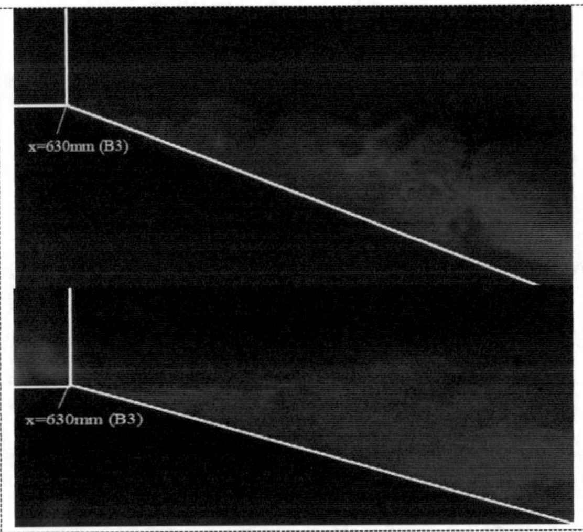

Figure 8. Separation suppression by VG (top) and PVG (bottom).

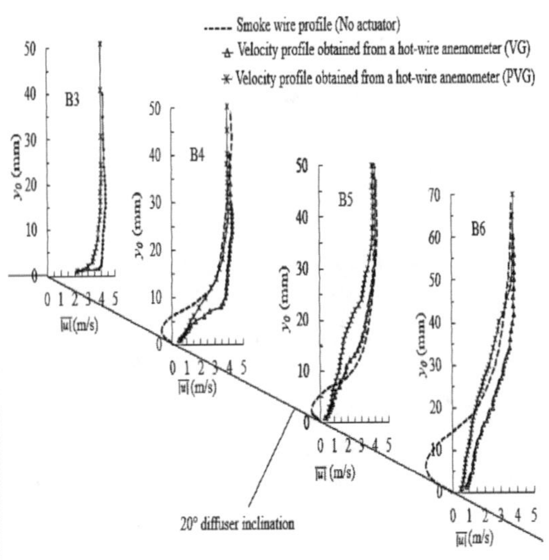

Figure 9. Velocity profiles by VG and PVG.

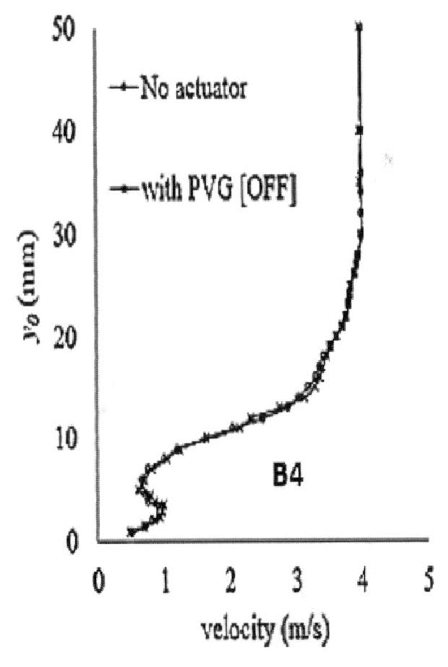

Figure 10. Impact of PVG in switching OFF condition on uncontrolled flow.

4. Conclusion

From the above discussion and pictorial representations, it is revealed that the new concept of flow control by plasma vortex generator is very promising as its effectiveness to suppress the separation in a turbulent boundary layer flow as like as conventional vortex generator. Moreover, it can be controlled as per requirement and in switching OFF condition, PVG doesn't have any impact on the flow, hence

IOP Conf. Series: Materials Science and Engineering **926** (2020) 012005 doi:10.1088/1757-899X/926/1/012005

the drag penalty is minimized that is a common phenomenon in uncontrolled flow situation if VG is used.

References

[1] Taylor H D, *United Aircraft Corporation Report 1947* **4012-3**

[2] Lin J C, 30th AIAA Fluid Dynamics *Conf.* Norfolk VA

[3] Lin J C and Howard F G 1990 *J. spacecraft* **27**

[4] Lin J C, Robinson S K, McGhee R J and Valarezo W 1994 *J. Aircraft* **31**

[5] Melin T, Crippa S, Holly M and Smidy M, 25th *Intl. congress of the aeronautical sciences*

[6] Culter D, Bradshaw P993 *Exp. in fluids* 393-401

[7] Satta F, Simoni D, Ubaldi M, Zunino P 2007 *J. Power and Energy* **221** 815

[8] Lengani D, Simoni D, Ubaldi M, Zunino P and Bertini F 2011 *Exp. Thermal and Fluid Science* **35** 1505

[9] Velte C M, Hansen M and Jonck K 2007 *J. Phy.* **75**

[10] Duriez T, Aider J and Wesfreid J E 2006 *ASME Joint U.S. - European Fluids Engineering Summer Meeting*

[11] Lin J C 2002 *Prog. Aerosp. Sci.* 389

[12] Mukut A N M M I, Mizunuma H, Obara H 2013 *J. Fluid Science and Technology* **8**

[13] Schlichting H 1979 *Boundary Layer Theory* Chapter XXI, 7th ed.

IOP Conf. Series: Materials Science and Engineering **926** (2020) 012006 doi:10.1088/1757-899X/926/1/012006

Application of Transfer Matrix Method in Calculating Flexible Deployment Mechanisms Modes

Min Wang[1,2], Qinda Liu[1,b], Tingting Zhao[1,c] and Jiaxing Liu[1,d]

[1] Institute of Intelligent Monitoring and Diagnosis, College of Mechanical Engineering and Applied Electronics Technology, Beijing University of Technology, Beijing 100124, China.
[2] Beijing Municipal Key Laboratory of EDM Technology, Beijing 100191, China

[a] wangm@bjut.edu.cn, [b] liuqinda@hotmail.com, [c] ztt9607@163.com, [d]1677486548@qq.com

Abstract: Flexible deployment mechanisms are widely used in the aerospace field. During the satellite antenna's attitude adjustment and deployment process, vibration will greatly affect the accuracy of the satellite antenna profile after deployment. The finite element method (FEM) has the disadvantages of high calculation order, slow calculation speed and difficult optimization in the process of establishing the dynamic model of the flexible deployment mechanism. This paper uses the principle of the transfer matrix method to propose a calculation method for the natural frequency and vibration mode of a flexible deployment mechanism, which is used to calculate the vibration characteristics of a radial rib deployment antenna. Finally, its accuracy is verified by the finite element method (FEM).

1. Introduction

With the development of satellite technology, the design requirements for large flexible self-deployed antennas are getting higher and higher. However, due to the large and complex structure of large-size flexible self-deployed antennas, they belong to a multi-body rigid-flexible coupling mechanism [1]. Coupling vibration is very prone to occur during posture and deployment, affecting the accuracy of the antenna's profile and its ability to communicate with the ground. By establishing its accurate damping dynamic model to analyze the natural frequency and mode shape of the deployed antenna, effective vibration control methods can be targeted to ensure the accuracy of the profile of the deployed antenna and increase the stability after deployment.

At present, most scholars usually use the finite element model method to establish the dynamic model of the flexible self-deploying truss structure. Masayoshi M, et al. Conducted a theoretical analysis of the dynamics and structural dynamics of petal-shaped, umbrella-shaped, and ring-column flexible deployment structures [2]. Zhang Chun et al. Used the finite element method to establish a finite element model of the antenna reflector structure in the working state [3]. Li Dongying et al. Solved the dynamic characteristics of a radial stiffener-film umbrella antenna structure by finite element method [4]. However, due to the complex structure of the large flexible self-deployment device and the large number of state variables, the calculation order of the finite element model is very high [5], the calculation speed is slow, and the finite element method has certain restrictions on the optimization after the dynamic model is established. This paper adopts the principle of transfer matrix method to establish the dynamic model of flexible deployment mechanism. The advantage of the transfer matrix method modeling is that the dimension of the transfer matrix does not increase with the increase of the degree of freedom of the system. When saving the computer, it is more suitable for dynamic modeling of large truss structures such as self-expanding devices, and it can also be more conveniently optimized. The antenna structure parameters and damping structure design parameters improve the design efficiency and reduce the calculation amount. This paper combines the equivalent structure of a radial rib flexible deployment antenna, establishes a transfer matrix to solve the vibration characteristics, and finally verifies the practicability and effectiveness of the method by the finite element method.

2. Establishment and solution of transfer matrix

The overall flexible expansion antenna transfer matrix can be established by establishing transfer matrices for different antenna components and combining them based on continuous conditions. The flexible deployment antenna can be equivalently formed by connecting a plurality of beams of different materials and different cross-sectional shapes, and some components at the node can be equivalent to adding a transfer function for concentrated mass.

This article introduces the calculation method by taking a radial rib-spread antenna as an example. As shown in Figure 1, the radial rib deployment antenna can be equivalent to a cantilever beam containing multiple complex components. The transfer matrix of different components can be discussed in sections, and all components are transferred via the point transfer matrix at the connection. The matrices are merged into a unified transfer matrix, and the overall transfer matrix is solved by boundary conditions.

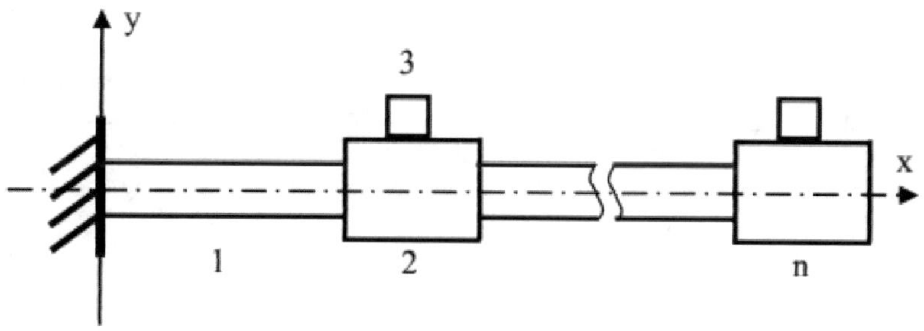

Figure 1. Simplified model of radial rib flexible deployment antenna

2.1. Transfer matrix of beam

Let $S = [y, \theta, M, Q]^T$be the state vector of the section, where y is the deflection, θ is the angle of rotation, M is the bending moment, and Q is the shear force. Based on the Euler beam theory and the uniformly distributed mass model, the free vibration differential equation of the beam is:

$$EI\frac{\partial^4 y}{\partial x^4} + \overline{m}\frac{\partial^2 y}{\partial t^2} = 0 \tag{1}$$

Where, EI is the bending stiffness, and $\overline{m}$ is the linear density.

The general solution of equation (1) can be obtained by using the separated variable method:

$$Y(x) = A_1 cosh\lambda x + A_2 sinh\lambda x + A_3 cos\lambda x + A_4 \sin\lambda x \tag{2}$$

Where, $\lambda = (\frac{\overline{m}w^2}{EI})^{-4}$, w is the natural circular frequency of the lateral vibration.

From the relationship between deflection, rotation angle, bending moment and shear force in material mechanics:

$$\theta_x = A_1 \lambda sinh\lambda x + A_2 \lambda cosh\lambda x - A_3 \lambda sin\lambda x + A_4 \lambda \cos\lambda x \tag{3}$$

$$M_z = A_1 EI\lambda^2 cosh\lambda x + A_2 EI\lambda^2 sinh\lambda x - A_3 EI\lambda^2 cos\lambda x - A_4 EI\lambda^2 \sin\lambda x \tag{4}$$

$$Q_y = A_1 EI\lambda^3 sinh\lambda x + A_2 EI\lambda^3 cosh\lambda x + A_3 EI\lambda^3 sin\lambda x - A_4 EI\lambda^3 \cos\lambda x \tag{5}$$

Where, A_1, A_2, A_3, A_4 is the undetermined coefficient determined by the boundary conditions at the left and right ends of the beam.

The above equation can be written by the Krylov function as the following equation:

$$Y(x) = L_1 S(x) + L_2 T(x) + L_3 U(x) + L_4 V(x) \tag{6}$$

$$\theta_x = \lambda(L_1 V(x) + L_2 S(x) + L_3 T(x) + L_4 U(x)) \tag{7}$$

$$M_z = -EI\lambda^2(L_1 U(x) + L_2 V(x) + L_3 S(x) + L_4 T(x)) \tag{8}$$

$$Q_y = -EI\lambda^3(L_1 T(x) + L_2 U(x) + L_3 V(x) + L_4 S(x)) \tag{9}$$

Where, the Krylov function is:

$$\begin{cases} S(x) = \frac{1}{2}[\cosh(x) + \cos(x)] \\ T(x) = \frac{1}{2}[\sinh(x) + \sin(x)] \\ U(x) = \frac{1}{2}[\cosh(x) - \cos(x)] \\ V(x) = \frac{1}{2}[\sinh(x) - \sin(x)] \end{cases} \tag{10}$$

L_1, L_2, L_3, L_4, is the undetermined coefficient determined by the boundary conditions at the left and right ends of the beam.

When $x = 0$,

$$Y(x) = Y_l,\ \theta_x = \theta_l,\ M_z = M_l,\ Q_y = Q_l \tag{11}$$

L_1, L_2, L_3, L_4 can be obtained:

$$L_1 = Y_1,\ L_2 = \frac{\theta_1}{\lambda},\ L_3 = -\frac{M_1}{EI_0\lambda^2},\ L_4 = -\frac{Q_1}{EI_0\lambda^3} \tag{12}$$

The transfer function of the left and right ends of the beam can be obtained as:

$$\begin{bmatrix} Y_r \\ \theta_r \\ M_r \\ Q_r \end{bmatrix} = \begin{bmatrix} S(\lambda l) & \frac{T(\lambda l)}{\lambda} & \frac{U(\lambda l)}{EI\lambda^2} & \frac{V(\lambda l)}{EI\lambda^3} \\ \lambda V(\lambda l) & S(\lambda l) & \frac{T(\lambda l)}{EI\lambda} & \frac{U(\lambda l)}{EI\lambda^2} \\ EI\lambda^2 U(\lambda l) & EI\lambda V(\lambda l) & S(\lambda l) & \frac{T(\lambda l)}{\lambda} \\ EI\lambda^3 T(\lambda l) & EI\lambda^2 U(\lambda l) & \lambda V(\lambda l) & S(\lambda l) \end{bmatrix} \begin{bmatrix} Y_l \\ \theta_l \\ M_l \\ Q_l \end{bmatrix} \tag{13}$$

The transfer matrix of the beam is:

$$Z_1 = \begin{bmatrix} S(\lambda l) & \frac{T(\lambda l)}{\lambda} & \frac{U(\lambda l)}{EI\lambda^2} & \frac{V(\lambda l)}{EI\lambda^3} \\ \lambda V(\lambda l) & S(\lambda l) & \frac{T(\lambda l)}{EI\lambda} & \frac{U(\lambda l)}{EI\lambda^2} \\ EI\lambda^2 U(\lambda l) & EI\lambda V(\lambda l) & S(\lambda l) & \frac{T(\lambda l)}{\lambda} \\ EI\lambda^3 T(\lambda l) & EI\lambda^2 U(\lambda l) & \lambda V(\lambda l) & S(\lambda l) \end{bmatrix} \tag{14}$$

2.2. Transfer Matrix of Concentrated Mass

For the transfer matrix of concentrated mass m, the relationship between the deflection, rotation angle, bending moment and shear-force of the left side and the right side at the mass point acting section is:

$$Y_r(x) = Y_l(x), \theta_r(x) = \theta_l(x), M_r(x) = M_l(x), Q_r(x) = mw^2 Y_l(x) + Q_l(x) \tag{15}$$

Written as a matrix:

IOP Conf. Series: Materials Science and Engineering **926** (2020) 012006 doi:10.1088/1757-899X/926/1/012006

$$\begin{bmatrix} Y_l \\ \theta_r \\ M_r \\ Q_r \end{bmatrix} = \begin{bmatrix} 1 & 0 & 0 & 0 \\ 0 & 1 & 0 & 0 \\ 0 & 0 & 1 & 0 \\ mw^2 & 0 & 0 & 1 \end{bmatrix} \begin{bmatrix} Y_l \\ \theta_l \\ M_l \\ Q_l \end{bmatrix} \tag{16}$$

Therefore, the transfer matrix of concentrated mass is:

$$Z_2 = \begin{bmatrix} 1 & 0 & 0 & 0 \\ 0 & 1 & 0 & 0 \\ 0 & 0 & 1 & 0 \\ mw^2 & 0 & 0 & 1 \end{bmatrix} \tag{17}$$

2.3. Continuous Condition

Because the joint of the structural rod and the connector obeys the continuous condition, that is, the state vectors of the left and right ends of the cross section of the connected parts are equal. So you can get the matrix form:

$$\begin{bmatrix} Y \\ \theta \\ M \\ Q \end{bmatrix}_i^l = \begin{bmatrix} Y \\ \theta \\ M \\ Q \end{bmatrix}_i^r \tag{18}$$

2.4. Transfer Matrix of Overall Structure

After the state transfer matrices of each segment of beam and concentrated mass are determined, the total transfer matrix relationship of the state vector transferred from the initial point state to the end position state vector is:

$$S_n = Z_n Z_{n-1} \cdots Z_1 S_0 = Z \cdot S_0 \tag{19}$$

Where,

$$Z = \begin{bmatrix} H_{11} & H_{12} & H_{13} & H_{14} \\ H_{21} & H_{22} & H_{23} & H_{24} \\ H_{31} & H_{32} & H_{33} & H_{34} \\ H_{41} & H_{42} & H_{43} & H_{44} \end{bmatrix} \tag{20}$$

2.5. Solution of Transfer Matrix

The boundary conditions of a radial rib-expanded antenna can be equivalent to a fixed end at one end and a free end, the displacement and rotation angle at the fixed end are zero, and the bending moment and shear force at the free end are zero. The initial point state vector S_0^L and the end point state vector S_N^R can be obtained as:

$$S_0^L = [0,0,\mathrm{M},\mathrm{Q}]^T, S_N^R = [\mathrm{y},\theta,0,0]^T, \tag{21}$$

The boundary conditions at both ends of the flexible expanded antenna are brought into equation (18) to obtain the transfer function. To make the transfer function solvable, let $|\,Z\,| = 0$, and the natural frequency w can be solved. Bringing the natural frequency back into the transfer function can solve the initial state vector. Through the initial point state vector and the transfer matrix of each beam, the state vectors of all points of the satellite antenna can be solved, and finally the modal shape of the flexible expanded antenna can be obtained.

3. Modal example of radical rib flexible deployment antenna

3.1. Calculation of Modal Features

In this paper, a radial rib antenna with a diameter of 9m is used as an example. It consists of 5 structural members and 5 connecting members. The structural parts and connecting parts are made of M60j carbon

IOP Conf. Series: Materials Science and Engineering **926** (2020) 012006 doi:10.1088/1757-899X/926/1/012006

fiber material, and their dimensions, cross-sectional areas and other physical parameters are shown in Table 1.

Table 1. Physical parameters of radial rib antenna structure

Physical parameters	Structural Members	Connector Members
Length/m	0.5	0.07
Diameter /mm	Φ25*1	Φ28*1.5
Density (kg/m^3)	1910	1910
Elastic Modulus /Pa	5.88e11	5.88e11

3.2. Calculation Results

By solving the physical parameters of the flexible deployment antenna using the method described above, the first third-order natural frequency and mode shape of the antenna deployment rib can be calculated. The results are shown in Table 2 and Figure 2.

Table 2. Calculation result of natural frequency

Frequency	Radial Rib Antenna Structure /*Hz*
First-order Natural Frequency	2.6
Second-order Natural Frequency	16.9
Third-order Natural Frequency	48.1

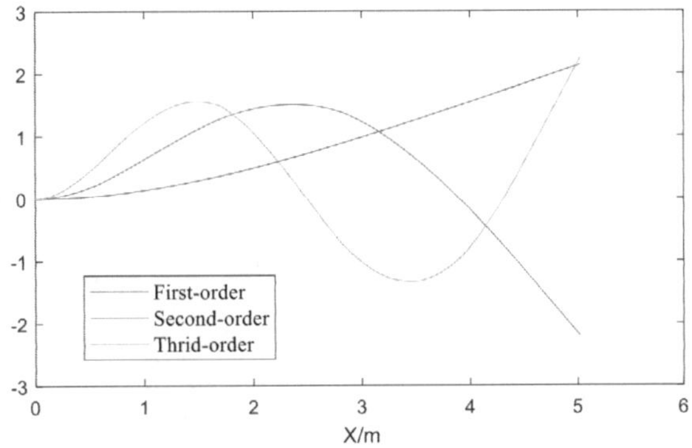

Figure 2. First third-order mode of radial rib antenna structure

4. Finite element certification

Based on the external dimensions of the radial rib flexible deployment antenna geometric model, a three-dimensional solid model is established, and the model is analyzed by frequency using the finite element method. The modeling results are shown in the Figure 3, and the analysis results of the finite element method and the calculation results of the method in this paper are shown in Table 3.

Figure 3. Finite element model of radial rib antenna structure

Table 3. Comparison of calculation results between the finite element method and the method of this paper

Frequency	Method of This Article/*Hz*	Finite Element Method /*Hz*	Error /%
First-order Frequency	2.6	2.5	3.84
Second-order Frequency	16.9	15.6	7.69
Third-order Frequency	48.1	43.6	9.35

As can be seen from Table 3, the method in this paper can accurately calculate the modal parameters of the radial rib flexible deployment antenna. Therefore, the mathematical model established in this paper has good practicability and effectiveness for calculating the vibration characteristics of flexible deployment mechanisms such as satellite antennas.

5. Conclusion

Based on the Euler-Bernoulli quantity theory, the transfer matrix method is used to derive the transfer matrix of the cantilever part of the radial rib flexible deployment mechanism. The state vectors of different positions of the deployment mechanism are solved, and the transfer relationship between different positions of the flexible deployment mechanism is established. The calculation examples show that the results of the transfer matrix in this paper are accurate and have faster calculation speed than the finite element method, which is more convenient to establish the dynamic model of the flexible self-deploying mechanism.

Acknowledgments

This work was supported by National Defense Equipment Pre-Research Foundation of China (Grant No.61402100103).

Reference

[1] Liu Rongqiang, Tian Dake, Deng Zongquan. Research Actuality and Prospect of Structure for Space Deployable Antenna [J]. Journal of Machine Design, 2010(09):4-13.

[2] Misawa M, Ogawa A. Analytical and Experimental Frequency Verification of Deployed Satellite Antennas[C]// Aiaa/ asme/ asce/ ahs/ asc Structures, Structural Dynamics, & Materials Conference. 2003.

[3] Zhang Chun, Wang Sanmin, Yuan Ru. Structural Design and Its Dynamics Analysis of Space Radial Deployable Antenna Reflector [J]. Mechanical Science and Technology, 2007(04):96-99.

[4] Li Dongying, Zhang Hua, Liu Hanwu, Dynamics Modeling and Numerical Simulation of Radial Stiffening Rib-membrane Umbrella Antenna Structure [J]. Journal of Dynamics and Control, 2017(2).

[5] Meguro A, Harada S, Watanabe M. Key technologies for high-accuracy large mesh antenna reflectors[J]. ACTA ASTRONAUTICA, 2003.

IOP Conf. Series: Materials Science and Engineering **926** (2020) 012007 doi:10.1088/1757-899X/926/1/012007

Research on Equivalent Thickness of Oil and Gas Spring Damping Valve

Yijie Chen[1,*], Xiaoling Han[1], Lang Xiang[2], Ning Zhao[1] and Jianjian Dai[1]

[1] China North Vehicle Research Institute, Beijing, China
[2] FAW Jiefang Automotive Co.LTD, Changchun, China

[*]Email: chenyijie1206@163.com

Abstract. Analytical algorithm for deflection analysis of superimposed throttle plate of oil and gas spring damping valve was studied, the theory of equivalent thickness of superposed valve plates was put forward, it can accurately calculate the equivalent thickness of a single valve plate corresponding to the superimposed valve plate groups, which greatly facilitates engineering calculations. After that, the boundary flow theory was used to derive the flow expression of the gap under the turbulent state. The relationship between the output force and displacement of the oil and gas spring was simulated and analyzed, compared with experimental data, the correctness of the mathematical model was verified.

1. Introduction

The damping device often uses superimposed valve plates of different thicknesses to generate damping force for oil throttling. This assembly method has the following advantages:1.Throttle valve plate is processed by elastic steel plates of different standard thicknesses, which is conducive to product standardization and serialization, and is easy to achieve mass production to reduce costs;2. It can reduce the concentration of stress and increase the service life of the damper valve by reducing the thickness of a single throttle valve plate; 3. By adjusting the thickness and number of superimposed valve plates, the purpose of changing the amount of deformation of the valve plate and adjusting the damping force can be achieved.[1-2]

In view of this situation, a theoretical study on the equivalent thickness of the valve plate is carried out, so that the same deformation amount is produced when different valve combinations are subjected to the same load, to ensure the stability of the damper valve opening, and to improve the design accuracy of the vibration reduction device.

2. Structure diagram

Figure 1 is simplified structural diagram of oil and gas spring. The throttle plate is installed at the end of the piston. During the reciprocating movement of the piston, pressure difference occurs between the upper and lower chambers to deform the valve plate, and oil flows through the annular gap between the deformed valve plate and the piston body. The nitrogen and oil are separated by floating piston, and the length of the gas chamber changes with the change of oil and gas spring stroke.

Figure 2 is assembly schematic diagram of throttle plate. In the figure 2, Δp is the pressure on the thin plate, $\delta_0$is the initial gap width, $\Delta\delta$ is the dynamic gap width after the throttle valve is deformed, r_a is the inner radius of the valve plate, r_b is the outer radius before the valve plate is deformed, and r is the outer

IOP Conf. Series: Materials Science and Engineering **926** (2020) 012007 doi:10.1088/1757-899X/926/1/012007

radius after the valve plate is deformed.

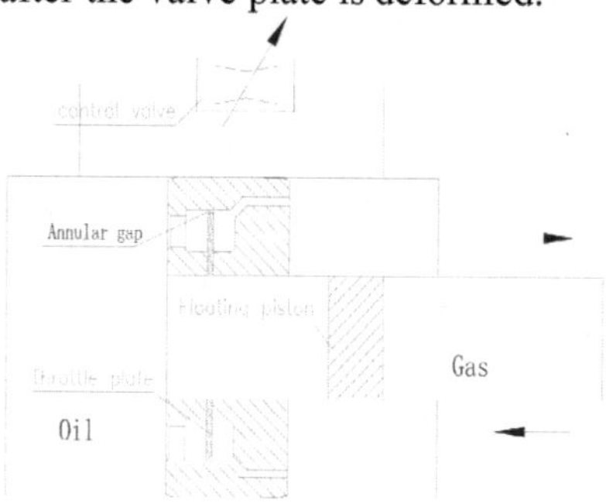

Figure 1. Oil and gas spring diagram

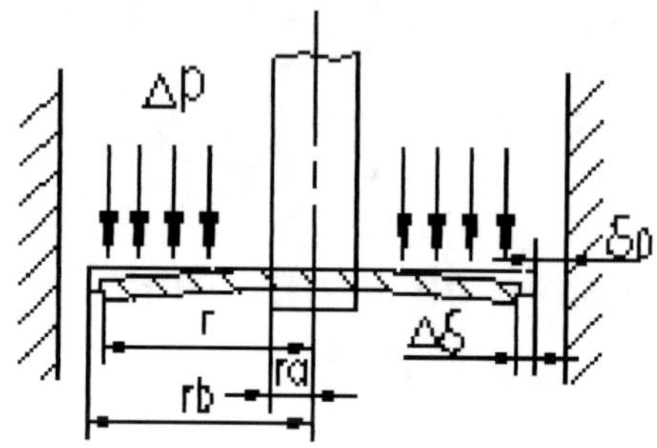

Figure 2. Schematic diagram of throttle valve deformation

3. Equivalent thickness of superposed throttle plate

The inner ring of the annular thin plate is fully constrained, and the free uniform load deformation of the outer ring is: [3]

$$f = C_6 \frac{\Delta p R^4}{E h^3} \tag{1}$$

Where f is the deformation of the outer edge of the throttle plate, R is the outer radius of the valve plate, C_6 is the deformation coefficient, E is the elastic modulus of the valve plate, and h is the thickness of the valve plate.

According to the parallel and split principle of superimposed valve plates, the equivalent thickness relationship of superimposed valve plates under different thicknesses can be derived:

$$h_e = \left(h_1^3 + h_2^3 + h_3^3 + \cdots + h_n^3\right)^{\frac{1}{3}} \tag{2}$$

Where h_i is the thickness of single valve plate, i=1,2, …, n.

And the equivalent thickness relationship of valve plate superposition under the same thickness:

$$h = h_p (n_p)^{\frac{1}{3}} \tag{3}$$

Where h_p is the thickness of single superimposed valve plate, and n_p is the number of superimposed valve plates.

If there are multiple groups of different thickness of valve plates, the equivalent thickness is:

$$h_e = \left(n_1 h_1^3 + n_2 h_2^3 + n_3 h_3^3 + \cdots + n_n h_n^3\right)^{\frac{1}{3}} \tag{4}$$

Where $n_1 \ldots n_n$ is the number of valve plates in different groups.

For example, from the above formulas, it can be estimated that under the same pressure, when 2 pieces of 4×10^{-4} m thick pieces or 3 pieces 3.5×10^{-4} m thick pieces are stacked together, the deformation is basically equivalent to the single piece of 5×10^{-4} m thick.

According to the corresponding relationship between the deformation size of the throttle plate, the dynamic gap width is derived:

$$\Delta\delta = (r_b - r_a) - \left[(r_b - r_a)^2 - f^2\right]^{\frac{1}{2}} \tag{5}$$

Then the total width of the gap is:

$$\delta = \delta_0 + \Delta\delta \tag{6}$$

Derive the outer radius of the valve plate after deformation:

IOP Conf. Series: Materials Science and Engineering **926** (2020) 012007 doi:10.1088/1757-899X/926/1/012007

$$r = r_a + \left[(r_b - r_a)^2 - f^2 \right]^{\frac{1}{2}} \tag{7}$$

In addition, the stress expression of the valve plate is: [4]

$$\sigma_{\max} = C_{yi} \frac{F}{2h^2} \tag{8}$$

Where the subscript is i=1,2,3, F is the total load under different stress modes.

It can be seen from the above derivation that the elastic modulus E is equal when the radial structure size and material properties of the superimposed valve plate are the same.

Sorted out:

$$(F : F_1 : F_2 : \cdots : F_n) = (h_e^3 : h_1^3 : h_2^3 : \cdots : h_n^3) \tag{9}$$

$$(\sigma_{\max e} : \sigma_{\max 1} : \sigma_{\max 2} : \cdots : \sigma_{\max n}) = (h_e : h_1 : h_2 : \cdots : h_n) \tag{10}$$

Where $\sigma_{\max e}$ is the maximum stress value of the valve plate equivalent thickness, $\sigma_{\max 1} \ldots \sigma_{\max n}$ are the maximum stress value of different valve plates.

Let the thickness relationship of the valve plates are:

$$\begin{gathered} h_1 > h_2 > \cdots > h_n \\ h_e > h_1 > h_2 > \cdots > h_n \\ F > F_1 > F_2 > \cdots > F_n \end{gathered} \tag{11}$$

Then

$$\sigma_{\max e} > \sigma_{\max 1} > \sigma_{\max 2} > \cdots > \sigma_{\max n} \tag{12}$$

It is not difficult to see that when the throttle plates of different thicknesses are stacked together, the thickest one is subjected to the largest partial pressure, and the stress value generated is also the largest compared to other valve plates. In other words, when valve plates of different thicknesses are superimposed, it is only necessary to check whether the thickest one of the plates exceeds the strength limit of the material, if its normal use no problem, the strength of the remaining valve plates can also be used.

4. Analytical calculation of gap flow

At present, the more commonly used is the gap flow analysis formula based on the laminar flow state: [5]

$$Q_{fc} = \frac{\Delta p B \delta^3}{12 \mu l} \tag{13}$$

Where μ is the dynamic viscosity of the fluid, l is the length of the gap, and B is the width of the gap.

In order to meet the design requirements of damping valves, the fluid flow boundary layer theory is used to derive the flow rate analysis formula of the gap in the turbulent state as follows.

$$Q_f = 2\pi r \left[\frac{\tau_w}{\mu} y_0^2 + A(y_1 - y_0) + 5\left(\frac{\tau_w}{\rho}\right)^{\frac{1}{2}} y_1 \ln y_1 - 10\left(\frac{\tau_w}{\rho}\right)^{\frac{1}{2}} y_0 \ln y_0 + 5\left(\frac{\tau_w}{\rho}\right)^{\frac{1}{2}} \frac{\delta}{2} \ln\frac{\delta}{2} + B\left(\frac{\delta}{2} - y_1\right) \right] \tag{14}$$

$$A = 10\left(\frac{\tau_w}{\rho}\right)^{\frac{1}{2}} \ln\left(\frac{\tau_w}{\rho}\right)^{\frac{1}{2}} - 16.1\left(\frac{\tau_w}{\rho}\right)^{\frac{1}{2}} - 10\left(\frac{\tau_w}{\rho}\right)^{\frac{1}{2}} \ln \nu \tag{15}$$

$$B = 6\left(\frac{\tau_w}{\rho}\right)^{\frac{1}{2}} - 5\left(\frac{\tau_w}{\rho}\right)^{\frac{1}{2}} \ln \nu + 5\left(\frac{\tau_w}{\rho}\right)^{\frac{1}{2}} \ln\left(\frac{\tau_w}{\rho}\right)^{\frac{1}{2}} \tag{16}$$

Where Q_f is the gap flow, τ_w is the wall shear stress, δ is the width of the single-sided gap, ρ is the fluid density, and the thickness of the viscous bottom layer is:

$$y_0 = 5\nu\left(\frac{\tau_w}{\rho}\right)^{-\frac{1}{2}} \tag{17}$$

The thickness of the transition layer is:

$$y_1 = 30\nu\left(\frac{\tau_w}{\rho}\right)^{\frac{1}{2}} \tag{18}$$

Where ν is the kinematic viscosity of the fluid.

Based on the principle that the pressure difference between the two ends of the gap is balanced with the frictional force of the fluid on the wall, the formula for the wall shear stress of the gap is derived:

$$\tau_w = \frac{\Delta p \delta}{2l} \tag{19}$$

Where Δp is the pressure difference across the gap.

Analytical formula for turbulent flow of annular gap is derived:

$$Q_f = 2\pi r\left[-127.8\nu + \left(2.5\ln\frac{\delta}{2} + 3 - 2.5\ln\nu + 2.5\ln\left(\frac{\Delta p\delta}{2l\rho}\right)^{\frac{1}{2}}\right)\delta\left(\frac{\Delta p\delta}{2l\rho}\right)^{\frac{1}{2}}\right] \tag{20}$$

$$A = 10\left(\frac{\tau_w}{\rho}\right)^{\frac{1}{2}} \ln\left(\frac{\tau_w}{\rho}\right)^{\frac{1}{2}} - 16.1\left(\frac{\tau_w}{\rho}\right)^{\frac{1}{2}} - 10\left(\frac{\tau_w}{\rho}\right)^{\frac{1}{2}} \ln\nu \tag{21}$$

$$B = 6\left(\frac{\tau_w}{\rho}\right)^{\frac{1}{2}} - 5\left(\frac{\tau_w}{\rho}\right)^{\frac{1}{2}} \ln \nu + 5\left(\frac{\tau_w}{\rho}\right)^{\frac{1}{2}} \ln\left(\frac{\tau_w}{\rho}\right)^{\frac{1}{2}} \tag{22}$$

Where Q_f is the gap flow, τ_w is the wall shear stress, δ is the width of the single-sided gap, ρ is the fluid density, and the thickness of the viscous bottom layer is:

$$y_0 = 5\nu\left(\frac{\tau_w}{\rho}\right)^{-\frac{1}{2}} \tag{23}$$

The thickness of the transition layer is

$$y_1 = 30\nu\left(\frac{\tau_w}{\rho}\right)^{-\frac{1}{2}} \tag{24}$$

Where ν is the kinematic viscosity of the fluid.

Based on the principle that the gap pressure difference is balanced with the frictional force of the fluid on the wall, the formula for the shear stress of the gap is derived:

$$\tau_w = \frac{\Delta p \delta}{2l} \tag{25}$$

Where Δp is the pressure difference across the gap.

Get the annular gap flow expression through the above:

$$Q_f = 2\pi r\left[-127.8v + \left(2.5\ln\frac{\delta}{2} + 3 - 2.5\ln v + 2.5\ln\left(\frac{\Delta p\delta}{2l\rho}\right)^{\frac{1}{2}}\right)\delta\left(\frac{\Delta p\delta}{2l\rho}\right)^{\frac{1}{2}}\right] \tag{26}$$

5. Mathematical model of oil and gas spring

The output force of the oil and gas spring is:[6]

$$F_z = F_d + F_t \tag{27}$$

Where F_z is the total output force of the oil and gas spring, F_d is the damping force, and F_t is the elastic force.

Simultaneously, the above equations can be used to solve the pressure difference across the gap. Also consider the local resistance loss of the oil passage:

$$\Delta p_j = \xi\frac{\rho\left(A_h V\right)^2}{2A_j^2} \tag{28}$$

Where ξ is the local loss coefficient, A_h is the effective working area of the piston, A_j is the cross-sectional area of the oil passage, and V is the working speed of the oil and gas spring.

The resistance value of the oil and gas spring is:

$$F_d = \left(\Delta p + \Delta p_j\right)A_h \tag{29}$$

The actual gas pressure expression is

$$p = \frac{RgTm_q}{V_q - m_q b} - \frac{am_q^2}{V_q^2} \tag{30}$$

Where p is the absolute gas pressure, R_g is the gas constant, T is the thermodynamic temperature, m_q is the gas mass, a, b is the gas constant, and the volume of the gas is:

$$V_q = L_j - \frac{A_{gw}s}{A_{gn}} \tag{31}$$

Where L_j is the length of the air chamber at the static equilibrium position, A_{gw} is the external cross-sectional area of the piston rod, A_{gn} is the internal cross-sectional area of the piston rod, and s is the displacement of the oil and gas spring.

Then the pressure of the oil and gas spring in the static equilibrium position is:

$$p_j = \frac{2m_d g i_j}{\pi D_{gw}^2} \tag{32}$$

Where m_d is the sprung mass of single wheel, i_j is the suspension lever ratio at the static equilibrium position, and D_{gw} is the outer diameter of the piston rod.

The elastic force of the oil and gas spring is:

$$F_t = pA_{gw} \tag{33}$$

6. Simulation and experimental research

Input sinusoidal excitation f_z=1 Hz, A_z=0.04 m, for external characteristic test of oil and gas spring, the test and simulation curves are shown in Figure 3, the force value pairs of each key point are shown in Table 1.

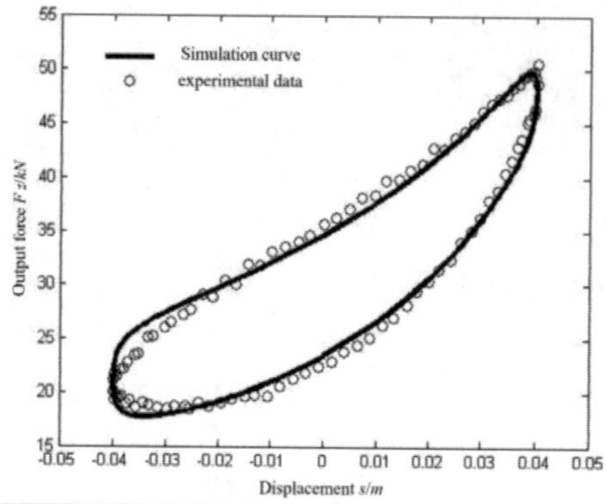

Figure 3. Comparison of test and simulation curves

Table 1. Comparison of force values at key points

Name	F_{zmax}	F_{zmin}	F_{zy0}	F_{zf0}
Test data(KN)	50.63	18.42	35.65	22.40
Simulation data (KN)	49.35	17.45	35.58	22.55
Data deviation(%)	2.5	5.3	0.2	0.7

In the table, F_{zmax} is the maximum output force of the oil and gas spring, F_{zmin} is the minimum output force, F_{zf0} is the return stroke output force at zero displacement, and F_{zy0} is the compression stroke output force at zero displacement.

It can be seen that the simulation curve basically agrees with the experimental data, which proves the correctness of the mathematical model and can provide a reference for the design of the oil and gas spring and its damping valve.

7. Conclusion

The mathematical model of the oil and gas spring was simulated by software MATLAB, compared with the external characteristic test data of the developed product, verified the correctness of the equivalent thickness rule for superimposed valve plates, provides convenient conditions for engineering design and theoretical analysis.

References

[1] Steven R, 2010 *SAE Paper.* **1** 09-48
[2] Nieto A J, Morales A L, Trapero J R, et al 2011 *J. Sound.Vibr.* **330** 1891-903
[3] Gimbun J 2006 *Powd. Technol.* **162** 126-32
[4] Daniel A, Mantarasa, Pablo L. 2012 *Inte. J. Vehi. Mech. Mobi* **50** 1563-84
[5] Li Z L, Li X Q, Chen X D.2014 *P. I. Mech. Eng. B-J. Eng.* **228** 1621-29
[6] Zuo L, Zhang P S 2013 *J. Vibr. Acou* **135** 48-65

IOP Conf. Series: Materials Science and Engineering **926** (2020) 012008 doi:10.1088/1757-899X/926/1/012008

Study on Calibration Technology and Uncertainty Evaluation of Asphalt Vacuum Decompression Capillary Viscometer

Liu Chang[1], Zhou Yishu[1], Ren Lishuo[1], Zhang Jinning[1,*] and Zhu Fenghua[1]

[1]Research Institute of Highway Ministry of Transport No. 8 Xitucheng Road, Haidian District, Beijing

[*]Corresponding author's E-mail address: jinning.zhang@foxmail.com

Abstract. 60 ℃ dynamic viscosity is one of the main indexes to evaluate the quality of asphalt products, and its influence on the quality of asphalt pavement is very important. 60 ℃ dynamic viscosity is usually measured by the asphalt vacuum decompression capillary viscometer (hereinafter referred to as the viscometer). The quality of the viscometer directly affects the test results of the asphalt dynamic viscosity. Therefore, it is urgent to study the detection method of viscometer to ensure the quality of the viscometer and the accuracy of the test results. In this paper, the calibration test method of viscometer is established, and the uncertainty evaluation is carried out, which provides a technical basis for ensuring the accuracy of viscometer.

1. Introduction

Asphalt pavement plays an extremely important role in our country's highway, both in use and quantity. 60 ℃ dynamic viscosity [1] is one of the main indexes of asphalt properties. The United States and Australia have used this index as the classification standard of road petroleum asphalt. According to JTG F40-2004, on the basis of properly improving the softening point index of A-grade asphalt, 60 ℃ dynamic viscosity has been added as the evaluation index of asphalt high temperature performance and has clear technical requirements and qualification standards [2]. After more than 10 years of use, it is shown that 60 ℃ dynamic viscosity is very important for the quality control of asphalt products [3], and has become one of the most widely used technical indexes in the evaluation of asphalt.

60 ℃ dynamic viscosity is measured by viscometer [4]. The quality of viscometer directly affects the test results of asphalt dynamic viscosity. At present, there are many problems with viscometer, such as many manufacturers, uneven quality, and the accuracy of viscometer calibration coefficient cannot be verified. Therefore, it is urgent to study the detection method of viscometer to ensure the quality of the viscometer and the accuracy of the test results, and provide technical support for the quality control of asphalt and asphalt pavement. In this paper, the calibration test method of viscometer is established, and the uncertainty evaluation [5-7] is carried out, which provides a technical basis for ensuring the accuracy of viscometer.

2. Calibration test method of viscometer

2.1. Introduction of viscometer

Viscometer [8] is used to determine dynamic viscosity of viscous petroleum asphalt, which is composed of vacuum decompression capillary viscometer and vacuum decompression system. Vacuum decompression system is composed of constant temperature and temperature measuring device, vacuum decompression device, timing device and controller. The structure of viscometer is shown in Figure 1.

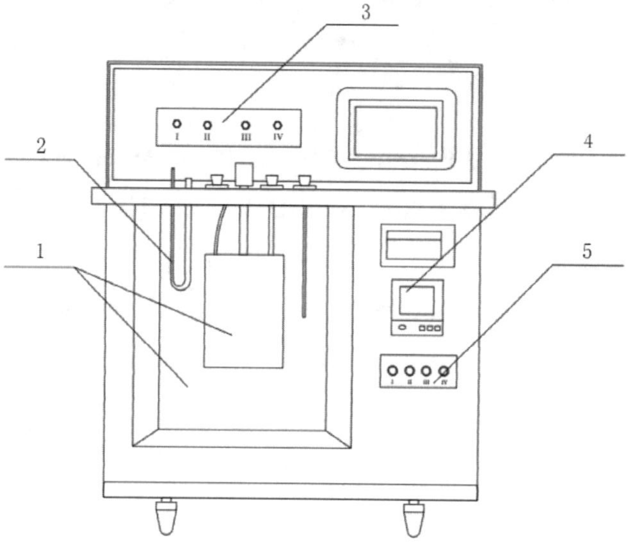

1--Constant temperature and temperature measuring device, 2--Vacuum decompression capillary viscometer, 3--Vacuum decompression device, 4--Controller, 5--Timing device.

Figure 1. The structure of viscometer

2.2. Design of calibration test scheme for viscometer

The calibration test scheme designed in this paper is as follows: first, calibrate [9] the viscometer coefficient. Then control the viscometer at a certain temperature and vacuum, and measure the time of the standard viscosity liquid flows through a certain height between the timing marks of the vacuum decompression capillary viscometer, the measured dynamic viscosity value can be converted by multiplying this time by the calibrated viscometer coefficient. By comparing the measured dynamic viscosity value with the standard one, the dynamic viscosity indication error of the viscometer can be calculated, so as to judge whether the measured value of the viscometer is accurate.

The spacing of timing marks is an important parameter to ensure the accuracy of the equipment. In this test, it is specified that the spacing of timing marks shall be 20 mm±0.6 mm. During the test, the temperature of the loading area of each vacuum decompression capillary viscometer shall be 60 °C±0.1 °C and the vacuum degree shall be 40 kPa±0.07 kPa. The indication error of timing device shall not exceed ±0.1 s. When the dynamic viscosity range is not more than 1000 Pa·s, the indication error of dynamic viscosity shall not exceed ±15%. When the dynamic viscosity range is 1000 Pa·s or above, the indication error of dynamic viscosity shall not exceed ±20%. The American canon standard viscosity liquid is used in the test.

2.3. Calibration test process of viscometer

2.3.1. Spacing of timing marks. Use a standard caliper to measure the spacing of each timing mark three times. Compare the measured value of the standard caliper with the spacing of timing marks of the viscometer, so as to judge whether the spacing of the timing marks of the viscometer meets the test requirements.

2.3.2. Temperature of constant temperature and temperature measuring device. Start the viscometer and set the water temperature to 60 °C, after the temperature is stable, insert the thermometer into the constant temperature and temperature measuring device, so that the thermometer is at the loading area of the vacuum decompression capillary viscometer. Read the temperature every 2 minutes after the data is stable, read a total of 9 data. Compare the value of the thermometer with the set value, it can judge whether the temperature of the loading area of each vacuum decompression capillary viscometer meets the test requirements.

2.3.3. Vacuum degree of vacuum system. Connect the precision pressure gauge to the vacuum device, read the pressure data every 3 minutes after the data is stable, read 3 data in total. Compare the value of the precision pressure gauge with the vacuum degree setting value, so as to judge whether the vacuum degree of the vacuum system meets the test requirements.

2.3.4. Timing device. Fix the fixture of the time calibration device on the time trigger button of the viscometer, and adjust the distance between the impact head of the fixture and the time trigger button, so that the equipment can work normally. Set a random time greater than 60 s on the time calibration device, start the viscometer and the time calibration device at the same time, read the time after the time calibration device is triggered. Compared the measured value of the time calibration device with the indicating value of the timing device of the viscometer, it can judge whether the timing device of the viscometer meets the test requirements.

2.3.5. Indication error of dynamic viscosity. According to the method specified in JTG E20, the dynamic viscosity value of standard viscosity liquid is measured by viscometer. By comparing the measured dynamic viscosity value with the standard dynamic viscosity value, the dynamic viscosity indication error of the viscometer can be calculated, so as to judge whether the measured value of the viscometer is accurate.

3. Uncertainty evaluation

3.1. Uncertainty of distance measurement results

3.1.1. Establishment of measurement model

$$\Delta = l - l_0 \tag{1}$$

In style:
Δ—Indication error of the spacing of timing marks, mm.
l—Measured value of the spacing of timing marks, mm.
l_0—Standard value of the spacing of timing marks, mm.

3.1.2. Evaluation of uncertainty components. (1) Uncertainty introduced by repeatability [10] u_1

Repeat the measurement for 3 times and the obtained data are: 20.20 mm, 20.10 mm and 20.12 mm. The difference between the maximum value and the minimum value is 0.1 mm. The uncertainty is calculated by the range method (3 times of measurement, the range coefficient is 1.69):

$$u_1 = \frac{0.1\ \text{mm}}{1.69\times\sqrt{3}} = 0.03\ \text{mm} \tag{2}$$

(2) Uncertainty introduced by the standard digital calipers u_2

According to the certificate of the standard digital caliper, U=0.02 mm, k=2, the uncertainty introduced by the standard digital caliper is:

$$u_2 = \frac{U}{k} = \frac{0.02\ \text{mm}}{2} = 0.01\ \text{mm} \tag{3}$$

3.1.3. Composite standard uncertainty u_c. The calculation results of the composite standard uncertainty are as follows:

$$u_c = \sqrt{u_1^2 + u_2^2} = \sqrt{0.03^2 + 0.01^2} = 0.03 \text{ mm} \tag{4}$$

3.1.4. Relative standard uncertainty u_{r1}

$$u_{r1} = 0.03 \text{ mm}/20.14 \text{ mm} \times 100\% = 0.15\% \tag{5}$$

3.2. Uncertainty of temperature measurement results

3.2.1. Establishment of measurement model

$$\Delta = T - T_0 \tag{6}$$

In style:
Δ—Indication error of temperature, °C.
T—Measured value of temperature, °C.
T_0—Standard value of temperature, °C.

3.2.2. Evaluation of uncertainty components. (1) Uncertainty introduced by repeatability u_1

Repeat the measurement for 3 times and the obtained data are: 59.92 °C, 60.08 °C and 60.10 °C. The difference between the maximum value and the minimum value is 0.18 °C. The uncertainty is calculated by the range method (3 times of measurement, the range coefficient is 1.69):

$$u_1 = \frac{0.18\ ^\circ\text{C}}{1.69 \times \sqrt{3}} = 0.06\ ^\circ\text{C} \tag{7}$$

(2) Uncertainty introduced by the standard thermometer u_2

According to the certificate of the standard thermometer, U=0.08 °C, k=2, the uncertainty introduced by the standard thermometer is:

$$u_2 = \frac{U}{k} = \frac{0.08\ ^\circ\text{C}}{2} = 0.04\ ^\circ\text{C} \tag{8}$$

3.2.3. Composite standard uncertainty u_c. The calculation results of the composite standard uncertainty are as follows:

$$u_c = \sqrt{u_1^2 + u_2^2} = \sqrt{0.06^2 + 0.04^2} = 0.07\ ^\circ\text{C} \tag{9}$$

3.2.4. Relative standard uncertainty u_{r2}. At 60 °C, the test results show that, the relative standard uncertainty of the influence of temperature on the dynamic viscosity of standard viscosity fluid is as follows:

$$u_{r2} = 0.33\% \tag{10}$$

3.3. Uncertainty of pressure measurement results

3.3.1. Establishment of measurement model

$$\Delta = P - P_0 \tag{11}$$

In style:
Δ—Indication error of pressure, kPa.
P—Measured value of pressure, kPa.
P_0—Standard value of pressure, kPa.

3.3.2. Evaluation of uncertainty components. (1) Uncertainty introduced by repeatability u_1

Repeat the measurement for 3 times and the obtained data are: 40.18 kPa, 40.07 kPa and 40.12 kPa. The difference between the maximum value and the minimum value is 0.11 kPa. The uncertainty is calculated by the range method (3 times of measurement, the range coefficient is 1.69):

$$u_1 = \frac{0.11\ \text{kPa}}{1.69\times\sqrt{3}} = 0.04\ \text{kPa} \tag{12}$$

(2) Uncertainty introduced by the standard digital pressure gauge u_2

According to the certificate of the standard digital manometer, MPE: ±0.03 kPa, assuming uniform distribution, k=$\sqrt{3}$, the uncertainty introduced by the standard digital manometer is:

$$u_2 = \frac{0.03\ \text{kPa}}{\sqrt{3}} = 0.02\ \text{kPa} \tag{13}$$

3.3.3. Composite standard uncertainty u_c. The calculation results of the composite standard uncertainty are as follows:

$$u_c = \sqrt{u_1^2 + u_2^2} = \sqrt{0.04^2 + 0.02^2} = 0.04\ \text{kPa} \tag{14}$$

3.3.4. Relative standard uncertainty u_{r3}

$$u_{r3} = 0.04\ \text{kPa}/40\ \text{kPa} \times 100\% = 0.1\% \tag{15}$$

3.4. Uncertainty of time measurement results

3.4.1. Establishment of measurement model

$$\Delta = t - t_0 \tag{16}$$

In style:

Δ—Indication error of the time calibration device, s.

t—Measured value of the time calibration device, s.

t_0—Standard value of the time calibration device, s.

3.4.2. Evaluation of uncertainty components. (1) Uncertainty introduced by repeatability u_1

Repeat the measurement for 3 times and the obtained data are: 90.01 s, 89.92 s and 90.03 s. The difference between the maximum value and the minimum value is 0.02 s. The uncertainty is calculated by the range method (3 times of measurement, the range coefficient is 1.69):

$$u_1 = \frac{0.11\ \text{s}}{1.69\times\sqrt{3}} = 0.04\ \text{s} \tag{17}$$

(2) Uncertainty introduced by the standard time calibration device u_2

According to the certificate of the standard time calibration device, U=0.01 s, k=2, the uncertainty introduced by the standard time calibration device is:

$$u_2 = \frac{U}{k} = \frac{0.01\ \text{s}}{2} = 0.01\ \text{s} \tag{18}$$

3.4.3. Composite standard uncertainty u_c. The calculation results of the composite standard uncertainty are as follows:

$$u_c = \sqrt{u_1^2 + u_2^2} = \sqrt{0.04^2 + 0.01^2} = 0.04\ \text{s} \tag{19}$$

3.4.4. Relative standard uncertainty u_{r4}

$$u_{r4} = 0.04\ \mathrm{s}/89.99\ \mathrm{s} \times 100\% = 0.04\% \quad (20)$$

3.5. Uncertainty of dynamic viscosity measurement results

3.5.1. Establishment of measurement model

$$\Delta = x - x_0 \quad (21)$$

In style:
Δ—Indication error of dynamic viscosity, Pa·s.
x—Measured value of dynamic viscosity, Pa·s.
x_0—Standard value of dynamic viscosity, Pa·s.

3.5.2. Evaluation of uncertainty components. (1) Uncertainty introduced by repeatability u_1
Repeat the measurement for 10 times and the obtained data are: 100.8904, 101.3155, 101.7406, 100.7487, 102.3074, 101.8823, 102.3074, 99.4734, 100.7487, 101.4572. The uncertainty is calculated:

$$u_1 = s(\bar{X}) = s(X)/\sqrt{n} = 0.9\% \quad (22)$$

(2) Uncertainty introduced by the standard viscosity liquid u_2
According to the certificate of the standard viscosity liquid, the maximum uncertainty of the standard viscosity liquid is 2%, then $u_2 = 2\%$.

3.5.3. Composite standard uncertainty u_c. The calculation results of the composite standard uncertainty are as follows:

$$u_c = \sqrt{u_{r1}^2 + u_{r2}^2 + u_{r3}^2 + u_{r4}^2 + u_1^2 + u_2^2} = 2.23\% \quad (23)$$

3.5.4. Composite extended uncertainty U. Take $\mathrm{k} = 2$,

$$U = ku_c = 2 \times 2.23\% = 4.46\% \approx 4.5\% \quad (24)$$

The expanded uncertainty of dynamic viscosity measurement results is: $U = 4.5\%,\ k = 2$.

4. Conclusion

In this paper, the parameters of viscometer are studied and determined. The spacing of timing marks shall be 20 mm±0.6 mm. The temperature of the loading area of each vacuum decompression capillary viscometer shall be 60 °C±0.1 °C. The vacuum degree shall be 40 kPa±0.07 kPa. The indication error of timing device shall not exceed ±0.1 s. When the dynamic viscosity range is not more than 1000 Pa·s, the indication error of dynamic viscosity shall not exceed ±15% and when the dynamic viscosity range is 1000 Pa·s or above, the indication error of dynamic viscosity shall not exceed ±20%.

According to the requirements of the above parameters, the calibration test method of viscometer is established and the uncertainty evaluation is carried out, which ensures the quality of the viscometer and the accuracy of the test results, and provides technical support for the quality control of asphalt and asphalt pavement. The research results of this paper lay a theoretical and experimental foundation for ensuring the accuracy of viscometer.

Reference

[1] Li L H, Geng H and Sun Y N 2010 *J. Journal of Building Materials.* Evaluation Method and Indicator for Viscosity of High-Viscosity Asphalt. 03 80-84.
[2] JTG F40-2004, Technical Specifications for Construction of Highway Asphalt Pavements.
[3] Zhu C R, Gao B M and Yao D H 2006 *J. Petroleum Asphalt.* Study on influencing factors of 60 °C dynamic viscosity of petroleum asphalt. 03 63-66.

[4] Yang J X, Mei L and Lu J 2010 *J. Petroleum Asphalt.* Comparative study on the test methods of 60 ℃ dynamic viscosity of asphalt. 01 15-20.
[5] JJF 1059.1-2012, Evaluation and Expression of Uncertainty in Measurement.
[6] Duan H and Luo Y H 2015 *J. Measurement and Testing Technology.* Analysis of Measurement Uncertainty and Error Detection. 42 39-41.
[7] GB/T 27411-2012, Commonly Used in Eesting Laboratories Uncertainty Evaluation Method and Representation.
[8] Wang L Z. 2005 *J. Chinese highway.* Measurement of asphalt viscosity. 02 128.
[9] Wang W, Chang X C and Cao F 2016 *J. Measurement and testing technology*. Calibration and Measurement Capability of Spindle Correction Factors of Rotational Viscometer. 07 62-63.
[10] Zhou S C, He J M and Yang J 2019 *J. Petroleum Asphalt.* Study on influencing factors of 60 ℃ dynamic viscosity test of asphalt. 04 42-46.

Deformation Analysis of Rotating Linear Guide

Cao Jiangping, Peng Yuanhui, Yang Chang, Li Jieji and Yang Li

National Institute of Measurement and Testing Technology, Cheng Du, China; Chengdu jade double road no. 10

1878281644@qq.com

Abstract. Rotating linear guide movement precision because of its flexible realization of two-dimensional space, and the advantages of compact structure, easy to install and debug, in more and more application in the industry and precision instruments, not only its application technology and rapid development in different areas, rotating linear guide itself material, structure and control technology have been rapidly increasing. Rotating linear guide application according to the use of the environment and the demand is different, have different requirements, but to control its deformation is the key to ensure the good operation to solve common technical problems. Rotating linear guide deformation is mainly composed of the weight of the structure itself and structure stiffness, support and fundamental cause, among them, the support (including support and support) and vehicle (including the distribution of vehicle quality, vehicle quality and vehicle movement way and the state) is a key factor, the deformation of the straight line guide rail and the trend of change over time, is crucial, especially in the condition of rotating linear guide. This paper focuses on the potential of several support state set of rotating linear guide bearing conditions of stress and deformation analysis, study its change trend, compared several different support condition state of linear guide.

1. Introduction

Linear guide is also called line rail, slide rail, linear guide, linear slide rail, often used to need to do a straight line reciprocating movement, which can realize higher bearing capacity than the straight line bearing, can absorb a certain amount of torque at the same time, but in the case of high load to ensure high precision linear motion. Practical applications, the linear guide is mainly used to support and guide the moving parts, in a specific position according to the given direction do reciprocating linear motion, a guide rail sliding friction, rolling friction guide rail, elastic friction guide, fluid friction guide several types, such as linear guide many with chromium bearing steel materials, such as common GCr15, also can consider to carburizing bearing steel, such as G20CrMo, to reduce the cracking and deformation, guarantee the stability and accuracy[1].

Usually in order to get good stability and greater carrying capacity, linear guide, generally with groups, the most common way of sliding table for four groups, in according to the requirement to choose the appropriate linear guide, precision of the composite guide mainly decided by the layout structure, installation and adjustment, the combination of the rotating guide due to the effect of gravity, layout structure, installation and calibration and the conventional fixed state, which affect the largest is the arrangement of structure, as shown in the diagram below a transverse vertical shaft adopt linear guide, different arrangement of structure, stress, deformation and the trend of the change over time. This article mainly discuss the layout structure of rotating linear guide pair of stress and deformation, the influence of the transverse structure of similar to figure 1 can rotate, and through the analysis of several kinds of condition of different performance to study the change trend.

Figure 1 combined linear guide rail pair

2. The layout structure comprehensive analysis

The combination of rotating linear guide vice layout structure includes three aspects: supporting form, rotation and bearing structure. Rotating state against doing straight line reciprocating movement load in the design process are usually as far as possible to reduce the lightest, and center of gravity is as far as possible near the center of linear guide rail pair, high to low as far as possible, after more than do these points, can play when bearing structure design and optimization of little space, so here don't do too much to discuss, in several rotation would focus on state support types influence on the combination of linear guide rail pair loading and deformation[2].

Combination according to the practical application of linear guide rail pair spin state change between horizontal and 60 ° inclined, choose levels, 30 ° inclined and 60 ° inclined three states are analyzed[3]. Combination among the support linear guide rail pair has, at both ends, and three forms "∏ type, figure 1 horizontal axis is the intermediate support way, on both ends of the support in short distance travel guide," ∏ type and its extension type "is often used in long distance guide, select 2800 mm long linear guide rail pair experiment analysis, choose the middle support, support on both ends and the ∏ type a third support three ways. Below is for the three rotating state and three kinds of support way, carry out specific analysis on nine kinds of state, assumes that the load bearing 1000 g, at 1.6 m, guide installed on 2800 mm long, 160 mm wide and 100 mm high hollow stent.

2.1 level + middle support

In accordance with the above preconditions, in horizontal state adopts middle support method, the stress and deformation was carried out on the linear guide rail pair is analyzed, the following figure 2: + middle support horizontal deformation analysis diagram shows, the linear guide rail pair deformation mainly produce at both ends.

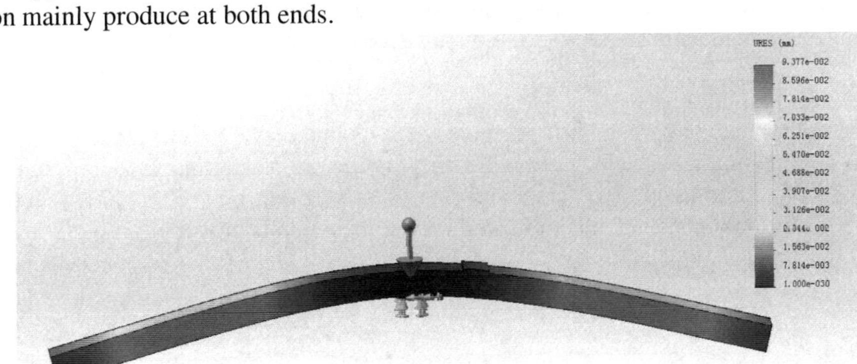

Figure 2 middle level + support deformation analysis diagram

IOP Conf. Series: Materials Science and Engineering **926** (2020) 012009 doi:10.1088/1757-899X/926/1/012009

2.2 level + support at both ends

In accordance with the above preconditions, in horizontal state adopts two-terminal bracing method, the stress and deformation was carried out on the linear guide rail pair is analyzed, the following figure 3: level + on both ends of the support deformation analysis diagram shows, the linear guide rail pair deformation mainly produce in the middle position.

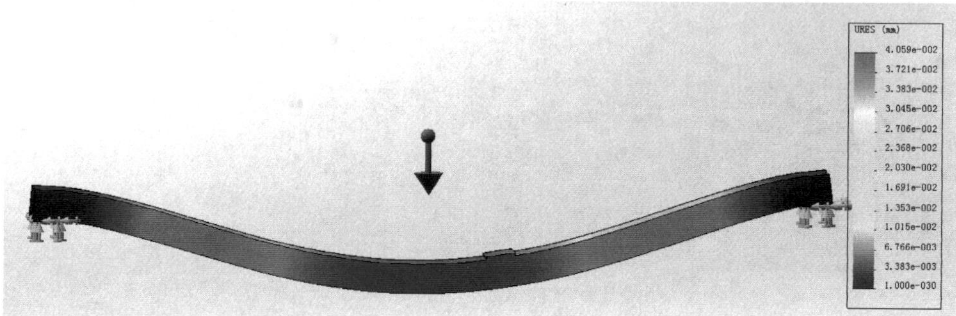

Figure 3 horizontal deformation analysis diagram + supported on both ends

2.3 level + 1/3 "∏ "type support

In accordance with the above preconditions, in horizontal state adopts "∏" type a third support method, the stress and deformation was carried out on the linear guide rail pair is analyzed, the following figure 4:1/3 +" ∏" type support horizontal deformation analysis diagram shows, the linear guide rail pair deformation mainly produce in the middle position.

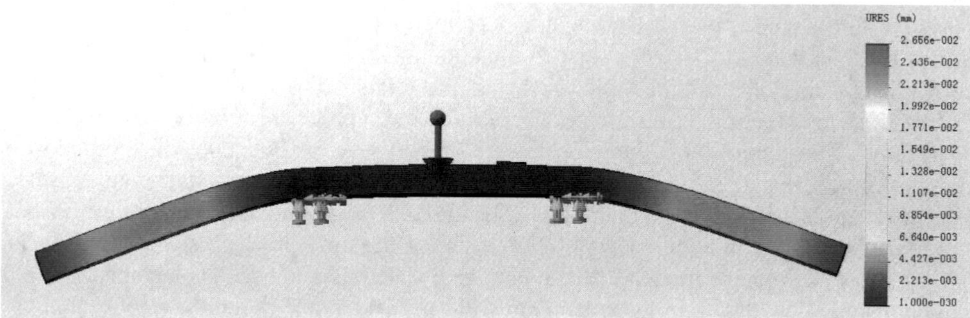

Figure 4 horizontal 1/3 + "∏" type support deformation analysis diagram

Through the analysis of the level in the state of three kinds of support way, found that the first way deformation approximately twice the second way, the second mode of deformation is about twice that of the third way, that in the condition of static level, "∏" type a third support better. When the linear guide rail pair tilts, stress state and deformation state whether there will be different, the following is to analyze in 30 ° inclined and 60 ° inclination of two kinds of condition.

2.4 30 ° inclined + middle support

In accordance with the above preconditions, under 30 ° inclined state adopts middle support method, the stress and deformation was carried out on the linear guide rail pair is analyzed, the following figure 5:30 ° inclined + middle support deformation analysis diagram shows, the linear guide rail pair deformation is mainly produced at both ends, and low deformation is bigger.

IOP Conf. Series: Materials Science and Engineering **926** (2020) 012009 doi:10.1088/1757-899X/926/1/012009

Figure 5 30 ° inclined + middle support deformation analysis diagram

2.5 30 ° inclined + support at both ends

In accordance with the above preconditions, under 30 ° inclined state adopts two-terminal bracing method, the stress and deformation was carried out on the linear guide rail pair is analyzed, the following figure 6:30 ° inclined + on both ends of the support deformation analysis diagram shows, the linear guide rail pair deformation mainly produce in the middle position.

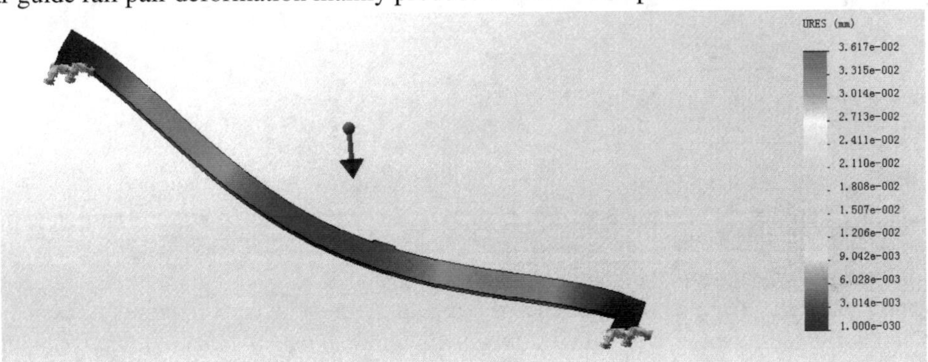

Figure 6 30 ° inclined + support deformation analysis diagram on both ends

2.6 30 ° inclined + 1/3 "∏" type support

In accordance with the above preconditions, under 30 ° inclined state adopts "∏ "type a third support method, the stress and deformation was carried out on the linear guide rail pair is analyzed, the following figure 7:30 ° inclined +" ∏ "type a third support deformation analysis diagram shows, the linear guide rail pair deformation is mainly produced at both ends.

Figure 7 30 ° inclined + 1/3 "∏ "type support deformation analysis diagram

Through the analysis of three kinds of support under 30 ° inclined state, found that the first way deformation is significantly higher than the second and the third way, the second mode of deformation is about 1.5 times that of the third way under the static state of 30 ° inclined, "∏" type a third support is better. Horizontal combination of the above three ways of data comparison, found that in the condition of 30 ° inclined, three ways of deformation is less than level condition, but the middle support way difference to be markedly less than the other two ways[4].

2.7 60 ° inclined + middle support

In accordance with the above preconditions, in 60 ° inclined state adopts middle support method, the stress and deformation was carried out on the linear guide rail pair is analyzed, the following figure 8:60 ° inclined + middle support deformation analysis diagram shows, the linear guide rail pair deformation is mainly produced at both ends, and low deformation is bigger.

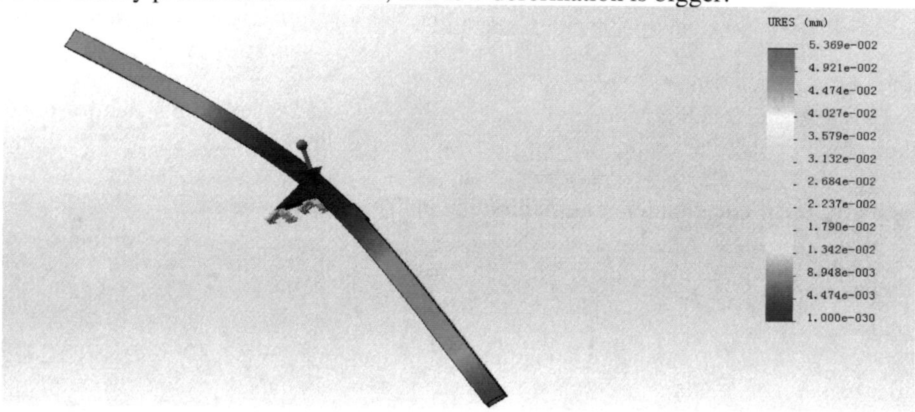

Figure 8. 60 ° inclined + middle support deformation analysis diagram

2.8 60 ° inclined + support at both ends

In accordance with the above preconditions, in 60 ° inclined state adopts two-terminal bracing method, the stress and deformation was carried out on the linear guide rail pair is analyzed, the following figure 9:60 ° inclined + on both ends of the support deformation analysis diagram shows, the linear guide rail pair deformation mainly produce in the middle position.

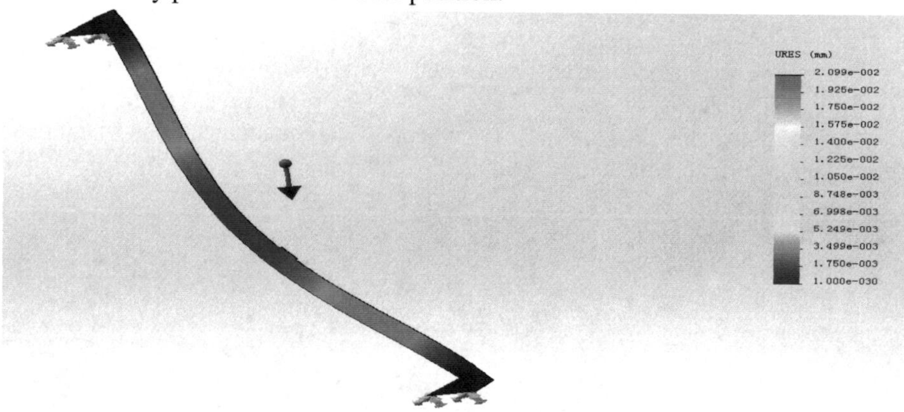

Figure 9. 60 ° inclined + support deformation analysis diagram on both ends

2.9 60 ° inclined + 1/3 "∏ type support

In accordance with the above preconditions, in 60 ° inclined state adopts "∏ type a third support method, the stress and deformation was carried out on the linear guide rail pair is analyzed, the following figure 10:60 ° inclined +" ∏ type a third support deformation analysis diagram shows, the linear guide rail pair deformation is mainly produced at both ends, and low deformation is bigger.

Figure 10. 60 ° inclined + 1/3 "∏ type support deformation analysis diagram

The above nine state maximum deformation comprehensive analysis and comparison, see Figure 11: the biggest deformation analysis table, can be seen from the table, "∏ type a third support" means in the same state minimum deformation, this is consistent with the traditional theory of the strut, with the increase of Angle, three kinds of support way cause deformation decrease gradually[5].

	The middle support	On both ends of the support	"∏ type support by one-third
level	9.377e-002	4.059 e-002	2.656 e-002
30 ° inclined	8.342 e-002	3.617 e-002	2.367 e-002
60 ° inclined	5.369 e-002	2.099 e-002	1.455 e-002

Figure 11. the biggest deformation analysis table

3. Conclusion

In the process of using rotating linear guide rail pair, in the case of position structure allows, give priority to "∏ type support way, specific pivot location determined comprehensively combined with bearing and linear guide rail pair[6]; At the same time, through the above analysis that the smaller the Angle, the greater the deformation, as far as possible when in use and stop the linear guide rail pair in a bigger perspective, is conducive to guarantee the stability and precision of the linear guide rail pair.

Reference

[1] Van-Cahn Tong;;Gung-ho Khim; Linear ball guide design optimization considering stiffness, friction force, and basic dynamic load rating using particle swarm optimization, Journal of Mechanical Science and Technology,2020,03

[2] Xu Huawei, Cao Jangling, Nondestructive testing technology in the application of the electronic components failure analysis, Electronic technology and software engineering,2019,01

[3] Song Xianchun, Fan Ming Zhen, Pre-tightening force on the roller linear guide pair of research on the effects of stiffness characteristics, Journal of Shandong construction university,2019.10

[4] Zhang Shoo, Shu Qulin, Weibull distribution in the application of the linear rolling guide vice accuracy stable analysis, Group technology and production modernization,2019.03

[5] JiaoJie, 14.5 m precision granite guide processing technology research, manufacturing technology and machine 2013, 06;

[6] Yao Weighing, a new measuring tool - granite slab, the aviation technology, 1982, 02.

IOP Conf. Series: Materials Science and Engineering **926** (2020) 012010 doi:10.1088/1757-899X/926/1/012010

The King of the New Generation Photovoltaic Technologies——Perovskite Solar Cells & the Opportunities and Challenges

Qianyu Chen

The School of Automobile and Traffic Engineering, Wuhan University of Technology, Wuhan, Hubei 430070, People's Republic of China

Chenqianyu121112@163.com

Abstract. Within the past several decades, photovoltaic technology is an emerging pioneer in the renewable energy field. Driven by the rapid decline in the price of photovoltaic products, the cost of photovoltaic power generation is becoming cheaper or even lower than the cost of thermal power generation. Among various photovoltaic technologies, organic-inorganic hybrid perovskite thin film solar cells are promising candidates for the next generation of solar cells in consideration of their outstanding advantages such as low-cost, easy manufacturing processing and high power conversion efficiencies. Perovskite solar cells stem from dye-sensitized solar cells. Within less than a decade of rigorous research and development in perovskite solar cells, the efficiency is boosted up to 25.2%. Aforementioned high PCE is mainly attributed to outstanding photovoltaic properties such as the long diffusion length of carriers, high optical absorption coefficient, excellent carrier mobility, etc. Meanwhile, the main barriers of commercialization for the perovskite solar cells are the poor stability of the devices, and the possible environmental pollution caused by lead. Herein, we briefly reviewed the opportunities and the challenges of this game changer in photovoltaic field. The development prospective are also discussed.

Key words: perovskite, solar cells, photovoltaic materials,

1. Introduction

With the gradual decrease of total amount of non-renewable fossil resources such as oil, coal and natural gas, the development and utilization of renewable and environmentally friendly new energy has become a major issue in the world. In the field of photovoltaics, Si-based photovoltaic devices have been making a great contribution, but their cost does not show an obvious advantage compared to the conventional energy. Compared to Si-based photovoltaic devices, perovskite solar cells can be fabricated at lower costs. Meanwhile, the power conversion efficiency (PCE) of perovskite solar cells has evolved from 3.8% [1] to 25.2% [2] which demonstrated a promising future in the related field. (Figure 1) In the past years, solar devices based on perovskite materials have been a superstar in the research institutes, and the number of publications and citations of related scientific and technological papers demonstrate a fierce growth trend. Since 2013, the number of papers published on perovskite solar cells has been growing continuously, and now it is about 3-4 thousand a year. Herein we mainly discuss the advantages and disadvantages of perovskite solar cell and the future prospect of it.

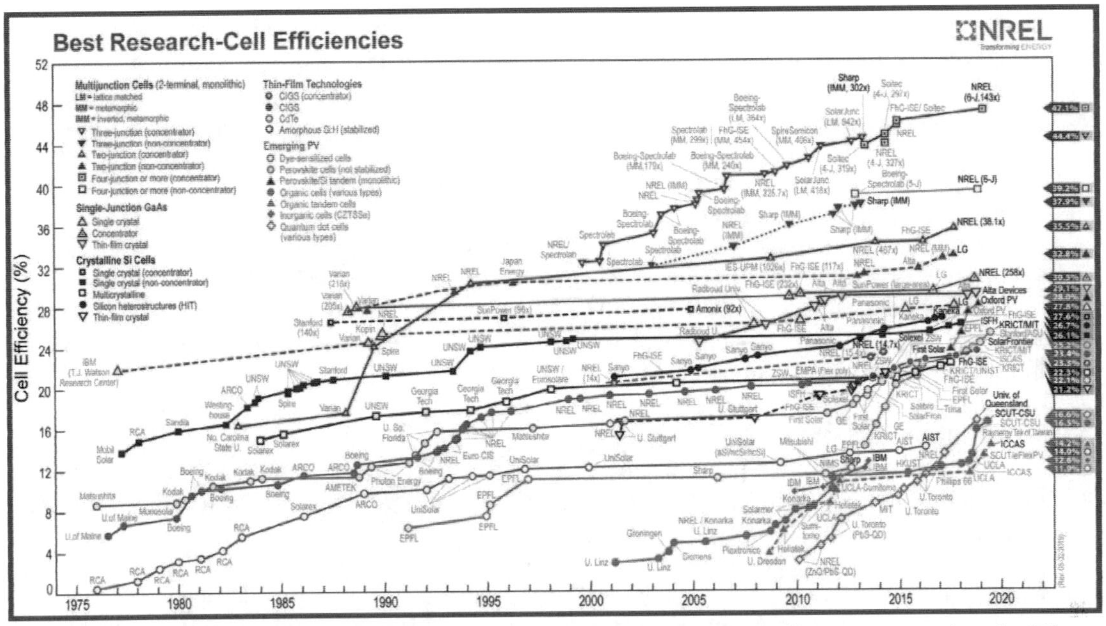

Figure 1. Trends in efficiency evolution for several different solar cells over recent decades[2] Copyright National Renewable Energy Laboratory

2.The opportunities of perovskite solar cells

Featuring extreme low cost and skyrocketing efficiency, organic-inorganic hybrid halide perovskite solar cells have emerged as the most promising superstar in PV technology field. There are several unique advantages of perovskite solar cells like the low fabrication expenses and mild device processing. To prepare the traditional silicon wafers, high vacuum and hot processing above 1000 °C are inevitable in several steps, whereas perovskite solar cells only need some mild chemistry processing in the commonly available ambient environment.

2.1 The organic - inorganic hybrid structure of perovskite materials

Perovskite solar cells have shown excellent photovoltaic properties in just a few years. Take other kinds of solar cells as references, studies indicate that the high efficiency of perovskite solar cells is mainly due to their organic-inorganic hybrid structure[3]. Such an organic-inorganic hybrid structure has several merits in some crucial properties of the materials, which is showed in the following passage.

2.1.1 Bandgap tuning of perovskite materials. Perovskites for photovoltaics have an intriguing characteristic that their band gap can be tuned by simple compositional substitution. Nevertheless, ordinary halide substitution cannot offer a way to reduce the band gap below 1.48 eV. There are two strategies to achieve lower band gap:

I. A site replacement in ABX_3 perovskite[4] This method can be combined with the B-site doping method. To conduct a further research on the influence of the A cation size in the perovskite materials, GE Eperon et al replaced the MA constituent with Cs and FA. The experimental results demonstrated that 1.48 eV could be obtained, allowing larger spectral absorption. It is further demonstrated that as the A cation increasing in ionic radius, the lattice would be expected to be larger and the bandgap would be reduced, leading to a red-shift in the absorption onset. Beyond MA and FA, organic cation guanidinium (CH_6N^{3+} GA) with a size of around 278 pm[5], which is larger than the size of FA^+(≈253 pm)[5]

II. Directly mutating the M-X bond. In spite of replacing the A site or B site cation, modifying the M-X bond can also tune perovskite band gap. According to the theoretical study, the band edges are adjudged by the orbitals of B-cation metals, therefore, the energy band structure will be affected. Likewise, band-gap running can be achieved by varying the halide of M-X bond. Jun Hong Noh et al fabricated inorganic-organic heterojunction solar cells with an entire range of $MAPb(I_{1-x}Br_x)_3$ When the x was changed from 0 to 1, the band gaps were changing between 1.5 and 2.3eV[1][6]. It was worthy of mention here that this kind of cells performed steadily when the x equals 0.2.

2.1.2 High carrier mobility of perovskite materials. Carrier mobility equals the ratio of the carrier moving velocity and the intensity of the applied electric field. Jongchul Lim et al [7]found that there were obvious differences in the remote carrier mobility of the composite films of perovskite fabricated by various preparation processes, such as solution based spin-coating and vacuum vapor deposition, in the range from 2.2 to 0.2 $cm^{-2}V^{-1}s^{-1}$.Compared to the polycrystalline perovskite films, perovskite single crystals showed better transport properties (≈60 $cm^2V^{-1}s^{-1}$) and smaller trap densities (≈1010cm^{-3})[8][9] All through the compositional range, $FA_{0.85}MA_{0.15}PbI_3$ and $FA_{0.15}MA_{0.85}PbI_3$ single crystals performed the optimum carrier mobility. Therefore, Huang et al[10]obtained mobility based on the Mott-Gurney law. They estimated the hole mobility and the electron mobility of $FA_{0.85}MA_{0.15}PbI_3$ perovskite single crystals, which is 5.31 $cm^2V^{-1}s^{-1}$ and 10.19 $cm^2V^{-1}s^{-1}$ while the two data for $FA_{0.15}MA_{0.85}PbI_3$ perovskite single crystals are 0.60 $cm^2V^{-1}s^{-1}$ and 0.77 $cm^2V^{-1}s^{-1}$, respectively. Above mentioned data indicates that the carrier mobility of $FA_{0.85}MA_{0.15}PbI_3$ is higher than $FA_{0.15}MA_{0.85}PbI_3$, which illustrates the high dependency of mobility and the organic cation ratio in the composition.

2.1.3 High optical absorption coefficient of perovskite materials. Absorption coefficient is one of the most important standards to evaluate the quality of an optical material. A high absorption coefficient usually means that only a very thin layer of optical material can absorb enough light, which can decrease the performance loss derived from the undesired charge recombination. In detail, $MAPbI_3$ has the absorption coefficient of larger than $3.0\times10^4 cm^{-1}$[11] in the visible region, which is over 10 times larger than that of silicon, indicating that the thickness of the absorption layer of perovskite is much smaller than that of crystalline silicon. In addition, numerous reports have pointed out that the thickness of organic and inorganic perovskite batteries is about 0.3-0.6μm, while the thickness of crystalline silicon solar cells is about 300μm. In that case, the fabrication investment of perovskite solar devices should be smaller than the conventional silicon solar wafers.[12][13]

2.1.4 Longer diffusion length of carriers. Samuel D. Strank et al reported the carrier diffusion lengths of $MAPbI_3$and $MAPbI_{3-x}Cl_x$ Both of them had over 1 micron diffusion length which further led to the high efficiency of perovskite solar device.[14] These experimental outputs can form important foundation for the research of perovskite solar cells.

2.2 Cost advantages of perovskite solar cells

In comparison with other kinds of photovoltaic technology, cost is an important advantage of perovskites. Song et al[15] developed a bottom-up cost model to evaluate the cost of perovskite solar cells modules fabricated with low-cost materials and processes. The manufacturing process which used a large area, low temperature deposition technique step by step was shown in the flowing diagram (Figure 2). The fabrication cost was estimated to be \$ 5.5~31.7/m^2 which was consisted of the perovskite cells process (\$1.2~6.8/m^2) and the related balance of module components (\$4.3~24.9) such as glass, frame and etc.

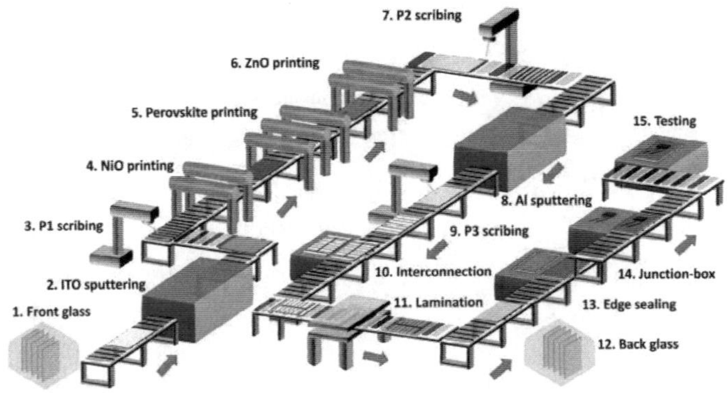

Figure 2. the manufacturing flow chat of perovskite solar cells modules. [15]. Copyright 2017 The Royal Society of Chemistry.

Cai et al[16] estimated the fabrication cost and showed a cost analysis of two representative kinds of perovskite modules. One of them was average efficient modules with mesoporous structure mainly based on cheap materials and the screen printing solution (module A), the other was highly efficient solar devices with accurate structure and high cost materials (module B), the whole calculated cost of materials for module A was 0.127 \$/W while the cost for module B was 0.102 \$/W.[17]

Later, Chang et al also analyzed the LCOE of a single-junction module. Compared to the LCOE of CdTe modules and water based c-Si modules respectively are \$86/m^{2}and \$136/m^2, They mainly researched the LCOE of three kinds of perovskite solar cell modules. Sequence A relies on gold evaporation for metallization, while sequence B substitute gold for silver, and sequence C is special for the removal of the sacrificial metal mask process for patterning the c-TiO_2 layer. The LCOE of these mentioned sequence are \$175/m^2, \$102/m^2, \$90/m^2, respectively. [18]

Nonetheless, different researchers adopt different ways to estimate the manufacturing cost of perovskite solar cells, all these results support the great potential of solar cells based on perovskite devices. The cost of different types of the PSCs is relatively lower than the traditional energy resources. The table 1 shows the four different research results about the cost of the perovskite solar cells.

Table 1. The information comparison of several perovskite solar cells modules from different researchers.[15][16][18][19]

Reference	Module assumption	Estimated cost	Device structure	Key layer function method
From [15]	PCE: 16% Lifetime: 30 years	3.5~4.9 US cents/kWh	Front glass/ITO/NiO/ $CH_3NH_3PbI_3$/ZnO/Al/Frame /junction-box	ITO and Al: industrial-scale magnetron sputtering, The charge transport layers(NiO and ZnO): screen printing followed by a post-deposition heat treatment,PbI_2: single-step screen printing, Perovskite layer : deposit by MAI mixture dissolved in a dimethylformamide (DMF) and dimethyl sulfoxide (DMSO) solution
From [16] Module A Based on a mesoporous structure	Assumed PCE:12% Assumed PCE:15%	4.9 US cents/kWh 4.2 US cents/kWh 3.5 US cents/kWh	Glass/TCO/Mesoporous scaffold perovskite layer/back electrode/encapsulation layer	TiO_2: Spray pyrolysis Scaffold layers: multiple screen printing Perovskite material: dip-coated within the mesoporous scaffold by moderate thermal annealing

From [16] Module B Based on a precise structure	Assumed PCE:20% (Assumed lifetime: 15 years)		Glass/TCO/Charge extraction layer/perovskite layer/Charge transport layer/Charge extraction/Metal electrode/Encapsulation layer	Metal electrodes: vacuum evaporation Production of patterns with lasers
From [18]	PCE: 18% Lifetime: 20 years	\$107/m^2 (about \$87 ~ 140/m^2)	Glass/FTO/TiO_2/Perovskite/ P3HT/Gold or Silver	TiO_2: spray pyrolysis, Pattern FTO and TiO_2: Scaffold print, Perovskite PbI_2: coat, PbI_2 Laser pattern, MAI: Dip coat, HTM(hole transport medium): laser pattern
From [19]Planar perovskite solar cells	PCE of 19% in the 2020s will be achieved	32.69 US\$/m^2	ITO/SnO_2/Perovskite/NiO/ Cu	Electron transport layer: low-temperature processing techniques such as screen printing and dip coating.

3. The challenges of perovskite solar cells

Although the perovskite solar cells have made an incredible progress in high efficiency and showed great advantages in the manufacturing cost, many barriers still remain on the way to perovskites' commercialization, among which the stability issue of the perovskite materials is a big trouble for the researchers especially under the influence of some external conditions such as oxygen, water, and ultraviolet light. In addition, the toxicity of lead is also a challenge to be overcome when considering the future feasible application.

3.1 Stability improvement of perovskite solar cells

The factors which can have impact on the stability of perovskite solar cells are listed as follows: the intrinsic chemical stability of perovskite solar cells, the stability of the solar energy device, the influence of external natural conditions. In addition, impacted by the electric field, the perovskite materials are not only easy to polarize. Finally, when the perovskite absorption layer and other functional layers are combined, the issue of instability is more serious. In terms of these issues, researchers mainly optimize the stability in the following two strategies. One is improving the intrinsic chemical stability of perovskite crystals, the other is avoiding the direct contact of perovskite absorber with those above-mentioned adverse factors. Here we mainly introduce four methods to alleviate these problems.

3.1.1 Additive engineering. Ionic liquid additives have previously been incorporated in the precursor in order to improve the long-term operational stability of the perovskite solar cells. Recently Bai et al employed a simple, widely used method, they added ionic liquids 1-butyl-3-methylimidazolium tetrafluoroborate($BMIMBF_4$)to the $(FA_{0.83}MA_{0.17})_{0.95}Cs_{0.05}Pb(I_{0.9}Br_{0.1})_3$ of the perovskite precursor. According to the experimental results, it was demonstrated that 0.3%mol doing could get the optimized result with a PCE of 19.8% under a light intensity of 105mW/m^2. Furthermore, the device could retain 85% of the initial efficiency after aging for 1885 hours[20]. Wang et al novelly proposed the trivalent and divalent europium ion pair that could alternatively oxidize Pb and reduce I defects at the same time. The modified solar cell could obtain a PCE of 21.52% with a certified efficiency of 20.52% with largely improved long-term durability. The devices could retain 92% and 89% of the champion PCE after continuous illumination or 1500 hours of heating at 85°C and 91% of the initial PCE after aging for 500 hours under maximum power point conditions.[21]

3.1.2 Compositional engineering of perovskite materials. Compositional engineering is another effective method to enhance the stability of devices based on perovskite materials. In this aspect, Zhou et al incorporated the CsI into the inorganic skeleton based on the precursor chemistry, which could greatly modify the film nucleation dynamics during the film formation. The results indicated that not only the efficiency of $FA_{0.9}Cs_{0.1}PbI_{3-x}Cl_X$ based device but also the stability were greatly improved[22].Furthermore, the total replacement of fragile organic species with stable inorganic cations would lead to all-inorganic perovskite devices. Lyubov A. Frolova et al systematically investigated the all-inorganic perovskite material $Cs_XPbI_2Br_X$, where x had a value range from 0.95 to 4. When the x equalled 1.2, the $Cs_{1.2}PbI_2Br_{1.2}$-based device had the best structural, device performance (the PCE is over 10%), as well as the optimized stability, in comparison with $CsPbI_3Br$, which could generate $CsPbBr_3$ and $CsPbI_3$ via a disproportionation reaction under the stress of ultraviolet irradiation or high temperature[23]

3.1.3 Interface modification of perovskite solar cells. Interfaces play a significant role in the functioning of perovskite voltaic devices. Therefore, it is essential to pay enough attention to the design and optimize interfaces between each functional layer to optimize the device stability. Yang et al fabricated the $PbSO_4$ and $Pb_3(PO_4)_2$ passivation layers by a simple chemical reaction of perovskite with SO_4^{2-} or PO_4^{3-}salts, after which a thin and dense inorganic lead-oxygen layer was generated on the top interface of perovskite. In comparison with the intermolecular forces, the strong chemical bond between perovskite and passivation materials could enhance the resistance ability towards light and moisture. Apart from that, it could also enhance the carrier lifetime and improve the efficiency of solar cells up to 21.1%. After aging for 1200 hours at 65 °C under maximum power point tracking, the modified device could retain 96.8% of the initial efficiency.[24] Besides, Bu et al demonstrated that using the phenylethylammonium iodide could reduce the iodide vacancy as well as the generation of superoxide. Experiment results showed that solar devices assembled with it could obtained better stability as well as enhanced the device PCE at the same time.[25].

3.1.4 Encapsulation technology for perovskite solar cells. Last but not least, encapsulation technology is also an effective method to improve the stability perovskite solar cells and protect them from environmental factors including moisture, oxygen and light. Ashraf Uddin et al summarized a series of different materials and technologies for encapsulation, and listed various situations where different packaging materials are appropriate. Until now, people have developed different interface layers as barriers to protect perovskite layers but they are not sufficient. By contrast, capsuling the devices with glass is the most efficient and effective method so far, which may be vastly applied in the perovskite device during its industrial manufacturing. In addition, many other materials such as Ethylene methyl acrylate, Ethylene vinyl acetate, Polyvinyl butyral, Thermoplastic polyurethane and etc., may be effective methods for device encapsulation as well.[26].

3.2 Possible environmental pollution caused by lead

Apart from the stability issue of solar cells based on perovskite materials, the impact on the environment is also one big challenge for the vast application of perovskite solar cells. For instance, when the perovskite solar modules are used outdoors, they may be corroded by natural factors such as rainfall, high winds, and fires, which can lead to the release of toxic lead from the damaged modules, which could further cause damage to natural environment. In that case, there are mainly two effective methods to reduce the lead pollution, encapsulation technology and doping method, namely partial replacement of toxic lead with Sn. What's more, Douglas Fabini et al had tried to quantify the possible upper limit of lead contamination during the lifetime of the perovskite solar cells modules. Based on the simulation result, it took 38 μg of lead to generate one kWh of electricity based on perovskite solar cells under the assumption that the lifetime and the efficiency are 25 years and 25% respectively.[27]

IOP Conf. Series: Materials Science and Engineering **926** (2020) 012010 doi:10.1088/1757-899X/926/1/012010

3.2.1 Encapsulation technology. As mentioned in the previous context, encapsulation can make a positive contribution to preventing lead leakage. Jiang et al investigated three packaging methods for perovskite solar devices under the simulation situations and recorded the amount of lead leakage under various weather simulations. The result showed that epoxy encapsulation could reduce lead leak rate by 375 times in comparison with glass encapsulation, which was mainly attributed to epoxy coating's self-healing performance and strong mechanical strength.[28]

3.2.2 Doping method. Partial replacement of lead with nontoxic tin is another effective strategy to reduce the possible pollution of perovskite solar devices. In this aspect, Hao et al did a pioneer work and adopted the tin-based materials ($CH_3NH_3SnI_3$) to fabricate perovskite solar cells. Compared with 1.5 eV bandgap of the conventional $MAPbI_3$, the tin-based one was only 1.3 eV, which showed better performance for sunlight harvesting. Besides, other properties like the mobility of carriers were better than the traditional one.[29] Further, Efat Jokar et al developed a Sn-based perovskite solar cells with the crystal structure of $FASnI_3$ doped different proportions of GA^+ When the ratio of FAI to GAI in the precursor was 8 to 2, the obtained solar cell demonstrated a initial PCE of 8.5%, after being stored in a glove box for two thousands hours, the efficiency continued to be improved to 9.6%.[30]

4. Conclusion

As one of the most efficient ways to use solar energy, solar cells have attracted worldwide attention. However, many kinds of solar cells cannot provide satisfactory device performance in consideration of those key factors such as cost, efficiency and stability. The organic and inorganic hybrid perovskite solar cells have rapidly rose to prominence within a few years and set off another research upsurge in the field of photovoltaics with good performance and a wide range of development prospects. The efficiency of perovskite solar cells has been up to 25.2% and it is still developing rapidly under the joint efforts of researchers. Herein we briefly reviewed the opportunities and challenges of the perovskite solar cells. The advantages mainly include the high efficiency in the solar performance, simple film preparation process, impressive physical properties and low fabrication cost, nevertheless, the disadvantages cover the poor stability, short lifetime and the toxicity of raw material etc.

5. Envisioned Future

Back in 2013, the perovskite solar cell was awarded as one of the ten scientific breakthroughs of the year by science journal. It is even being expected to realize its commercialization in the following decades and become a new generation of mainstream solar cells beyond monocrystalline silicon cells. It is worthy to mention here that there is still space to further reduce the module costs of perovskites solar cells such as the replacement of expensive electrode materials and the optimization of the film formation method in the large area device production. Even so, lots of obstacles to the large scale commercialization still remain and numerous technological challenges need to be overcome urgently. We believe that in the near future, perovskite-based clean and sustainable energy will occupy the mainstream of photovoltaic field, and perovskite solar cells will make great contributes to the human society.

Reference

[1] Kojima, A., Teshima, K., Shirai, Y., & Miyasaka, T. (2009). Organometal halide perovskites as visible-light sensitizers for photovoltaic cells. Journal of the American Chemical Society, 131(17), 6050-6051.

[2] NREL Best Research-Cell Efficiencies.https://www.nrel.gov/pv/cell-efficiency.html (2019)

[3] Loi, M. A., & Hummelen, J. C. (2013). Hybrid solar cells: perovskites under the sun. Nature materials, 12(12), 1087-1089.

[4] Geng, W., Zhang, L., Zhang, Y. N., Lau, W. M., & Liu, L. M. (2014). First-principles study of lead iodide perovskite tetragonal and orthorhombic phases for photovoltaics. The Journal of Physical Chemistry C, 118(34), 19565-19571.

[5] Kieslich, G., Sun, S., & Cheetham, A. K. (2015). An extended tolerance factor approach for organic–inorganic perovskites. Chemical science, 6(6), 3430-3433.
[6] Kim, H. S., Lee, C. R., Im, J. H., Lee, K. B., Moehl, T., Marchioro, A., ... & Grätzel, M. (2012). Lead iodide perovskite sensitized all-solid-state submicron thin film mesoscopic solar cell with efficiency exceeding 9%. Scientific reports, 2, 591.
[7] Lim, J., Hörantner, M. T., Sakai, N., Ball, J. M., Mahesh, S., Noel, N. K., ... & Wenger, B. (2019). Elucidating the long-range charge carrier mobility in metal halide perovskite thin films. Energy & Environmental Science, 12(1), 169-176.
[8] Dong, Q., Fang, Y., Shao, Y., Mulligan, P., Qiu, J., & Cao, L., et al. (2015). Solar cells. electron-hole diffusion lengths > 175 Î¼m in solution-grown ch3nh3pbi3 single crystals. Science,347(6225), 967-70.
[9] Shi, D., Adinolfi, V., Comin, R., Yuan, M., Alarousu, E., Buin, A., ... & Losovyj, Y. (2015). Low trap-state density and long carrier diffusion in organolead trihalide perovskite single crystals. Science, 347(6221), 519-522.
[10] Huang, Y., Li, L., Liu, Z., Jiao, H., He, Y., Wang, X., ... & Zhou, H. (2017). The intrinsic properties of FA (1− x) MA x PbI 3 perovskite single crystals. Journal of Materials Chemistry A, 5(18), 8537-8544.
[11] Huang, J., Yuan, Y., Shao, Y., & Yan, Y. (2017). Understanding the physical properties of hybrid perovskites for photovoltaic applications. Nature Reviews Materials, 2(7), 1-19.
[12] Xiao, Z., Dong, Q., Bi, C., Shao, Y., Yuan, Y., & Huang, J. (2014). Solvent annealing of perovskite-induced crystal growth for photovoltaic-device efficiency enhancement. Advanced Materials, 26(37), 6503-6509.
[13] Lee, M. M., Teuscher, J., Miyasaka, T., Murakami, T. N., & Snaith, H. J. (2012). Efficient hybrid solar cells based on meso-superstructured organometal halide perovskites. Science, 338(6107), 643-647.
[14] Stranks, S. D., Eperon, G. E., Grancini, G., Menelaou, C., Alcocer, M. J., Leijtens, T., ... & Snaith, H. J. (2013). Electron-hole diffusion lengths exceeding 1 micrometer in an organometal trihalide perovskite absorber. Science, 342(6156), 341-344.
[15] Song, Z., McElvany, C. L., Phillips, A. B., Celik, I., Krantz, P. W., Watthage, S. C., ... & Heben, M. J. (2017). A technoeconomic analysis of perovskite solar module manufacturing with low-cost materials and techniques. Energy & Environmental Science, 10(6), 1297-1305.
[16] Cai, Molang, Wu, Yongzhen, Chen, Han, Yang, Xudong, Qiang, Yinghuai, & Han, Liyuan. . Cost-performance analysis of perovskite solar modules. Advanced Science, 1600269.
[17] Becker, C., Amkreutz, D., Sontheimer, T., Preidel, V., Lockau, D., Haschke, J., ... & Steffens, S. (2013). Polycrystalline silicon thin-film solar cells: Status and perspectives. Solar Energy Materials and Solar Cells, 119, 112-123.
[18] Chang N L, Yi Ho-Baillie A W, Basore P A, et al. A manufacturing cost estimation method with uncertainty analysis and its application to perovskite on glass photovoltaic modules[J]. Progress in Photovoltaics: Research and Applications, 2017, 25(5): 390-405.
[19] Li, Z., Zhao, Y., Wang, X., Sun, Y., Zhao, Z., Li, Y., ... & Chen, Q. (2018). Cost analysis of perovskite tandem photovoltaics. Joule, 2(8), 1559-1572.
[20] Bai, S., Da, P., Li, C., Wang, Z., Yuan, Z., Fu, F., ... & Huettner, S. (2019). Planar perovskite solar cells with long-term stability using ionic liquid additives. Nature, 571(7764), 245-250.
[21] Wang, L., Zhou, H., Hu, J., Huang, B., Sun, M., Dong, B., ... & Xu, Z. (2019). A Eu3+-Eu2+ ion redox shuttle imparts operational durability to Pb-I perovskite solar cells. Science, 363(6424), 265-270.
[22] Zhou, N., Shen, Y., Zhang, Y., Xu, Z., Zheng, G., Li, L., ... & Zhou, H. (2017). CsI Pre-Intercalation in the Inorganic Framework for Efficient and Stable FA1− x CsxPbI3 (Cl) Perovskite Solar Cells. Small, 13(23), 1700484.
[23] Frolova, L. A., Chang, Q., Luchkin, S. Y., Zhao, D., Akbulatov, A. F., Dremova, N. N., ... & Troshin, P. A. (2019). Efficient and stable all-inorganic perovskite solar cells based on

nonstoichiometric Cs x PbI 2 Br x (x> 1) alloys. Journal of Materials Chemistry C, 7(18), 5314-5323.
[24] Yang, S., Chen, S., Mosconi, E., Fang, Y., Xiao, X., Wang, C., ... & De Angelis, F. (2019). Stabilizing halide perovskite surfaces for solar cell operation with wide-bandgap lead oxysalts. Science, 365(6452), 473-478.
[25] Bu, X., Westbrook, R. J., Lanzetta, L., Ding, D., Chotchuangchutchaval, T., Aristidou, N., & Haque, S. A. (2019). Surface Passivation of Perovskite Films via Iodide Salt Coatings for Enhanced Stability of Organic Lead Halide Perovskite Solar Cells. Solar RRL, 3(2), 1800282.
[26] Uddin, A., Upama, M. B., Yi, H., & Duan, L. (2019). Encapsulation of organic and perovskite solar cells: a review. Coatings, 9(2), 65.
[27] Fabini, D. (2015). Quantifying the potential for lead pollution from halide perovskite photovoltaics.
[28] Jiang, Y., Qiu, L., Juarez-Perez, E. J., Ono, L. K., Hu, Z., Liu, Z., ... & Qi, Y. (2019). Reduction of lead leakage from damaged lead halide perovskite solar modules using self-healing polymer-based encapsulation. Nature Energy, 4(7), 585-593.
[29] Hao, F., Stoumpos, C. C., Cao, D. H., Chang, R. P., & Kanatzidis, M. G. (2014). Lead-free solid-state organic–inorganic halide perovskite solar cells. Nature photonics, 8(6), 489.
[30] Jokar, E., Chien, C. H., Tsai, C. M., Fathi, A., & Diau, E. W. G. (2019). Robust Tin-Based Perovskite Solar Cells with Hybrid Organic Cations to Attain Efficiency Approaching 10%. Advanced Materials, 31(2), 1804835.

IOP Conf. Series: Materials Science and Engineering **926** (2020) 012011 doi:10.1088/1757-899X/926/1/012011

Porous Graphene for Sea Water Desalination Considering the Effects of Fluorine/nitrogen Modification: A Molecular Dynamic Study

Zhang Fa

College of Chemistry and Chemical Engineering, Chongqing University

zhangfajason@163.com

Abstract. Sea water desalination becomes more and more important as the consumption of fresh water. Forward osmosis (FO) is a novel technology for sea water or brackish water desalination, where a most important device, semi-permeable membrane, are required low resistance, high selection and inexpensive. In this study, based on molecular dynamic simulations, we explored the performance of porous graphene as the semi-permeable membrane for sea water desalination. Fluorine (F) and nitrogen (N) are adopted to optimize the property of graphene pore. We found that although pure pore have highest water flux (indicating lower resistance), N modified pore has the best selection due to the high electronegativity of N atoms. The about 60 L/cm^2/h water flux and 100% solute rejection ratio confirm the graphene with N modified pores is good candidate as a semi-permeable membrane for sea water desalination.

1. Introduction

Mixture separation, such as sea water desalination and gas purification, plays an important role in many industrial processes [1-5]. Compared to conventional distillation, solvent adsorption, and solid adsorption, membrane separation system has many advantages, such as energy conservation, high flexibility, and no moving parts and exotic chemicals, especially for small-scale applications [6]. Many carbon materials and their analogues, such as graphene, graphene oxide, fluorine-modified graphene and hexagonal boron nitride (h-BN) have been widely explored as promising membranes for sea water desalination and gas separation [7-8].The mechanism of membrane separation is the different permeation fluxes for mixtures due to their differences in the adsorption, diffusion, potential energy, and even the size ratio of pore and molecule.

Graphene, as a 2D material with high thermal stability, low dielectric constant and high mechanical strength, exhibits good impermeability and preferential adsorption properties to many molecules and ions, and is deemed as very promising sea water desalination membrane [1-2]. What's more, the adsorption capacity, interaction energy and selection of graphene to molecules can be significantly improved after the oxidation, fluorination and ion coating treatments, etc. [9-10]. For example, using some other atoms to modify the pore properties can significantly enhance the selection of water molecules and ions [11, 12].

In this work, molecular dynamic simulations are performed to explore the application of porous graphene for sea water desalination under a forward osmosis (FO) technology. To enhance the selection the graphene pore, fluorine (F) and nitrogen (N) atoms are used at the pore edge. A uniform

IOP Conf. Series: Materials Science and Engineering **926** (2020) 012011 doi:10.1088/1757-899X/926/1/012011

pore diameter and ion concentration in sea water are chosen to compare the effects of pure pore, F modified pore and N modified pore. The fluxes of water and ions with time are obtained and the solute rejection ratios are quantitatively discussed. Last, the adsorption energies of water molecule and several ions are calculated for the three pores to explain the difference in fluxes and solute rejection ratios.

2. Methods

The establishment of models and calculation are based on the Materials Studio version 7.0 (MS) of Accelrys. We visualize the structures by using the Materials Visualizer in MS, the sea water filling is by the Packing function in Amorphous Cell module, the geometry optimization and dynamic simulations are by the Forcite module. The COMPASS forcefield is used in all the simulations. Van der Waals interactions and Ewald electrostatic interactions were applied with a cutoff distance of 9.5 Å. COMPASS forcefield was developed by an ab initio calculation, and therefore this forcefield can accurately predict many condensed phase properties with covalent models and ionic models, such as graphene and seawater [13-14].

The front, side views of computational model and pores in graphene are shown in Figure 1. The sheet areas of both prefect graphene and graphene with pore are 24.6×25.6 Å2. The pores in this study are with a diameter of ~14.7 Å and N atoms and F atoms are used to modify the pores as the edge atom can significantly affect the molecule permeance [11]. The computational domain in z-axis is 60 Å. Periodic boundary conditions are applied in all directions. Two graphenes with periodic condition divide the region to two parts. The left part of the pore graphene contains fresh water, but the right part with seawater. Sea water part are with 540 H_2O, 27 Cl-, 14 Na+ and 13 K+, which makes the density of ~1.05 g/cm^3 and salting concentration of ~2.8 mol/L, and the fresh water part only includes 540 H_2O.

The prefect and pored graphenes are set as fixed and movable, respectively. Before the dynamic simulation, the system is optimized using the Smart algorithm with a force tolerance of 0.5 kJ/mol/Å, then a Langevin thermostat is used to control the temperature at 298 K. With a 1 fs time step, a 10 ps NVT (constant number of atoms, N; constant volume, V; constant temperature, T) dynamics is run to equilibrate the systems. Next, an NVE (constant number of atoms, N; constant volume, V; constant energy, E) is run more than 1 ns for the systems.

To quantify the flux of water, JW, and solute rejection ratio, R, we define JW and R as,

$$J_W = \frac{\frac{18g}{mol\rho}N_W(t)/N_A}{tS}, R = \frac{(N_{Na}+N_K)R_m}{N_W+(N_{Na}+N_K)R_m}, \quad (1)$$

where ρ is the density of water, NW(t) is the net number of water molecule through the pore, NA is avogadro's number, NNa and NK are net number of Na+ and K+, and S is the area of a graphene sheet. Rm=20 is the molar ratio of salt and water in the seawater side.

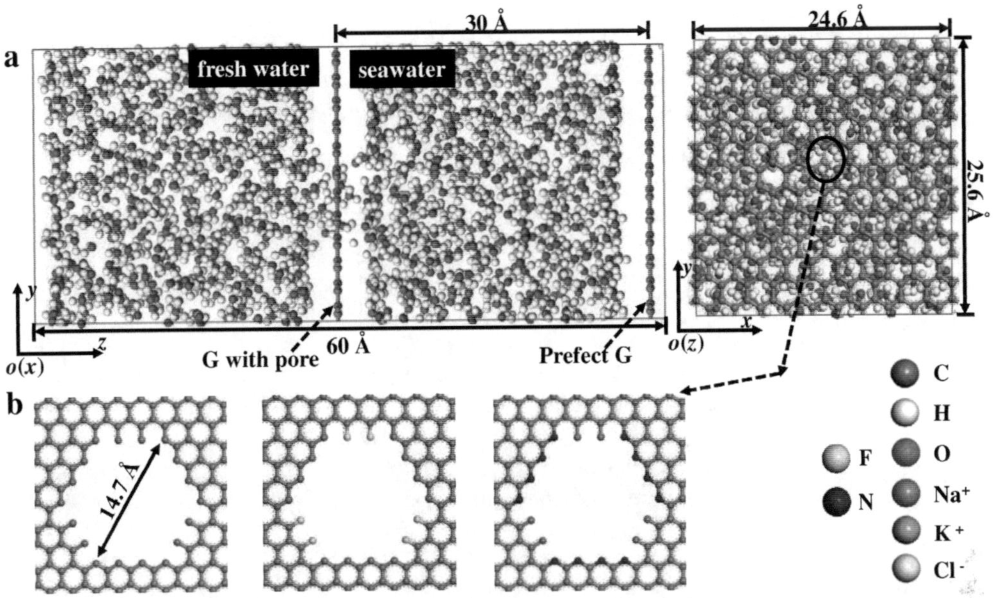

Figure 1. Graphical representation of calculation model. (a) are the front view and side view of the modeling region, where two parts are divided by two graphenes and one part with fresh water and the other with seawater. The middle graphene with a pore, and the pore structure are displayed in (b), where N and F atoms are used to modify the graphene pores.

3. Results and discussions

If the pored graphene has solute rejection ability, the osmotic pressure (salinity in seawater is higher) can drive the fresh water into the seawater part. As the pored graphene is movable in the simulation, the equivalent pressure in the two parts can be guaranteed in the molecule migration process [15]. Figure 2 shows the net number changes of (a) water molecule and (b) Na+ and K+ ions in fresh water part with the increase of time for the systems of pure pore, F modified pore and N modified pore. We found the water number in fresh part is reduced and the Na+ and K+ and Cl- ions are increasing, which is consistent with the forward osmosis.

In detail, with the increase of time, the changes of water number in 200-1000 ps are almost linear in all the three cases. That is due to the stable salinity difference between the seawater and fresh water parts in the simulation, although there is slight variation in concentration. It can also be found that Nw of the system with pure pore are higher than those with F and N modified pores, even the pore diameters are same. The reason might be the destruction of electroneutral C-C bond by F and N atoms. In the pure pore edge, C atoms are zero valent. However, F and N atoms have higher electronegativity, thus F and N atoms in pore edge are with negative valent [16-17], which may enhance the adsorption of the pore to water molecules. Similar trends of NNa+NK vs. t are found in Fig. 2(b). However, the values of NNa+NK are much smaller than Nw, which implying a solute rejection effects of the pores.

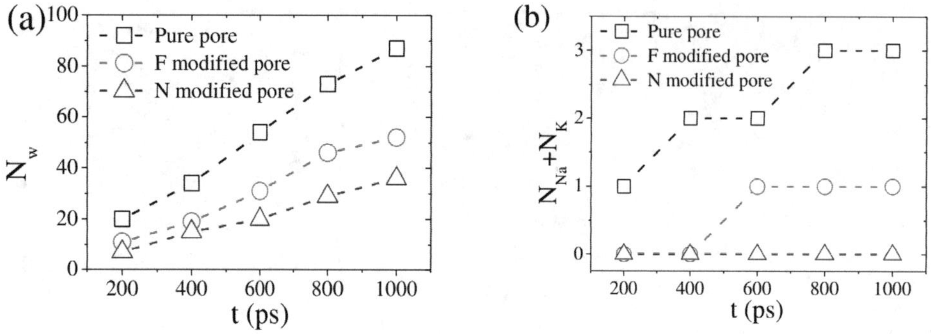

Figure 2. Net number changes of (a) water molecule and (b) Na+ and K+ ions in fresh water part with the increase of time for the systems of pure pore, F modified pore and N modified pore.

According to Eq. (1), we obtained the flux of water changing with time for the systems of pure pore, F modified pore and N modified pore, and the solute rejection ratio for the three pores, which are shown in Fig. 3 (a) and (b). Jw for pure pore, F and N modified pores are about 150, 90 and 60 $L/cm^2/h$, indicating lower resistance of those pores. Meanwhile, the significant difference means the F and N modifications have obvious effects on the water permeation. Other than flux, solute rejection ratio, R, is also an important parameter for sea water desalination application. As shown in Fig. 3(b), R of pure pore is only ~0.6, representing a very weak solute rejection effect. However, R of F modified pore is larger than 0.7, and that of N modified case is 1 in all the simulation time. The use of F and N atoms remarkably increased solute rejection ratio, especially for the N modified case.

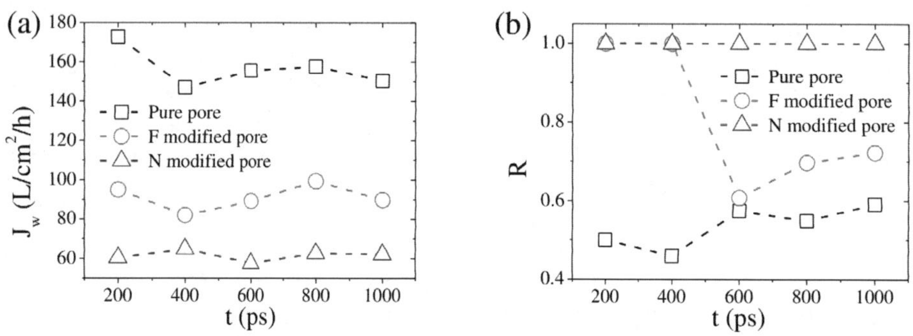

Figure 3. (a) Flux of water changing with time for the systems of pure pore, F modified pore and N modified pore, and (b) solute rejection ratio for the three pores.

To explain the effects of N and F modification to pores, we calculated the adsorption energy, Ea, of water molecule, Na+ and K+ in the pure pore, F modified pore and N modified pore, the approximate adsorption structure and Ea in the unit of kcal/mol are shown in Fig. 4. When bonding with C atoms, the valence of N is less than F by 0.35 [12], i.e. N atom is more negatively charged than F atom. Then, when molecules or ions through the pores, the charged N and F atoms have electrostatic force because the polar H_2O and +1 valences of Na+ and K+. Electrostatic effects is clearly shown in the adsorption energy values. For instance, Ea of pure pore for H_2O, Na+ and K+ are only -2.42~-1.22 kcal/mol due to the zero valent C atoms. However, the existence of F atoms increases Ea (see the absolute values) to -17.6~-13.7 kcal/mol. More remarkably, N atoms increase the Ea to -29.1~-48.6 kcal/mol, much higher than pure and F modified pores.

Higher Ea indicates higher energy barrier for molecules and ions through the pore. The values of Ea can explain Fig. 1 and Fig. 2 (a), where Nw, NNa+NK and Jw of pure pore are higher than F and N modified cases. Then, comparing Ea of H_2O, Na+ and K+ in a same pore can find the reason of selective rejection. In pure and F modified pores, Ea of H_2O are similar with Na+ and K+, thus inducing lower R as shown in Fig. 3 (b). However, the high negative valences of N atoms significantly enhances Ea of Na+ and K+, and the Ea improvement for H_2O is lower. The large gap between ions and H_2O lead to the 100% solute rejection ratio for N modified pore. Our study indicates that pure graphene pore is not good candidate for water desalination, but using high electronegative atoms, such as N, modifying the pore edge can significantly improve the performance of graphene pore.

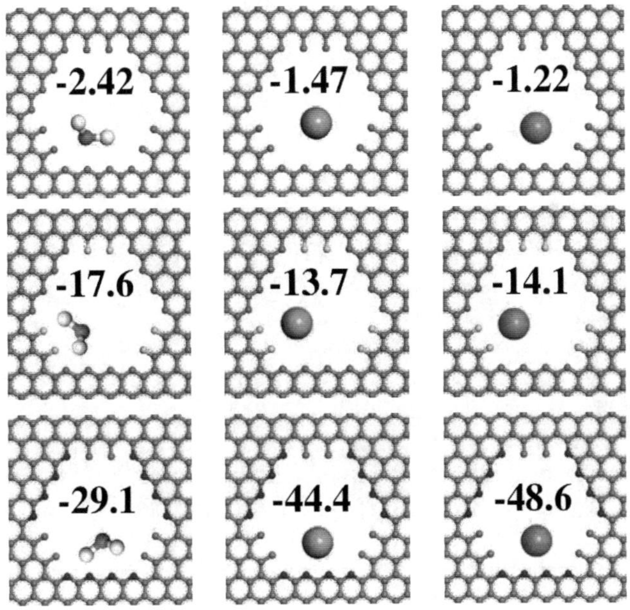

Figure 4. The adsorption energy, Ea, of water molecule, Na+ and K+ in the pure pore, F modified pore and N modified pore (kcal/mol), which is obtained by calculating the difference of the total energies of optimized structure and the structure of molecule or ion far from the pore. The negative sign of Ea suggests that the adsorption process is exothermic.

4. Conclusion

In this study, we explored the possible application of porous graphene for sea water desalination using molecular dynamic simulations on the Forcite module of Materials Studio. COMPASS forcefield is used in all the simulations due to the satisfactory accuracy. The pore in graphene is 14.7 Å and F and N atoms are adopted as the pore edge to modify the chemical property. Water flux and solute rejection ratio are calculated and compared for the pure, F modified and N modified pores. Results show pure pore has the highest flux (~150 $L/cm^2/h$), and F modified and N modified pores only with fluxes about 90 and 60 $L/cm^2/h$. Although flux of pure pore is favorable, the solute rejection ratio is only ~0.6, much lower than ~0.7 and 1 of F and N modified cases. The adsorption energies of H_2O, Na+ and K+ in the three pores are obtained to explain the differences in flux and solute rejection ratio. Due to the negative valences of F and N, the adsorption energies in F and N modified pores are higher than the pure pore. Na+ and K+ have greater adsorption energies than H_2O in the N modified pore, which induces a higher energy barrier for Na+ and K+ through the N pore. Our study demonstrate that N modified porous graphene is a good candidate for sea water desalination.

References

[1] Surwade, S. P., Smirnov, S. N., Vlassiouk, I. V., Unocic, R. R., Veith, G. M., Dai, S., & Mahurin, S. M. (2015). Water desalination using nanoporous single-layer graphene. Nature nanotechnology, 10(5), 459-464.

IOP Conf. Series: Materials Science and Engineering **926** (2020) 012011 doi:10.1088/1757-899X/926/1/012011

[2] Celebi, K., Buchheim, J., Wyss, R. M., Droudian, A., Gasser, P., Shorubalko, I., ... & Park, H. G. (2014). Ultimate permeation across atomically thin porous graphene. Science, 344(6181), 289-292.

[3] Yang, H.; Xu, Z.; Fan, M.; Gupta, R.; Slimane, R. B.; Bland, A. E.; Wright, I. Progress in Carbon Dioxide Separation and Capture: A Review. J. Environ. Sci. 2008, 20, 14-27.

[4] D'Alessandro, D. M.; Smit, B.; Long, J. R. Carbon Dioxide Capture: Prospects for New Materials. Angew. Chem. Int. Ed. 2010, 49, 6058-6082.

[5] Boutilier, M. S.; Sun, C.; O'Hern, S. C.; Au, H.; Hadjiconstantinou, N. G.; Karnik, R. Implications of permeation through intrinsic defects in graphene on the design of defect-tolerant membranes for gas separation. Acs Nano, 2014, 8, 841-849.

[6] Baker, R. W. Future Directions of Membrane Gas Separation Technology. Ind. Eng. Chem. Res. 2002, 41,1393-1411.

[7] Khakpay, A.; Rahmani, F.; Nouranian, S.; Scovazzo, P. Molecular Insights on the CH_4/CO_2 Separation in Nanoporous Graphene and Graphene Oxide Separation Platforms: Adsorbents versus Membranes, J. Phys. Chem. C 2017, 121, 12308-12320.

[8] You, Y., Sahajwalla, V., Yoshimura, M., & Joshi, R. K. (2016). Graphene and graphene oxide for desalination. Nanoscale, 8(1), 117-119.

[9] Kim, H. W.; Yoon, H. W.; Yoon, S.-M.; Yoo, B. M.; Ahn, B. K.; Cho, Y. H.; Shin, H. J.; Yang, H.; Paik, U.; Kwon, S.; et al. Selective Gas Transport through Few-Layered Graphene and Graphene Oxide Membranes. Science 2013, 342, 91-95.

[10] Sun, P.; Zhu, M.; Wang, K.; Zhong, M.; Wei, J.; Wu, D.; Xu, Z.; Zhu, H. Selective Ion Penetration of Graphene Oxide Membranes. ACS Nano 2013, 7, 428-437.

[11] Wu, T., Xue, Q., Ling, C., Shan, M., Liu, Z., Tao, Y., & Li, X. (2014). Fluorine-modified porous graphene as membrane for CO2/N2 separation: molecular dynamic and first-principles simulations. The Journal of Physical Chemistry C, 118(14), 7369-7376.

[12] Gai, J. G., Gong, X. L., Wang, W. W., Zhang, X., & Kang, W. L. (2014). An ultrafast water transport forward osmosis membrane: porous graphene. Journal of Materials Chemistry A, 2(11), 4023-4028.

[13] Mijović, J., & Zhang, H. (2003). Local dynamics and molecular origin of polymer network–water interactions as studied by broadband dielectric relaxation spectroscopy, FTIR, and molecular simulations. Macromolecules, 36(4), 1279-1288.

[14] Mijović, J., & Zhang, H. (2004). Molecular dynamics simulation study of motions and interactions of water in a polymer network. The Journal of Physical Chemistry B, 108(8), 2557-2563.

[15] Dahanayaka, M., Liu, B., Hu, Z., Pei, Q. X., Chen, Z., Law, A. W. K., & Zhou, K. (2017). Graphene membranes with nanoslits for seawater desalination via forward osmosis. Physical Chemistry Chemical Physics, 19(45), 30551-30561.

[16] Jiang, D. E., Cooper, V. R., & Dai, S. (2009). Porous graphene as the ultimate membrane for gas separation. Nano letters, 9(12), 4019-4024.

[17] Schrier, J. (2010). Helium separation using porous graphene membranes. The Journal of Physical Chemistry Letters, 1(15), 2284-2287.

IOP Conf. Series: Materials Science and Engineering **926** (2020) 012012 doi:10.1088/1757-899X/926/1/012012

The Development of Elderly Care Robot and Current Challenges for Functional Design

Jiaying Wang

Industrial robot technology, Lingnan Institute of Technology, GangZhou, China.

Wangjiaying85@163.com

Abstract. The emergence and development of robot industry is one of the important achievements in the 20th century. After years of development, the elderly care robot can only complete some mobile auxiliary functions at the earliest, and now it can complete many functions such as detection, reminder, chat, etc. Even so, the elderly care robot is still in the direction of more functions and more wisdom. So, what is the function of elderly care robot? What should be paid attention to in the design and use of elderly care robot? Based on the needs of the elderly, this paper discusses the functions of the elderly care robot and the practical factors that the elderly care robot should consider on the premise of meeting the interests of the elderly and not violating the ethics.

1. Introduction

With the development of technology, medical technology and people's life have been greatly improved. Many countries are faced with the problem of "population aging", in which more elderly means more medical, social and nursing services. If only the traditional way including build more nursing homes, equipment and employ more nurses were adopted, it would be hard to solve this problem. As the "aging population" will not only bring more old people but also result in the lack of young labor force due to lower of fertility rate.

For example, in 2018, the elderly aged 65 and above accounted for about 12.5% in the United States, which is expected to reach 20.7% in 2050. In terms of medical demand, according to the current growth rate, the number of Americans living in nursing homes will reach 3 million in 2030. According to the green paper of EU population, the total population of EU will reach 468.7 million by 2030, and the gap of labor population will reach 20.8 million. At that time, two labor population (i.e. population aged 15-64) will support one nonlabor population (over 65 years old)[1].

The above two data show that the core of the current problem lies in the imbalance of population structure, the increasing number of the elderly and the serious lack of young labor force, which means that not only the nursing problems of the elderly cannot be solved, but also the cost of living of the elderly can be further improved[2]. Japan is currently the most aging country in the world. It is expected that by 2060, Japan will become a society with 39.9% of the elderly population. And many welfare and care facilities are often understaffed. The number of nursing homes can not meet the number of applications for more than 524000 elderly people. Instead of building new facilities, the Japanese government began to support and encourage home-based care. Under these circumstances, the assistive technologies, which support elderly people using technologies, attract great attention[3].

With the increase of the number of elderly people, people have more urgent needs for it, so these assistive technologies are attracting more and more attention, but it is not a new thing. In the mid-1970s, France and Germany started the research and development of service robots. In 1987, the "handy" rehabilitation robot was developed in the UK. In 1990, Engelberg's "nurse assistant" robot was put into use[4]. After entering the 21st century, the application of assistive technology for the elderly has been expanding, from the former daily life of the elderly to health management, social care, and other aspects. For example, the "Pearl" robot developed by Carnegie Mellon University in the United States can remind the elderly to take medicine and monitor their health status, and the "huggable" robot can ask for hugs from the elderly.[5].

To sum up, we can see that the elderly care technology has become a relatively mature field. It is true that today's elderly assistant equipment can meet the needs of the elderly through different types of auxiliary equipment, however, these are only assistant machines, not a robot. They need the elderly to use the remote control to manually manipulate or shout a word of speech using voice control. In addition, these machines are difficult to associate this sentence, observe the actual situation and combine the elderly's habits to analyze the needs of the elderly, when they can't find the remote control or the instruction given out due to memory loss lacks a certain word. Such machines only do assistant work, and can't achieve the effect of "nursing", it's true that these assistive devices can make care workers more relaxed in many places, but the elderly still need care of care workers for a long time. The difference between machinery and robot is Artificial intelligence (AI) AI technology is a key which is regarded as a structure consisting of five components namely Knowledge, the basic component and Comprehension, Perception, Communication, as well as Learning[6]. They can give robots the ability to understand and remember the old people's habits, making intelligent nursing robots possible.

2. Development of elderly care robot

2.1. Definition of robot

The word "robot" first appeared in R. U. R: Rossum s universal robots, written by Czech writer Karel capuk in 1920. After that, there were many explanations for the word "robot"[7]. According to B. Roth of Stanford University in the United States, "a robot is a kind of machine that works together with human, animal or other machines, including automatic and semi-automatic types.". According to B. espiau of CNSR: National Council of scientific research in France, robotics is a knowledge that takes the use of systems that can achieve the following goals as the research object, that is, it can complete pre-determined tasks according to the information of sensors. According to K.J. Waldron of the United States, "robot is a kind of machine with more degrees of freedom, which is used for decision-making and control by an active system with interface with mechanism"[8]. Up to now, there is no strict definition of the word "robot". As a kind of robot, there is no strict definition of service robot. The International Federation of robotics, after collecting and sorting out various data, gives a preliminary definition: "service robot is a kind of semi-autonomous or fully autonomous robot, which can complete the service work intended for human health, but does not include the equipment engaged in production". Elderly care robot is one of the service robots designed for the elderly[9].

2.2. The development of robots

At the beginning, service robots were mainly designed to meet the needs of special groups of people. In the 1970s, France's "Spartacus" operator project and Germany's "Heidelberg" project aimed at helping patients with high paraplegia opened the door of nursing robots.[4] In 1982, the Netherlands developed an experimental manipulator "RSI" which can feed and turn over books on a tea tray, providing experience and basis for the 1984 wheelchair manipulator "Manus"[10].In 1987, a prototype of "handy1" rehabilitation robot developed by UK Mike topping helped 11-year-old boys with cerebral palsy finish eating independently. Since then, the "handy1" rehabilitation robot has been developing and innovating continuously to provide convenience for people in need[11]. In 1985, Engelberger began to develop " Rona（Robotic Nursing Assistant）" robot, which was put into production and sale in 1990 and used in dozens of hospitals around the world. Once programmed, it can accomplish many tasks at any time[12]. The most representative robots in the United States are "Helpmate"[5]. The robots listed above are shown in Figure 1 and the specific functions of the robots listed above are shown in Table 1.

Figure 1. Different types of elderly care robot

Table 1. Different elderly care robots

Robots	Characteristics
Manus	Simple tasks such as feeding, turn over a book, etc.
Handy1	Feast, feed, wash, shave, brush, etc.
Rona	Delivery of medical equipment and equipment, meals for patients, medical records, reports and letters, medicines, test samples and test results.
HelpMate	Independent walking in complex environment, such as opening door and taking elevator

After entering the 21st century, the development direction began to expand. In 2005, Toshiba company of Japan developed a companion robot for the elderly, which realized both image and speech recognition. The elderly can control it and let the robot follow up[13]. "Paro", developed by the Japan Institute of industrial technology, is a health care robot with the function of psychological assistant therapy. It can communicate with the elderly and play an emotional role in their lives[14]. In 2015, "Pepper", developed by Softbank group of Japan and Aldebaran robotics of France, can comprehensively consider the surrounding environment and actively respond. The robot is equipped with speech recognition technology, joint technology to present beautiful posture, and emotion recognition technology to analyze expression and tone, which can communicate with human beings[15]. In 2017, inventor Stephen von rump introduced "Giraff" through video at the Health Council in minster. It enables the interaction of nursing staff, the elderly and their families in-home service and nursing service. The elderly can keep in touch with nursing staff, family and friends, and their loneliness is reduced. It has a sensor network that can detect the internal and external environment of the family, as well as the state and activity of the elderly, and feedback to the nursing staff, so as to help the nursing staff to serve the elderly better and more efficiently without rushing through the homes of the elderly[16-18]. The robots listed above are shown in Figure 2.

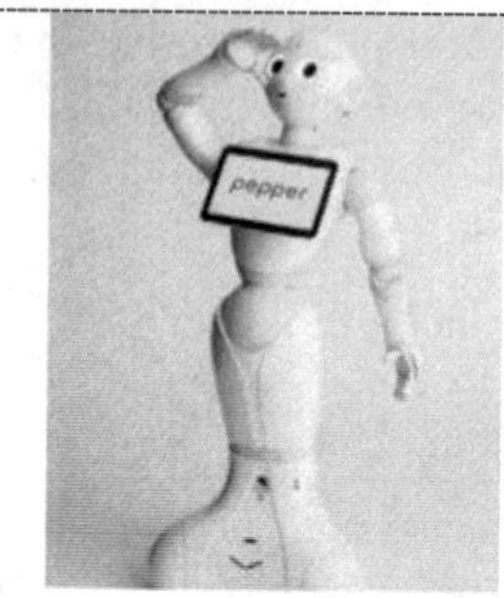

Paro Giraff Pepper

Figure 2. More recent elderly care robots

With the continuous expansion of the functions of the development of nursing robots, from action assistance, operation assistance [9]to health management, social care, etc., to communication, it can analyze the expressions and emotions of the elderly and maintain the contact between the elderly and society. Now, the development of nursing robots is still moving towards a more mature and intelligent direction without stopping.[19]

3. Elderly care robot functional design based on needs theory

The elderly care robot is designed for the elderly, that is to say, it should meet the needs of the elderly. In order to carry out the functional design of the elderly care robot, we must clarify the needs of the elderly. According to Maslow's hierarchy of needs theory (Abraham Harold Maslow, 1943)[20], human needs are divided into five levels: (1) Physiological needs; (2) Security needs; (3) Social needs; (4) Respected needs; (5) Self-realization needs. Through the above five levels of exploration, we can realize the targeted functional design of elderly care robots[21]. The above five points are shown in Figure 3.

(1) Physiological needs. Physiological needs are the lowest but the highest in the demand level, such as eating, hygiene, moving, dressing, etc. This layer is the most basic survival needs of the elderly, and the elderly care robot should first consider meeting the level of the elderly[21]. Before the 21st century, most of the basic design directions of elderly care robots were carried out at this level, for example, "Manus"[10].

(2) Security needs. Safety needs refer to the personal safety, property safety and health safety of the elderly.[22] As old people's memory would decline, it results in forgetting to close the door, window, gas, some electrical appliances, etc., which might be harmful to their lives and property. In addition, many old people will get lost, mislaid things, forget to take medicine and so on, which will cause health and safety risks. The design of the elderly robot on this floor is also a key point. It includes using sensors at home to detect gas concentration, door and window conditions, electrical conditions, etc[23], as well as reminding them to taking medicine and testing the health status of the elderly. Robots with this layer also represents a necessary design direction, such as "Jiaolong" of Shanghai Jiaotong University[24] and "Pearl" of Carnegie Mellon University in the United States[5].

(3) Social needs. This layer mainly refers to communication, emotional communication and entertainment. The elderly's physical function and mobility are reduced, while their social needs are still strong, and they are still eager to be loved and needed. Family communication, peer communication and nurse communication all contribute to the mental health of the elderly. Meanwhile, certain entertainment or music can improve the mentality of the elderly. In the design of this layer, we also need to consider that the elderly may be difficult to use the intelligent system, so the communication function is more humanized. The representative robots in this layer are "giraff"[17] with communication as the research direction and "Paro"[14] with emotional communication as the research direction.

(4) Respected needs. Everyone has the need to be recognized and respected. The respect of the elderly can never be ignored. The feeling of being respected mainly comes from two aspects: namely, the

approval from others and the feeling from the heart[21]. In the design of this layer, we should consider the unique needs of the elderly. For example, whether the elderly prefer personification[25], whether the elderly care about if robots will affect pets[26], etc., will affect the elderly's sense of respect[5].

(5) Self-fulfilling needs. This layer is the highest level, which refers to the needs that people's ideas and pursuits can achieve. After retirement for some time, the elderly usually have the needs of serving the society again, educating the third generation of families[27], etc., The elderly care robot can also try to pay attention to this part, and become a medium for the elderly to convey the outside world and introduce themselves. To accomplish anything meaningful will greatly meet the needs of self-realization of the elderly.

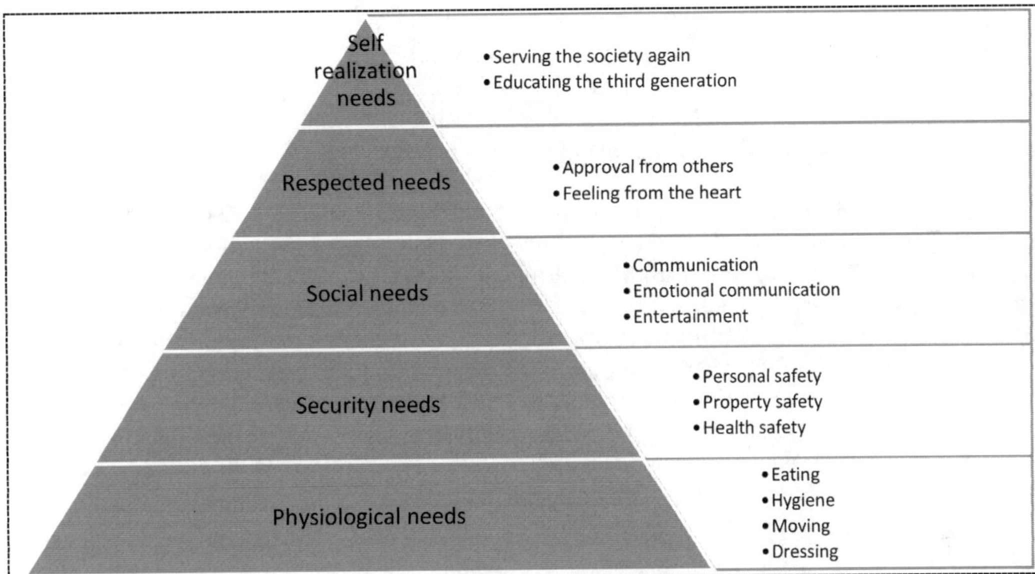

Figure 3. The needs theory for elderly care robot functional design

4. Practical problems to be considered in the design process

In order to meet the needs of the elderly, the elderly care robot itself must have many functions, some functions will indeed have many benefits, but in the process of design or use, there will inevitably be some practical problems.

4.1. Cost

This refers to the manufacturing cost of the elderly care robot itself and the purchase cost of consumers. Because each elderly has their own preferences, the robot needs personalized design[28], which will inevitably increase the cost of the robot. In addition, we should consider the economic situation of the elderly in different regions[2]. At present, due to the high cost, it is still difficult for the elderly care robot to enter the ordinary family[4].

4.2. Ethical issues

In the process of using robots, how to protect the benefits of the elderly and do not violate the ethical values of research is a very important part, that is, the physical and psychological welfare of the elderly is as important as the welfare of other people[29]. Specifically, this could be explained in the following different aspects, in detail.

4.2.1. The amount of social contact of the elderly is reduced. To some extent, the elderly care robot is to make the elderly stay at home longer and more safely, which will inevitably affect the amount of social activities of the elderly at home[3]. Research shows that the elderly with high social participation have higher quality of life and lower level of physiological and cognitive disorders[30].

4.2.2. Rights and interests of the elderly. How many functions should elderly care robots open to the elderly? Can offspring or nursing workers use more functions or restrict the elderly to use some functions? Although some functions may need to be banned in some cases, this will still affect the rights and interests of the elderly to a certain extent.

4.2.3. Violation of privacy. In the process of nursing, the elderly care robot will inevitably encounter some privacy problems on working. For instance, when the bathing robot helps the elderly to bathe or the changing robot changes clothes for the elderly, it will involve body privacy. Besides, some robots can detect the security of the house, their detectors often detect and store the situation in the house of the elderly, which involves property privacy. To some extent, this weakens the initiative of the elderly[29]. Research shows that more than 72% of the elderly are willing to accept home-based monitoring, and 60% of them are concerned about privacy and security issues in the monitoring process[31], which shows that the elderly pay attention to privacy.

4.2.4. Responsibility attribution, human rights, and morality are inextricably linked, but more from the perspective of legislation, rather than practical ethics[29]. Due to different national policies and laws and regulations, the division of responsibility is different. At present, there is not a sound responsibility system for the elderly care robot in many countries[4]. For example, if the elderly refuse to take medicine assisted by the nursing robot, who is responsible for the consequences, the manufacturer or the elderly themselves? Additionally, if the old people pour too much emotion into the nursing robot, and the new psychological diseases of the old people are caused by it[32], how to divide the cost of treatment? If these problems are only viewed from the perspective of morality and human values, there will be totally different views from different perspectives. If there is no sound legal system to divide them, it is difficult to determine the responsibility.

5. Conclusion

This paper discussed the functional design of the elderly care robot and the practical factors to be considered in the design. The research on elderly care robot has been developing for many years. Currently, the elderly care robot can accomplish many tasks. Whether it is reminder, detection, call or mobile assistance, it can almost meet most of the needs of the elderly. However, with more and more functions, the scope of functions is more and more wide, and many problems also follow, including the lack of practicability of some functions, privacy issues of the elderly, etc.

Some impractical and unnecessary functions not only make the system more complex, but also increase the cost. Hence, it is necessary to establish a theory to carry out the function design of the elderly care robot. The function design based on the needs of the elderly is reviewed considering that the elderly care robot is emerging for the elderly care. The needs theory determines many practical and necessary functions, at the same time. Specifically, the design of these functions should consider the feelings, privacy, and rights of the elderly.

Acknowledgments

I would like to thank Dr. Liu Fan for his guidance and help in the research and thesis writing. I also want to thank teacher Zhou, teacher Li and teacher Zhao for their concern and guidance.

References

[1] CINR. http://www.chinairn.com/scfx/20181112/113414456.shtml (accessed.
[2] Reinhardt U E, "Does the aging of the population really drive the demand for health care?," *Health Aff (Millwood),* vol. 22, no. 6, pp. 27-39, Nov-Dec 2003, doi: 10.1377/hlthaff.22.6.27.

[3] Tokunaga S, Horiuchi H, Tamamizu K, Saiki S, Nakamura M, and Yasuda K, "Deploying service integration agent for personalized smart elderly care," 2016.
[4] Maoni Z and Qiufan L, "Enlightment from development of foreign care robots for the age," *School of Nursing, New Campus of Zhengzhou University, Zhengzhou 450000, China,* no. 3.
[5] Ke L, Junqing X, and Guilan K, "Functional requirements of elderly health care robots," *China information industry,* no. 03, pp. 90-93, 2018.
[6] Davydov O I. and Platonov A K, "Robot and Artificial Intelligence. Technocratic approach," p. 112, 2017.
[7] Wowen W, "The development of robot in the world," *Defense Technology,* no. 1, pp. 70-75, 2001.
[8] McGraw H 1986, "Standard Handbook of Mechine Design[M]."
[9] Lanlan Y, "Ergonomical Design and Application of Free-old Robotic Bed," School of Mechanical Engineering Automation Beihang University, Beijing, China, 2011.
[10] Rosier J C, Woerden J. A. Van, Kolk L. W. Van Der, Driessen B. J. F and Bruyn P. M, "Rehabilitation robotics: the MANUS concept," in *Advanced Robotics, 1991. 'Robots in Unstructured Environments', 91 ICAR., Fifth International Conference on,* 1991.
[11] Tingting Z and Jing C, "The application of the elderly care service robot in China," *Management observation,* no. 10, pp. 139-141.
[12] I. t. s. robot. http://ishare.iask.sina.com.cn/f/34hOfoUhzsh.html (accessed.
[13] Koga T, Suzuki K, Hirokawa J, Ogawa H and Matsuhira N, "ApriAlpha V3 - sharp ear robot - and the omni-directional auditory process," 2007.
[14] Sabanovic S, Bennett C C, Chang W L and Huber L, "PARO robot affects diverse interaction modalities in group sensory therapy for older adults with dementia," *IEEE International Conference on Rehabilitation Robotics: [proceedings],* vol. 2013, pp. 1-6, 2013.
[15] Amit P and Gelin R, "A Mass-Produced Sociable Humanoid Robot: Pepper: The First Machine of Its Kind," *IEEE Robotics & Automation Magazine,* pp. 1-1.
[16] Portugal D, Trindade P, Christodoulou E, Samaras G andDias J, "On the development of a service robot for social interaction with the elderly," 2016.
[17] Coradeschi S, Cesta A, Cortellessa G, Coraci L and Ötslund B, *GiraffPlus: A System for Monitoring Activities and Physiological Parameters and Promoting Social Interaction for Elderly.* 2014.
[18] T. h. o. E. r. i. S. r. Valley. http://www.sohu.com/a/250668324_693803 (accessed.
[19] Doukas C, Metsis V, Becker Eric, Le Zhengyi and Maglogiannis I, "Digital cities of the future: Extending @home assistive technologies for the elderly and the disabled," *Telematics & Informatics,* vol. 28, no. 3, pp. 176-190, 2011.
[20] Anderson A, "Maslow's Hierarchy of Needs," vol. 67, no. 2, pp. 172–178, 2014.
[21] Yue S, "Research on the elderly nursing robot design based on design affairology," Yanshan University.
[22] Sinoff G and Perla W, "Anxiety disorder and accompanying subjective memory loss in the elderly as a predictor of future cognitive decline," *International Journal of Geriatric Psychiatry,* vol. 18, no. 10, pp. 951-959.
[23] Chifor B C, Bica I, Patriciu V. V and Pop F, "A security authorization scheme for smart home Internet of Things devices," *Future Generation Computer Systems,* p. S0167739X17311020.
[24] Bo Wen L, "Jiaolong service robot project," Knowledge of modern physics, no. 3, p. 2.
[25] Duffy B R, "Anthropomorphism and the social robot," *Robotics & Autonomous Systems,* vol. 42, no. 3-4, pp. 177-190.
[26] Scopelliti M, "If I had a robot at home... peoples' representation of domestic robots," *Designing A More Inclusive World,* 2006.
[27] Ya L, Yun Li Y, Chao Y, Hong J, Jun Xiang L, Jun Hui Z and Dong Fu C , "A study on the mental health status and influencing factors of the elderly," *Journal of Community Medicine,* no. 13, pp. 61-62.

IOP Conf. Series: Materials Science and Engineering **926** (2020) 012012 doi:10.1088/1757-899X/926/1/012012

[28] Ho Seok A, IHan K, Chandan D, Rebecca S, Ngaire K, Kathy P, Elizabeth B and Bruce A M, "Design of a Kiosk Type Healthcare Robot System for Older People in Private and Public Places," 2014.
[29] Sharkey A and Sharkey N, *Granny and the robots: ethical issues in robot care for the elderly.* 2012, pp. 27-40.
[30] Thomas P A, "Trajectories of Social Engagement and Mortality in Late Life," vol. 52, no. 4, pp. 430-443, 2011.
[31] Boise L, Wild K, Mattek N, Ruhl M, Dodge H H and Kaye J, "Willingness of older adults to share data and privacy concerns after exposure to unobtrusive in-home monitoring," *Gerontechnology,* vol. 11, no. 3.
[32] Frennert S and ?stlund B, "Review: Seven Matters of Concern of Social Robots and Older People," *International Journal of Social Robotics,* vol. 6, no. 2, pp. 299-310.

IOP Conf. Series: Materials Science and Engineering **926** (2020) 012013 doi:10.1088/1757-899X/926/1/012013

Calculation of Energy Parameters of LNG Power Plant with Utilization its Cold Energy

Dmitry Uglanov*, Olga Tremkina, Dmitry Sarmin, Olga Manakova and Alexey Lopatin

Samara University, Department of Heat Engineering and Heat Engines, 443086, Moskovskoye shosse, 34, Samara, Russia

*dmitry.uglanov@mail.ru

Abstract. This paper proposes the possibility of using a cryogenic power plant as part of a hybrid engine, which can be used in the system of urban transport infrastructure. The model of this cryogenic power plant consists of main contour and auxiliary contour of power plants. The parameters of a hybrid cryogenic power plant were evaluated and the characteristics of the main contour of a hybrid power plant driven by a piston expander and a turbo expander with heat input from both the ambient and the exhaust gases of a diesel engine were calculated. Liquid methane is used as the main fuel. The optimal operation cycle of the auxiliary contour of power plant itself has been determined. For this purpose the main parameters of Rankin and Brayton cycles have been analyzed - consumption, power, efficiency. Besides, the most suitable cryogenic fuels for efficient operation were determined and the schemes of the plants were proposed. The calculations were performed using calculation algorithms, examples of which are presented in this article.

1. Introduction

Reducing the growth of urban air pollution and greenhouse gas emissions from vehicle exhaust pipes is the main motivation for the development of alternative transport technologies that are not dependent on burning fossil fuels. Technological solutions such as the creation of a hybrid engine will eliminate the emission of automobile combustion products in areas with low air quality, as well as reduce the net amount of pollutants emitted into the environment. A thorough review of the overall environmental impact, efficiency and capacity is necessary to assess the ability of any new transport system to meet the needs of society [1].

This paper proposes a possible solution to the problem for urban transport infrastructure. It is proposed to use a hybrid cryogenic power plant, which consists of two contours: the main and auxiliary. The fuel in the main contour is methane, in the auxiliary contour - cryogenic fuel, the most effective for the system. It is proposed to implement four possible schemes of a power plant:

1. Cryogenic power plant driven by a piston expander;
2. Cryogenic power plant driven by a piston expander with the supply of heat of exhaust gases;
3. Cryogenic power plant driven by a turboexpander;
4. Cryogenic power plant driven by a turboexpander with the supply of heat of exhaust gases.

IOP Conf. Series: Materials Science and Engineering **926** (2020) 012013 doi:10.1088/1757-899X/926/1/012013

Nomenclature

G_{CH4}	Consumption (kg/sec)	$F_w(\varphi)$	Total cylinder surface area (m²)
p	Pressure (Pa)	F_p	Piston area (m²)
T	Temperature (K)	F_h	Inner area of the cylinder head (m²)
d	Cylinder diameter (m)	t_r	Piston ring thickness (m)
φ	Crank angle (°)	F_c	Clearance surface area (m²)
$V_{cyl}(\varphi)$	Cylinder working area volume (m³)	γ	Adiabatic exponent
V_c	A measure of the minimum gap between piston and cylinder head (m³)	m	Mass flow rate of the working fluid (kg/sec)
V_h	Volume inside the expansion chamber that is not occluded by the piston (m³)	h	Enthalpy (kJ/kg)
n	RPM	Q_1	Heat input (kJ)
$l(\varphi)$	Length of stroke	N	Power (kW)
R	Ratio of connecting rod to crank radius	N_s	Specific power (kW)
$Q(\varphi)$	Heat transfer (kJ)	η	Thermodynamic efficiency
K_x	Heat transfer coefficient (W/m² K)	η_T	Turboexpander efficiency
δ_w	Wall thickness (m)	C_p^g	Isobaric heat capacity (J/kg)
λ_w	Coefficient of thermal conductivity (W/ mK)	π_T	Expansion ratio of the turbo expander
α_x	Heat dissipation coefficient (W/ m² K)	η_m	Mechanical efficiency
Nu	Nusselt number	N_T	Turboexpander power (kW)
W_m, W_a,	Power of the main and auxiliary circuit	W_{liq}	Liquefied gas single stage heat difference
H_{T0}	Heat transfer turbine expander (kJ/kg)	H_{m0}	Single stage heat difference (kJ/kg)
H_{0T}	Available heat loss (kJ/kg)	H_T	Calculation of the expansion effect of one kilogram of the substance used in a turboexpander (kJ/kg)
ρ_z	Gas density behind the last stage (kg/m³)	z	number of stages of turboexpander
ΔP	Pressure drop (Pa)	α_T	Heat recovery coefficient
π_T	Expansion ratio of the turbo expander	c	Velocity (m/sec)

2. Initial data

In the developed hybrid power plant it is proposed to use methane at cryogenic temperatures. An open contour of a power plant using low-potential heat of a cryoproduct is the most simple and economical [2]. Figure 1 illustrates the T-s diagram of the Rankine cycle. The Rankine cycle consists of several sequential processes: adiabatic compression in a pump 1-2, heating to a boil 2-3, boiling process 3-4, superheat 4-5, adiabatic expansion 5-5 ', additional superheat 5-7 [3].

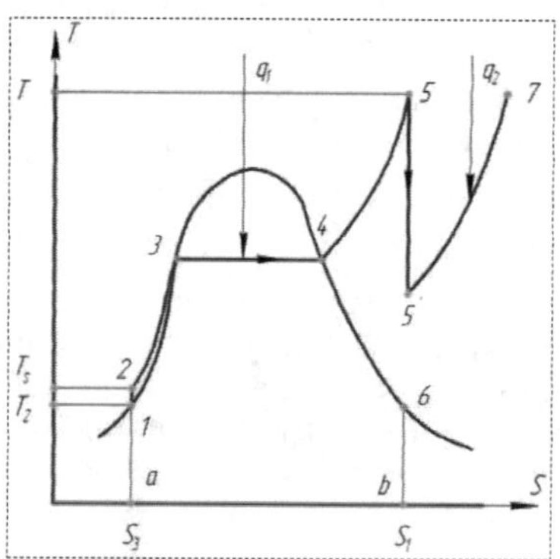

Figure 1. T-s diagram

The advantage of the presented scheme for the implementation of the Rankine cycle is the low consumption of the working fluid and the possibility operating of the cycle at high degrees of pressure increase π_k.

In this power plant, the parameters of the coolant at point 1: $p_1 = 0.5$ MPa; $T_1 = 139K$.

Since the ambient temperature varies on average from 240 to 280 K depending on the time of the year and day cycles are considered taking into account three variants of temperature values at point 5, with ΔT = 10 °.

3 variants of the Rankine cycle:

$p_{2,3,4,5} = 3288$ kPa – pressure at points 2, 3, 4, 5; $T_{3,4} = 180\ K$ – temperature in points 3, 4.

$p_{2,3,4,5} = 3854$ kPa – pressure at points 2, 3, 4, 5; $T_{3,4} = 185\ K$ – temperature in points 3, 4.

$p_{2,3,4,5} = 4641$ kPa – pressure at points 2, 3, 4, 5; $T_{3,4} = 190{,}55\ K$ – temperature in points 3, 4.

Based on the calculations of the four options for the Rankine cycle, the most suitable option is selected that matches the declared capacity N=265 kW, G_{CH4}=0.017 kg/sec.

3. Calculation of the parameters of the main contour of cryogenic power plant driven by a piston expander

The layout of cryogenic power plant driven by a piston expander with heat input from the ambient is shown in Figure 2 (A), and cryogenic power plant driven by a piston expander with the heat supply of exhaust gases shown in Figure 2 (B).

Cryogenic fluid (liquid methane) from tank 1 with the help of pump 2 enters heat exchanger 3. Next gaseous enters piston expander 5 (Figure 2), where a polytropic expansion process occurs through heat inflows through the walls of the cylinder and piston (1≤n≤ k) [4].

The auxiliary contour of a power plant also includes: a heat exchanger-evaporator 11, pump 10, turbine 12 and a cryogenic working fluid heat exchanger 3. The energy obtained in the combustion process is transmitted to the electric generator 13 and is accumulated in the battery 16. The energy from the auxiliary contour turbine 12 and expander 5 of the main contour is transmitted to the generator 13 and 14 and goes to the drive.

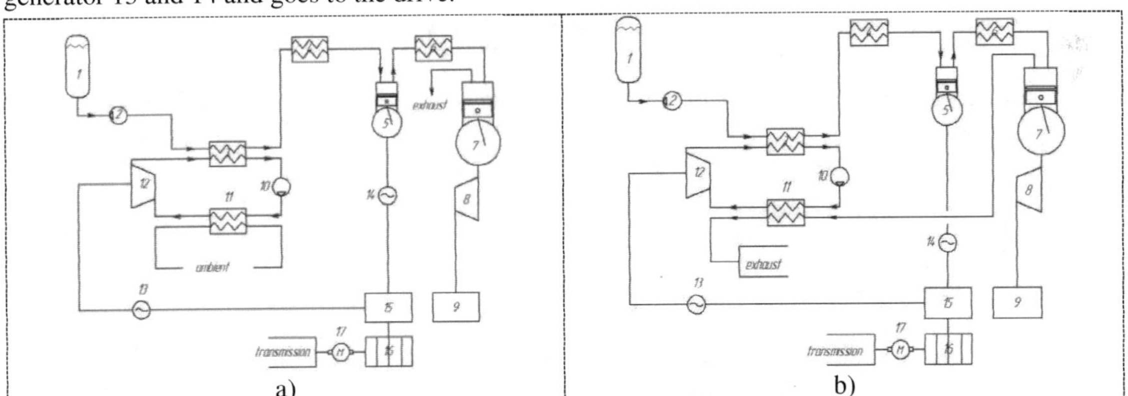

Figure 2. Diagram of a cryogenic power plant driven by a piston expander
a) with heat supply from the environment b) with the supply of heat of exhaust gases
1 - tank with liquid methane, 2- pump, 3- evaporator heat exchanger, 4 - additional heat exchanger, 5 - piston expander, 6- heat exchanger-heater, 7- internal combustion engine, 8-gearbox, 9- mover, 10 - pump, 11 - heat exchanger, 12- turbine, 13,14- generator, 15- controller, 16 - battery, 17- electric motor.

Initial data: G_{LCH4}=0.017 kg/sec, p_{input} = 0.5 MPa, T_{input} = 290 K.

The piston expander is a cylinder with a diameter is d = 260 mm, outlet gas pressure is 0.5 MPa and a temperature is 290 K.

The calculations were carried out and the geometric characteristics of the expander were determined in accordance with the methods [5].

The calculation algorithm is presented in Figure 3.

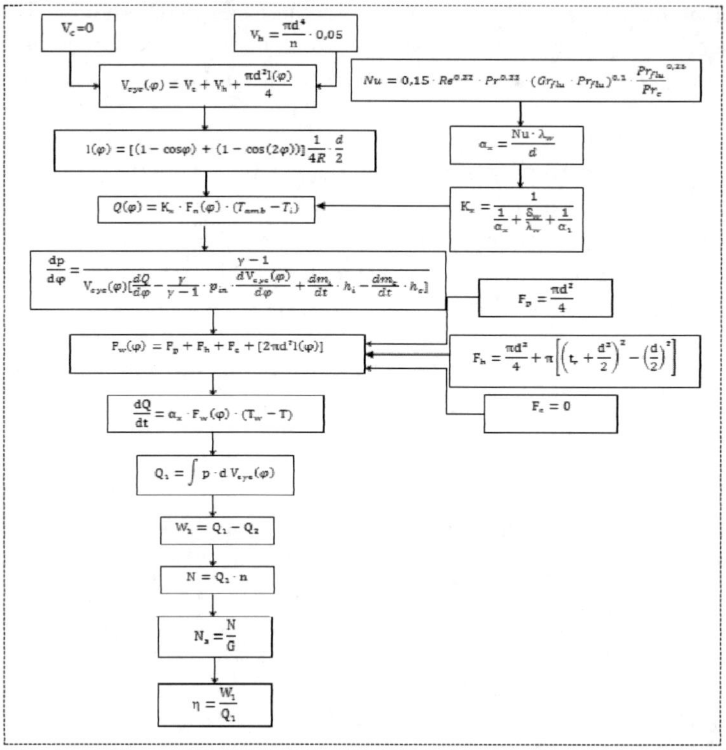

Figure 3. The calculation algorithm of a cryogenic power plant driven by a piston expander

4. Calculation of the parameters of the main contour of cryogenic power plant driven by a turboexpander

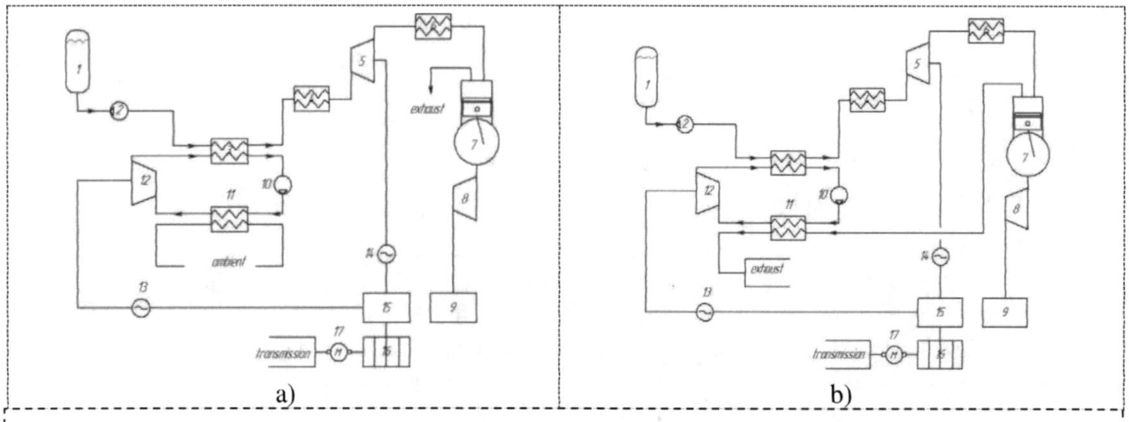

Figure 4. Diagram of a cryogenic power plant driven by a turboexpander

a) with heat supply from the environment b)with the supply of heat of exhaust gases

1 - tank with liquid methane, 2- pump, 3- evaporator heat exchanger, 4 - additional heat exchanger, 5 - turboexpander, 6- heat exchanger-heater, 7- internal combustion engine, 8-gearbox, 9- mover, 10 - pump, 11 - heat exchanger, 12- turbine, 13,14- generator, 15- controller, 16 - battery, 17- electric motor

IOP Conf. Series: Materials Science and Engineering **926** (2020) 012013 doi:10.1088/1757-899X/926/1/012013

The layout of cryogenic power plant driven by a turboexpander with heat input from the ambient is shown in Figure 4 (A), and cryogenic power plant driven by a turboexpander with the heat supply of exhaust gases shown in Figure 4 (B). The principle of operation of a power plant driven by a turboexpander is similar to the principle of a power plant driven by a piston expander. The calculation algorithm is presented in Figure 5.

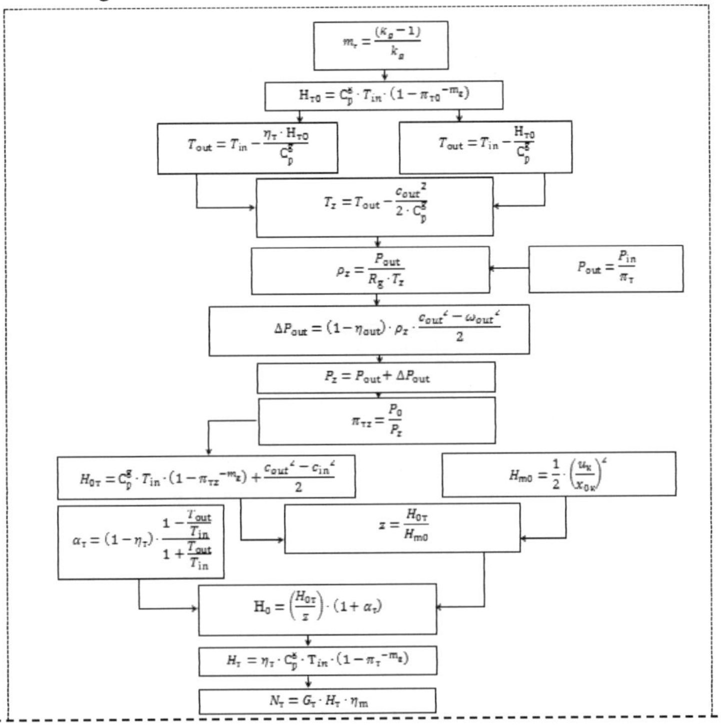

Figure 5. The calculation algorithm of a cryogenic power plant driven by a turboexpander

5. The choice of optimal parameters of the power plant of the auxiliary contour

It is proposed to use a heat engine operating according to the Rankine cycle [6] or Brayton cycle [7] as the power plant for the auxiliary circuit. The auxiliary contour of the power plant (Figure 6) can operate on various cryogenic fuels: methane, ethane, krypton, xenon.

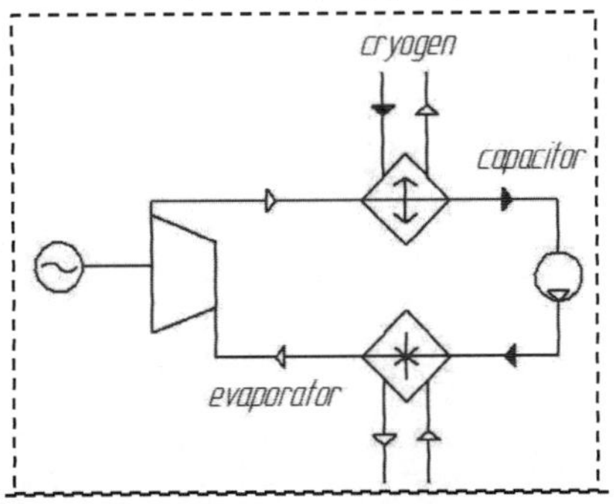

Figure 6. Diagram of the auxiliary contour of the power plant

An estimate of the amount of additional electrical energy received is given in tables 1 and 2.

IOP Conf. Series: Materials Science and Engineering **926** (2020) 012013 doi:10.1088/1757-899X/926/1/012013

Table 1. Estimation of the amount of received additional electric energy

	Rankine cycle					
	q_1, kJ/kg	q_2, kJ/kg	L, kJ/kg	G, kg/s	N, kW	ηt, %
Methane (p=0.6 MPa)	795	-595	395	0.015	11.58	25
Ethane	598	-500	500	0.017	10.37	16
Krypton	122	-97	97	0.089	10.90	20
Xenon	92	-83	83	0.104	9.61	10
Methane (p=0.7 MPa)	770	-600	400	0.014	11.13	22
	Brayton cycle					
	πk = 2					
	q_1, kJ/kg	q_2, kJ/kg	L, kJ/kg	G, kg/s	N, kW	ηt, %
Methane (p=0.6 MPa)	260	220	40	0.039	1.6	15
Ethane	118	110	8	0.079	0.6	7
Krypton	23.5	18	5.5	0.482	2.6	23
Xenon	3	2.5	0.5	3.468	1.7	17
Methane (p=0.7 MPa)	255	202	53	0.043	2.3	21
	πk = 2.5					
	q_1, kJ/kg	q_2, kJ/kg	L, kJ/kg	G, kg/s	N, kW	ηt, %
Methane (p=0.6 MPa)	222	175	47	0.050	2.3	21
Ethane	92	80	12	0.108	1.3	13
Krypton	13	10	3	0.867	2.6	23
Xenon	2.5	2	0.5	4.335	2.2	20
Methane (p=0.7 MPa)	215	147	68	0.059	4.0	32
	πk = 5					
	q_1, kJ/kg	q_2, kJ/kg	L, kJ/kg	G, kg/s	N, kW	ηt, %
Methane (p=0.6 MPa)	170	102	68	0.085	5.8	40
Ethane	81	65	16	0.133	2.1	20
Krypton	3	2.5	0.5	3.468	1.7	17
Xenon	-	-	-	-	-	-
Methane (p=0.7 MPa)	165	107	58	0.081	4.7	35
	πk = 10					
	q_1, kJ/kg	q_2, kJ/kg	L, kJ/kg	G, kg/s	N, kW	ηt, %
Methane (p=0.6 MPa)	85	40	45	0.217	9.8	53
Ethane	50	30	20	0.289	5.8	40
Krypton	-	-	-	-	-	-
Xenon	-	-	-	-	-	-
Methane (p=0.7 MPa)	95	47	48	0.184	8.9	51

Conclusion: Based on the data obtained, it can be concluded that it is most efficient to use an auxiliary contour of the power plant with heat supply from the environment operating according to the

IOP Conf. Series: Materials Science and Engineering **926** (2020) 012013 doi:10.1088/1757-899X/926/1/012013

Rankine cycle, using liquid methane as a working substance at a pressure is 0.6 MPa. Power N = 11.6 kW, efficiency is 20% at a flow rate is G = 0.015 kg / s.

Table 2. Estimation of the amount of additional electric energy received by the cryogenic power plant with the supply of heat of exhaust gases.

	Rankine cycle					
	q_1, kJ/kg	q_2, kJ/kg	L, kJ/kg	G, kg/s	N, kW	ηt, %
Methane (p=0.6 MPa)	1398.64	-445.14	445.14	0.019	27.2	68%
Ethane	1160.09	-434.76	4344.76	0.020	23.1	63%
Krypton	196.50	-97.18	97.18	0.089	17.5	51%
Xenon	136.29	-86.60	86.6	0.100	13.6	36%
Methane (p=0.7 MPa)	1616.42	-438.95	438.95	0.020	31.9	73%
	Brayton cycle					
	πk = 2					
	q_1, kJ/kg	q_2, kJ/kg	L, kJ/kg	G, kg/s	N, kW	ηt, %
Methane (p=0.6 MPa)	1040.6	787.7	252.9	0.011	2.75	24%
Ethane	795.0	559.0	236.1	0.015	3.62	30%
Krypton	103.2	84.4	18.8	0.102	1.91	18%
Xenon	56.0	46.4	9.5	0.185	1.76	17%
Methane (p=0.7 MPa)	1031.9	789.7	242.2	0.011	2.63	23%
	πk = 2.5					
	q_1, kJ/kg	q_2, kJ/kg	L, kJ/kg	G, kg/s	N, kW	ηt, %
Methane (p=0.6 MPa)	1040.6	602.6	418.4	0.014	5.95	41%
Ethane	776.0	487.7	288.3	0.018	5.06	37%
Krypton	99.8	67.1	32.6	0.128	4.17	33%
Ксенон	52.8	34.6	18.2	0.248	4.51	35%
Methane (p=0.7 MPa)	1013.8	701.5	312.3	0.012	3.81	31%
	πk = 5					
	q_1, kJ/kg	q_2, kJ/kg	L, kJ/kg	G, kg/s	N, kW	ηt, %
Methane (p=0.6 MPa)	949.5	505.4	444.1	0.017	7.53	47%
Ethane	703.2	314.8	388.5	0.027	10.57	55%
Krypton	90.7	57.3	33.3	0.149	4.98	37%
Xenon	44.7	28.1	16.6	0.305	5.06	37%
Methane (p=0.7 MPa)	940.7	498.5	442.2	0.017	7.60	47%
	πk = 10					
	q_1, kJ/kg	q_2, kJ/kg	L, kJ/kg	G, kg/s	N, kW	ηt, %
Methane (p=0.6 MPa)	861.2	353.1	508.1	0.024	12.33	59%
Ethane	604.1	192.3	411.9	0.045	18.35	68%
Krypton	81.9	40.9	41.0	0.209	8.58	50%
Xenon	35.5	17.5	18.0	0.489	8.80	51%
Methane (p=0.7 MPa)	847.1	345.6	501.4	0.025	12.43	59%

Conclusion: Based on the data obtained, it can be concluded that it is most efficient to use an auxiliary contour of the power plant with heat supply from exhaust gases operating according to the Rankine cycle, using liquid methane as a working fluid at a pressure is 0.7 MPa. Power N = 31.9 kW, efficiency is 73% at flow rate is G = 0.020 kg / s.

6. Comparison of the energy efficiency of hybrid power plants

Table 3 presents the main comparative parameters of cryogenic plants driven by a piston expander and a turboexpander.

Table 3. Comparison of energy parameters of cryogenic power plants

	N_{en}, kW	N_{mc}, kW	N_{ac}, kW	N_{Σ}, kW	η	η_{Σ}
Cryogenic power plant driven by a piston expander	265	1.1	11.6	277.7	0.39	34.5%
Cryogenic power plant driven by a turboexpander	265	3.75	11.6	280.35	0.67	35 %
Cryogenic power plant driven by a piston expander with the supply of heat of exhaust gases	265	1.1	31.9	298	0.39	37.1%
Cryogenic power plant driven by a turboexpander with the supply of heat of exhaust gases.	265	3.75	31.9	300.65	0.67	37.4%

Where N_{mc}, N_{ac} и N_{Σ} – installation power of the main and auxiliary circuits and total power, respectively.

$$\eta_{\Sigma} = \frac{Q_1 \cdot \eta_T}{Q_2} \cdot \eta_m \cdot \eta_g \cdot \eta_b \quad (1)$$

We can be seen from the point of view of calculating the efficiency of the energy parameters of power plants, the most efficient is a power plant driven by a turboexpander with an efficiency of 67%, however, the turboexpander has structural difficulties, vibrations and large weight and size characteristics. The efficiency of the power plant driven by a piston expander is 39%, the advantage of a piston expander is its relatively low cost and proven technology for its manufacture.

7. Conclusion

Based on the studies, the following conclusions:

A. It is proposed to use a hybrid power plant for driving a bus, consisting of a power plant of the main contour driven by a turboexpander and a power plant of the auxiliary contour, which will solve the environmental problems of large cities without compromising market competitiveness in terms of efficiency.

B. The calculation results show that it is most efficient to use power plant with a supply of heat of exhaust gases operating according to the Rankine cycle using liquid methane as a working fluid at a pressure is 0.7 MPa, G_{CH4} = 0.020 kg / s, N_{acpp} = 31.9 kW and η = 78%. As well as plants with a supply of ambient heat operating on the Rankine cycle using liquid methane as a working fluid at a pressure is 0.6 MPa, G_{C2H6} = 0.015 kg / s, N_{acpp} = 11.6 kW and η = 20%.

C. Analysis of the calculations shows that it is most efficient to use power plants of the auxiliary contour operating according to the Rankine cycle with the supply of heat of exhaust gases using liquid methane or liquid ethane as a working fluid. From the point of view of economy, methane is the most suitable working fluid for the auxiliary contour of power plant.

D. A comparative analysis of power plants showed that the most optimal power plant is a power plant driven by a turboexpander N = 265 kW, η = 61%. The total power of the power plant is N = 300.65 kW.

IOP Conf. Series: Materials Science and Engineering **926** (2020) 012013 doi:10.1088/1757-899X/926/1/012013

Acknowledgement
The results obtained by using equipment of collective use center «INTERDEPARTMENTAL EDUCATIONAL-INDUSTRIAL RESEARCH CENTER of CAM-TECHNOLOGIES», with financial support from Ministry of Education and Science (project # FSSS-2020-0019).

References
[1] C. Knowlen, A.T. Mattick, A.P. Bruckner and A. Hertzberg. High Efficiency Energy Conversion Systems for Liquid Nitrogen Automobiles / SAE Technical paper series August 11 – 13, 1998, Costa Mesa, California.
[2] Uglanov, D. A. Efficiency Criteria and Comparative Analysis of Combined Energy Plants Utilizing LNG Cold Energy [Text] / D. A. Uglanov, E. V. Blagin, A. A. Shimanov, A. I. Dovgyallo // Procedia Engineering. - 2016. - Vol. 152. - PP. 219-225.
[3] Richard T Jacobsen, Steven G. Penoncello, Eric W. Lommon Thermodynamic Properties of cryogenic fluids / The international cryogenics monograph series, 1997, New York.
[4] Arkharov, A. M. Cryogenic systems [Text] / A. M. Arkharov – M.: mechanical engineering, 1987. – P. 320.
[5] Tereshchenko, O.V. The choice of the optimal scheme of a hybrid cryogenic power plant [Text] / O.V. Tereshchenko, D.A. Uglanov // Scientific and technical journal «Pumps. Turbines. Systems.» - 2019. - No. 32. - No. ISSN 2226-177X. - P. 30-37.
[6] Tereshchenko, O. V. Investigation of the possibility of utilization of low-potential heat by means of installations operating on the Rankin cycle [Text] / O. V. Tereshchenko, E. V. Blagin / / Lukachevsky readings-2017: sat.nauch. Trudov / Samara University, - 2017. - PP. 7-12.
[7] Tereshchenko O. V. Investigation of the possibility of utilization of low-potential heat using installations operating on the Brayton cycle [Text] / O. V. Tereshchenko, A.V. Zaika, E. V. Blagin / / Lukachevsky readings-2017: sat.nauch. trudov / Samara University, - 2017. - PP. 58-62.

IOP Conf. Series: Materials Science and Engineering **926** (2020) 012014 doi:10.1088/1757-899X/926/1/012014

Research on Variable Sweepback Angle of Winglet Driven by Shape Memory Alloy

Zhang Wei[1,2], Zhang Chao[1], Liu Bingfei[1,*], Zhang Boli[3] and Wang Liwen[1,2]

[1] College of Aeronautical Engineering, Civil Aviation University of China, Jinbei Road 2898, Dongli District, Tianjin 300300, China
[2] CAAC Aviation Special Ground Equipment Research Base, Jinbei Road 2898, Dongli District, Tianjin 300300, China
[3] Basic Experiment Center, Civil Aviation University of China, Jinbei Road 2898, Dongli District, Tianjin 300300, China

[*] E-mail: bfliu@cauc.edu.cn

Abstract. Because of its shape memory effect, shape memory alloy (SMA), as a driver, provides more possibilities for the design and optimization of winglets of variable sweepback angle. In view of three different flight states of a large civil aircraft, i.e. take-off, cruise and landing, the effects of different winglets of variable sweepback angle on the aerodynamic performance of the aircraft are analyzed by computational fluid dynamics (CFD), and then the optimal sweepback angles are 35°, 45° and 50° respectively. Based on the change of the sweepback angle of winglets in different flight states, a variable sweepback winglet structure based on SMA wire is designed. Combined with SMA constitutive relation, SMA driving mechanism is designed by the coupling method of force, heat and strain. The mechanism can automatically adjust the swept angle of winglets according to different flight conditions, so as to improve the lift drag ratio of wings.

1. Introduction

With the increasing shortage of oil resources in the world, it has become more and more important to improve the aerodynamic efficiency of aircraft to reduce fuel consumption, and drag reduction technology is an effective way to improve the aerodynamic efficiency. For large aircraft, the induced resistance accounts for about 40% of the total resistance in cruise state and 60% in low-speed and high angle of attack flight such as take-off or landing [1-2]. According to Breguet relationship, the reduction of induced drag can improve lift drag ratio, reduce fuel consumption rate and reduce the quality of the aircraft, thus increasing the range of the aircraft and reducing the flight cost. However, the current design of the winglet is only based on the cruise condition [3], and the drag reduction effect is not obvious for the take-off and landing conditions with larger proportion of induced drag.

In order to optimize the aerodynamic performance of winglets under various flight conditions, the design of variable sweepback angle winglets came into being. The design can change the shape of winglets under different flight conditions, so as to obtain better drag reduction effect. Airbus and the University of Bristol have jointly developed a winglet based on a motor-driven variable tilt angle. This winglet can increase the lift-drag ratio by about 3% during take-off [4]. At present, there are few researches on variant winglets in China. Si L of Northwest Polytechnic University put forward a kind of variant winglet design scheme of the steerable trailing edge rudder. The numerical simulation shows that the aerodynamic performance can be improved by deflecting the trailing edge [5]. Li W and others from Nanjing University of Aeronautics and Astronautics proposed a research scheme of variable tilt angle winglet with SMA spring actuator [6]. This solution can automatically adjust the tilt angle of the winglet according to different flight conditions, and optimize the aircraft's drag characteristics in real time.

This paper takes a large civil aircraft as the research object. The influence of leading-edge sweepback angle of winglet on the aerodynamic performance of aircraft in different flight stages is analyzed by computational fluid dynamics (CFD). On this basis, a small wing with variable leading-edge sweepback angle driven by SMA wire is proposed. Combining with the constitutive theory of SMA, the SMA driving mechanism is designed by the coupling method of force, heat and strain. The research work of this paper can provide some reference for the research of variant winglets and the corresponding driving technology in the future.

2. Aerodynamic performance analysis of winglet

2.1. Research model

In this paper, the wing of a large civil aircraft with winglets is taken as the research object, and the simplified model is shown in figure 1.

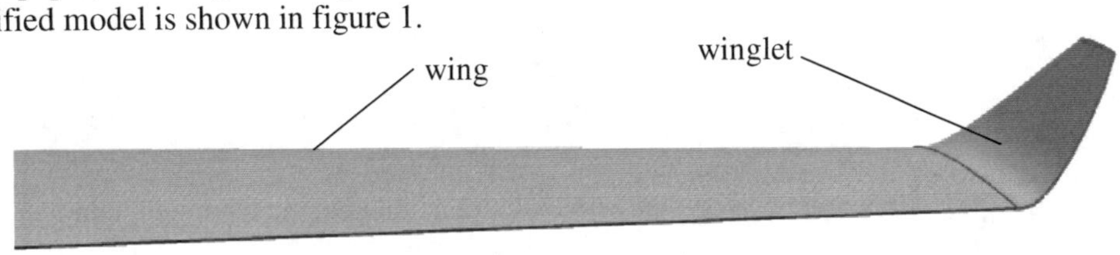

Figure 1. Research model.

2.2. Aerodynamic performance analysis

In this paper, it is assumed that that during the take-off state, the angle of attack is13°, the velocity of free flow is 0.4 Ma, and the temperature of free flow is 300 K; In the cruise state, the angle of attack is 4°, the free flow velocity is 0.8 Ma, and the free flow temperature is 250 K; in the landing state, the angle of attack is 8°, the free flow velocity is 0.4 Ma, and the free flow temperature is 300 K. In this paper, it is assumed that in each flight state, the leading-edge sweepback angle of the winglet changes in the range of 10 ~ 60 degrees with a step of 5 degrees. So, there are 11 different winglets in each flight state.

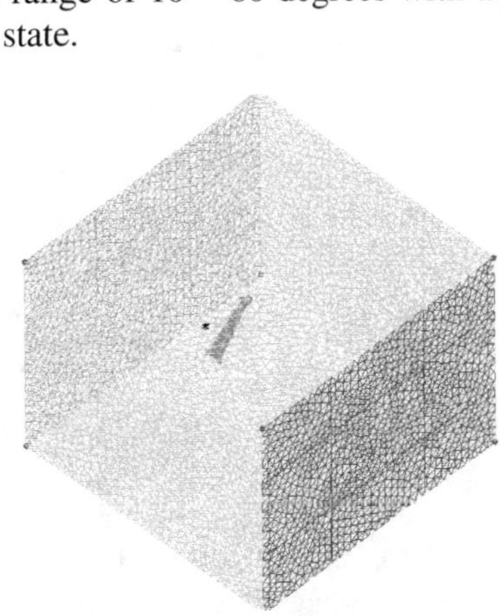

Figure 2. The flow field grid diagram of wing

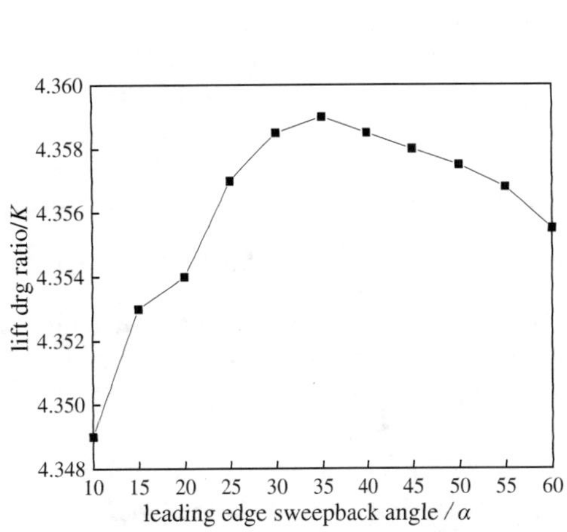

Figure 3. The relation curves of lift drag ratio and leading-edge sweepback angle

The numerical calculation was performed by commercial calculation software ANSYS. The numerical simulation calculation only considered the wing and the winglet and ignored the fuselage and other components. The density-based implicit solver is used and the turbulence model is selected from the Spalart-Allmaras model. Numerical calculations are performed on the wing in each state. The wing outflow field grid is shown in figure 2.

In this paper, the lift drag ratio *K* is taken as the optimization object to analyze the aerodynamic performance of the wing. The relation curves of lift drag ratio *K* and leading-edge sweepback angle α of winglet in takeoff and landing state are shown in figure. 3 and figure. 4. The curve of lift drag ratio and leading-edge sweepback angle α of winglet in cruise state is shown in figure 5.

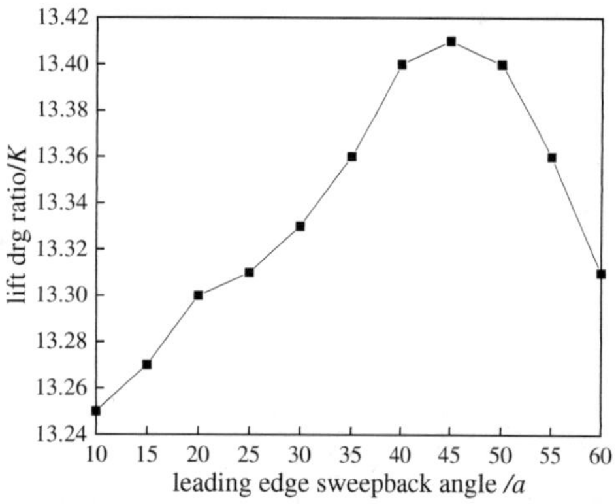

Figure 4. The relation curves of lift drag ratio and leading-edge sweepback angle

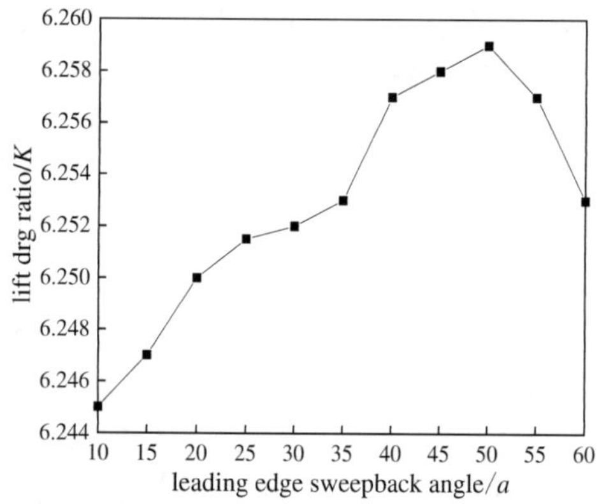

Figure 5. The relation curves of lift drag ratio and leading-edge sweepback angle

As can be seen from figure 3, with the increase of leading-edge sweepback angle of winglet, the overall trend of lift drag ratio of wing increases first and then decreases in take-off and landing flight. The same is true of figure 4. Because in the high angle of attack flight, such as take-off and landing, the pressure difference resistance and induced resistance are the main parts of aircraft resistance, and the winglet can effectively reduce the induced drag of the aircraft. With the increase of the leading-edge sweepback angle, the wingtip fossa of the winglet and the wingtip fossa of the wing are getting closer and closer, which can further reduce the overall strength of the tail socket, and then increase the lift-to-drag ratio. It can be seen from figure.5 that the lift-to-drag ratio of the wing also increases and then decreases during cruise. Because in the initial stage, changing the sweepback angle of the winglet can effectively reduce the energy loss and improve the lift drag ratio of the wing. However, when the sweepback angle is too large, it will cause stall and reduce the lift drag ratio of the wing.

In summary, the change of the lift-drag ratio with the sweepback angle of the leading edge of the winglet is generally similar under different flight conditions. However, the sweepback angles corresponding to the maximum lift-drag ratios in different states are not consistent. It can be seen from figure.3, figure.4 and figure.5 that the optimal sweepback angles corresponding to the maximum lift drag ratios of take-off, cruise and landing states are 35°, 45°and 50° respectively.

3. Design of driving mechanism of variable angle of sweepback of winglet

3.1. SMA drive mechanism requirements

The design of the variable sweepback angle winglet drive mechanism focuses on how to reasonably arrange the drive mechanism in the small space of the winglet. Moreover, because the surface of winglet has a large aerodynamic load, the driving mechanism should produce enough driving force to overcome the aerodynamic load. The driving method of the traditional hydraulic mechanical system is more complicated, and the weight and volume are relatively large. Therefore, a drive system represented by a new type of smart material must be selected as the drive actuator of the variable

IOP Conf. Series: Materials Science and Engineering **926** (2020) 012014 doi:10.1088/1757-899X/926/1/012014

sweepback angle winglet. Current research on SMA shows that SMA materials have unique shape memory effects and super elasticity, and they can be very effectively used as fast-response, light-weight, and high-force actuators. Therefore, it is very suitable for the application of variable sweepback angle winglets [7-9].

3.2. The Selection of actuator for SMA driving mechanism

SMA actuator can be activated through temperature change by using SMA wires, SMA belts, SMA torque tubes and other forms, therefore, the selection of SMA actuator needs to be considered according to the specific situation. The variable sweepback angle winglet scheme proposed in this paper only requires deformation in a single direction, and considering the control complexity, SMA wire is selected as the actuator. Such an actuator has the advantages of smaller volume, simple control and larger displacement. It can be designed to be installed on the winglet and will not affect the normal flight of the aircraft.

3.3. The design of SMA drive mechanism

In this paper, the variable sweepback angle winglet driving mechanism is composed of winglet matrix and SMA actuator, the simplified model is shown in figure 6. SMA actuators can be connected to the point inside the winglet after cooling below the martensite temperature. When the wires are heated above the initial temperature of austenite, they begin to shrink to their memory length. In this process, the transformation recovery force is generated and the matrix of winglets is deformed, thus changing the sweepback angle of winglets. When the temperature is below the completion temperature of martensite, the SMA actuator returns to the original state, so that the winglet sweepback angle returns to the original shape.

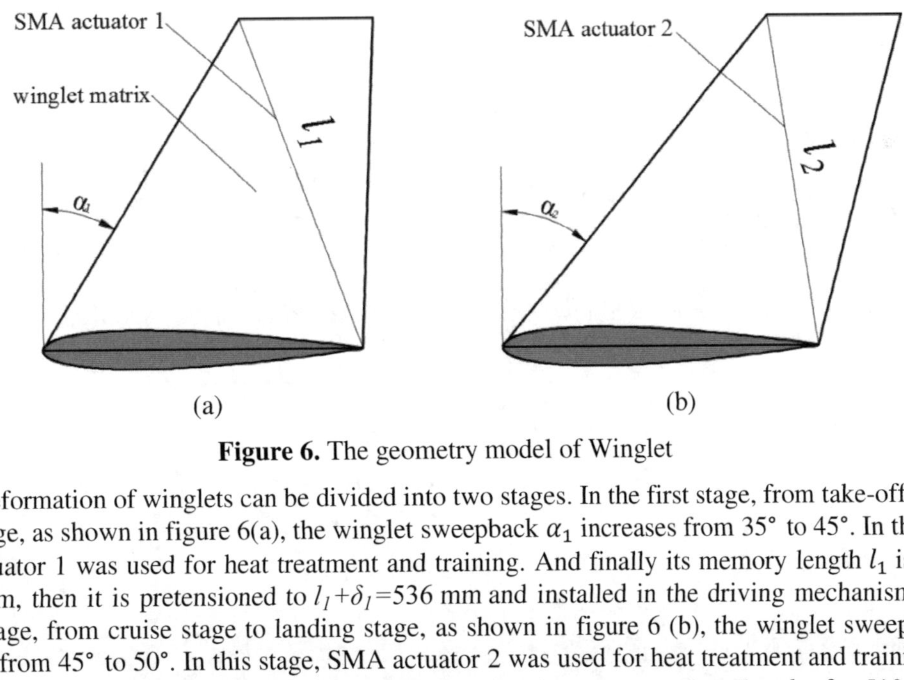

Figure 6. The geometry model of Winglet

The deformation of winglets can be divided into two stages. In the first stage, from take-off stage to cruise stage, as shown in figure 6(a), the winglet sweepback α_1 increases from 35° to 45°. In this stage, SMA actuator 1 was used for heat treatment and training. And finally its memory length l_1 is trained to 510 mm, then it is pretensioned to $l_1+\delta_1$=536 mm and installed in the driving mechanism. In the second stage, from cruise stage to landing stage, as shown in figure 6 (b), the winglet sweepback α_2 increases from 45° to 50°. In this stage, SMA actuator 2 was used for heat treatment and training. And finally its memory length l_2 was trained to 480 mm, then it was pretensioned to $l_2+\delta_2$=510 mm and installed in the driving mechanism. Electric heating is used to control the SMA drive mechanism. First, wrap the resistance wire around the surface of the SMA wire. Then, the outermost layer is covered with an elastic insulating skin, so that it is not affected by the external environment during the flight. By controlling the on-off button installed on the control panel, the SMA drive mechanism is turned on and off to control the sweepback angle of the leading edge of the winglet.

4. The realization of variable sweepback winglet driving mechanism

4.1. SMA Constitutive Model

The SMA material model selected in this paper is only subjected to axial tensile force, and the model has a larger axial dimension than a radial dimension, so only the axial deformation is considered. For the one-dimensional case, the constitutive relationship of the SMA can be finally obtained.

$$\begin{cases} \varepsilon = \varepsilon^{te} + \varepsilon^{t} = \sigma S + \alpha\left(T - T_0\right) + H\xi \\ \xi = \begin{cases} \dfrac{1}{2}\left\{\cos\left[a^M\left(T - M^f\right) - \dfrac{a^M}{C^M H}\left(\sigma H + \dfrac{\sigma^2\left(S^M - S^A\right)}{2} + \sigma\left(\alpha^M - \alpha^A\right)\left(T - T_0\right)\right)\right] + 1\right\}, & \dot{\xi} > 0 \\ \dfrac{1}{2}\left\{\cos\left[a^A\left(T - A^s\right) - \dfrac{a^A}{C^M H}\left(\sigma H + \dfrac{\sigma^2\left(S^M - S^A\right)}{2} + \sigma\left(\alpha^M - \alpha^A\right)\left(T - T_0\right)\right)\right] + 1\right\}, & \dot{\xi} < 0 \end{cases} \end{cases} \tag{1}$$

In the formula, σ is the axial tensile stress, H is the maximum transformation strain, ξ is the martensite volume fraction. $S=1/E$ is the elastic compliance, $E=E^A+\xi\left(E^M-E^A\right)$ is the elastic modulus, and E^A is the elastic modulus of austenite, E^M is the elastic modulus of martensite, $\alpha = \alpha^A+\xi\left(\alpha^M-\alpha^A\right)$ is the coefficient of thermal expansion, α^A is the coefficient of thermal expansion of austenite, and α^M is the coefficient of thermal expansion of martensite, T is the temperature, and T_0 is the initial temperature, C^M is martensite stress influence coefficient.

4.2. Numerical calculation

In this paper, a summary of the material properties of simulated SMA is shown in table 1 [10].

Table 1. SMA material parameters

M^f	M^s	A^s	A^f	E^A	E^M	α^A	H	C^M
-88°C	-55°C	0.2°C	28.4°C	70.6Gpa	32Gpa	3.5×10^{-5}/°C	0.045	0.593

According to the SMA theoretical model (1) and the material parameters in table 1, the stress-strain curve at 30°C is calculated, as shown in figure 7. It can be seen from the figure that with the increase of stress, the strain will increase correspondingly. When the stress reaches 800MPa, the maximum deformation of SMA wire is 6.7%.

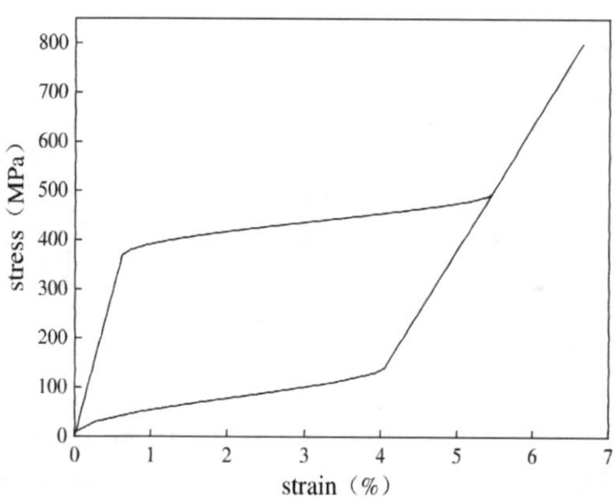

Figure 7. The curve of relationship between stress and strain

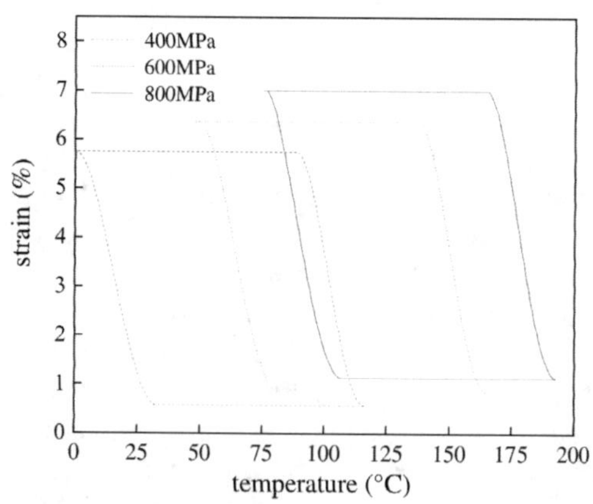

Figure 8. The temperature-strain curves under different stress

Based on the stress-strain relationship in figure 7, the temperature-strain relationship of the SMA wire was calculated under the pre-stress of 400MPa, 600MPa, and 800MPa under the thermal cycling condition of 0 °C to 200 °C. The numerical calculation results are shown in figure 8. The temperature control interval and achievable strain of the SMA wire under different stresses can be obtained from figure 8, and the numerical calculation results are shown in table 2.

Table 2. Temperature control interval and achievable strain of SMA wire under different stresses.

Strain (MPa)	400	600	800
Temperature (°C)	0~120	55~165	80~194
Stress (%)	5.2	5.5	5.8

It can be seen from table 2 that the deformation of the SMA wire gradually increases with increasing stress. It is known from Section 3.3 that during the first stage of winglet deformation, the SMA wire deformation amount is $\delta_1/l_1 = 5.1\%$, so under 400MPa stress, the control temperature range is 0 ~ 120 °C, which can meet the design requirements. In the second stage of winglet deformation, the amount of SMA wire deformation is $\delta_2/l_2 = 5.8\%$, so under 800MPa stress, the control temperature range is 80 ~ 194°C, which can meet the design requirements.

5. Conclusion

In this paper, the influence of different sweepback angle winglets on the aerodynamic performance of the aircraft are analyzed through computational fluid dynamics (CFD), and the optimal sweepback angle under different flight conditions is determined. Based on this, a new type of sweepback angle winglet structure scheme based on shape memory alloy drive is proposed. The scheme can automatically adjust the sweepback angle of winglet according to different flight conditions, so as to optimize the lift drag ratio of wings. The optimization theory and design can be extended to more wings, which lays a foundation for further optimization of aerodynamic efficiency of wings.

Acknowledgments
The authors acknowledge the financial support of National Natural Science Foundation of China (No. 11502284).

References

[1] Li W M, Bai J Q and Zhu J 2008 *J. Aviation computing technology*. (06) 36-39
[2] Ma H D and Cui E J 2007 *J. Mechanics in Engineering*. 29 (2)
[3] Jiang Y Q 2009 *Design of Aircraft Wingtips* (Beijing: Aviation Industry Press)
[4] Qian G P, Liu P Q, Zhang D W, Wu Y 2009 J. *Civil Aircraft Design and Research*. (S1) 30-33
[5] Si L and Wang H P 2010 *J. Aeronautical Computing Technology*. 40(3)
[6] Li W, Xiong K, Chen H, Zhang X, Su Y G and Ren Z Y 2012 *J. Journal of Aeronautics*. 33 (1), 22- 33
[7] BO Z and Lagoudas D C 1999 *J. International Journal of Engineering Science*. 37(9) 1089-1140
[8] Tanaka K 1986 *J. Res Mechanics*. 18(2) 251-263
[9] Liang C, Rogers C A 1990 *J. Journal of Intelligent Material Systems and Structures*. 1(2) 207-234
[10] Liu B F, Jin S Y, Chen K Y, Wang F S and Du C Z 2019 *J. Journal of Alloys and Compounds*. 797 1142-1150

IOP Conf. Series: Materials Science and Engineering **926** (2020) 012015 doi:10.1088/1757-899X/926/1/012015

Similarity Model Development of Spar Floating Wind Turbine for Vibration Experimental Study

W.Q. Huang, E.M. He and J.J. Yang

School of Aeronautics, Northwestern Polytechnical University, Xi'an, China

E-mail address: 18838250210@163.com

Abstract. As the fixed foundation wind turbine no longer meets the design and cost requirements in deep sea power plant, and the offshore floating wind turbine (FOWT) suitable has been developed rapidly. The spar FOWT model is established in ANSYS-AQWA, and the influences of spectral peak frequency and incident angle for irregular waves on the vibration responses of FOWT are studied under five load cases. Under the guidance of similar theory, 1/150 scaled model has been simulated under typical load cases. The results show that the peak frequency of the spectrum has a significant effect on the vibration responses of FOWT, the incident angle also has a slight effect. By comparing the maximum and standard deviation (SD) of the tower top fore-aft displacement (TTDspFA) with the physical model and the experimental model, the minimum errors are 3.60% and 3.56%, respectively. This work will lay a foundation for experimental study on vibration of the floating wind turbine.

1. Introduction

As the main method of utilizing wind energy, wind power generation has become one of the commercialized power generation methods in new energy technology [1]. Offshore wind power has become an important direction for the development of the global wind power industry due to its unparalleled advantages in onshore wind power. Offshore wind farms are far from residential areas on land, which can eliminate the impact on the lives of land residents during operation, and can install larger-scale wind turbines to form wind farms, and fully convert offshore wind energy into electrical energy [2-3]. Therefore, the research on FOWT has important practical significance.

At present, Sultania et al. [4] used numerical simulation methods to study the motion responses performance of the spar floating wind turbine platform under extreme marine environment. Jonkman et al. [5] developed HydroDyn, a hydrodynamic calculation module embedded in FAST, established a fully coupled calculation model for FOWT, and analysed the dynamic response characteristics of 5MW barge floating wind turbines under different wind and wave loads.

Nielsen et al. [6] studied the motion responses of the spar floating wind turbine platform system by combining numerical simulation and 1/47 scale model method. Utsunomiya et al. [7-8] further studied the motion responses of the floating wind turbine system under wave load by using the methods of 1/22.5 and 1/100 model experiments. Ishihara et al. [9] conducted a wind-wave experiment on a semi-submersible floating system using a 1/150 scale model to effectively verify the motion response performance of the platform system.

The influences of spectral peak frequency and incident angle for irregular waves on the vibration responses of FOWT are studied in this paper. For the experimental work, 1/150 scaled model has been simulated. The content of this paper is as follows: spar FOWT model is introduced in section 2. The influences of wave load on FOWT vibration responses are introduced in section 3. The similarity theory is introduced in section 4. The simulation comparison results of the physical model and the experimental model are in section 5. The last section presents the conclusions of these simulations.

2. Floating offshore wind turbine model

At present, four types of design schemes for floating-based wind turbine support platforms are adopted: the spar floating platform of the European Hywind project, Tension Leg Platform developed by the National Renewable Energy Laboratory and MIT, Barge floating platform and the Dutch three-float structure, etc. [10]. Compared with the other three platforms, the spar floating platform has obvious comprehensive advantages in terms of load performance, cost, and structural form. It seems to be the most suitable support way for deep waters [11]. The spar FOWT is selected as the research object in this paper, and the wind turbine parameters are shown in Table 1 [12-13].

Table 1. Parameters of spar FOWT.

Parameter	Parameter value
Rating	5MW
Rotor Orientation, Configuration	Upwind, Three blades
Rotor, Hub Diameter	126m, 3m
Hub Height	90m
Cut-In, Rated, Cut-Out Wind Speed	3m/s, 11.4m/s, 25m/s
Rotor Mass	110000kg
Nacelle Mass	240000kg
Tower Mass	347460kg
Platform Mass (Including Ballast)	7466330kg
Coordinate Location of Overall CM	(-0.25m, 0, -78.7m)

The rigid-body platform DOFs include translational surge, sway, and heave motions and rotational roll, pitch, and yaw motions. Positive surge is defined along the positive X-axis, sway is along the Y-axis, and heave is along the Z-axis. Positive roll is defined about the positive X-axis, pitch is about the Y-axis, and yaw is about the Z-axis. The illustrations of structural model and finite element model are shown in Figure 1.

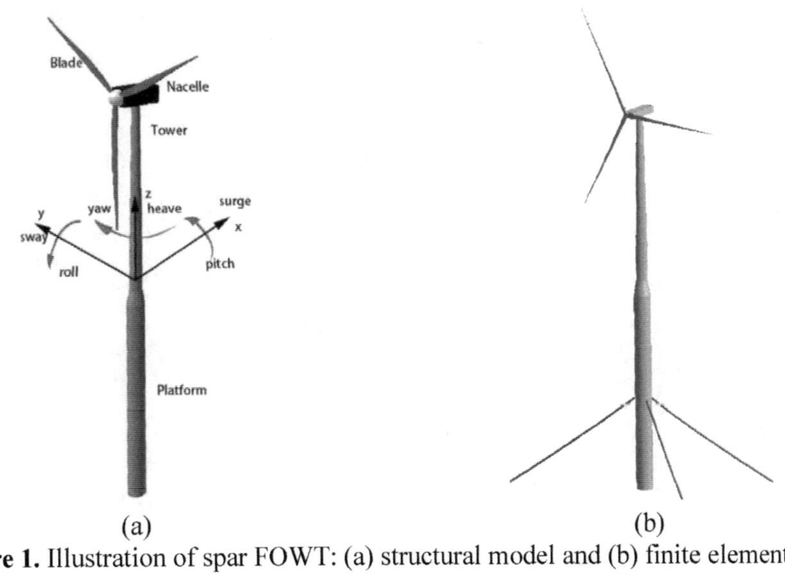

(a) (b)

Figure 1. Illustration of spar FOWT: (a) structural model and (b) finite element model.

3. Influences of wave load on FOWT vibration responses

The marine environment at the former Stevenson Weather Station (located at 61° 20′ N latitude, 0° 0′ E longitude) is extremely harsh. This reference site is chosen for its fairly extreme wind and wave conditions, with the implication that if the results of the loads analysis are favourable, the floating wind turbine system under consideration will be applicable at almost any site around the world. [14]. Five typical load cases are shown in Table 2. The load cases 1-5 represent the cut-in wind speed, below rated speed, rated speed, higher than rated speed and cut-out speed, respectively [14].

Table 2. Five typical load cases.

Load case number	1	2	3	4	5
Hub-height mean wind speed/(m/s)	4	8	12	18	24
Significant wave height/m	1.7	2	2.6	4.0	5.6

The influences of wave spectrum peak frequency and incident angle on wind turbine vibration responses are studied in this section. The three spectral peak frequencies are 0.034Hz, 0.08Hz and 0.12Hz, where 0.034Hz is the critical natural frequency of the wind turbine platform. When the FOWT is working in the marine environment, the external environmental load is random, and the direction of the waves cannot be predicted. Several wave incident angles of 0°, 30°, 60° and 90° are selected for analysis at the wave spectrum peak frequency of 0.034Hz. The NPD wind spectrum is selected to simulate wind, and the Jonswap spectrum is selected to simulate irregular waves.

In order to study the influences of wave spectrum peak frequency, it is assumed that the incident wave angle and the wind direction are both 0°. Figure 2 shows the influences of different spectrum peak frequencies on the FOWT vibration responses under five load cases. The SD of the TTDspFA and the SD of the platform heave displacement (PFDspH) are used as the evaluation criteria.

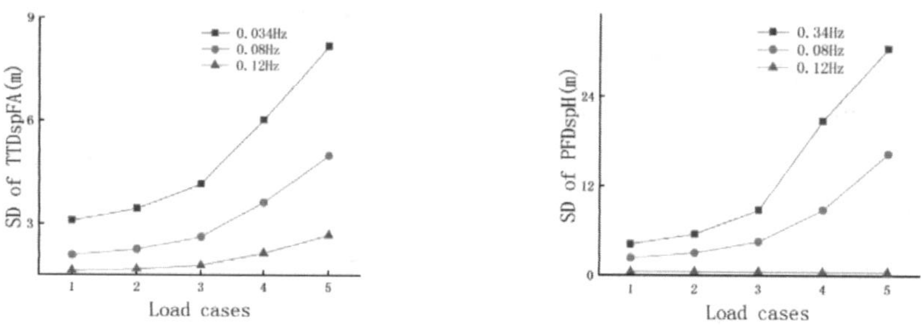

(a) (b)

Figure 2. Effects of wave spectrum peak frequency on spar FOWT vibration responses: (a) SD of TTDspFA and (b) SD of PFDspH.

Figure 2 illustrates TTDspFA and PFDspH gradually increase as the marine environment becomes more and more severe. When the peak frequency of the spectrum is 0.12 Hz, PFDspH does not change obviously. The vibration responses of the FOWT are greatest when the spectral peak frequency is 0.034 Hz because 0.034 Hz is close to the critical natural frequency of the FOWT.

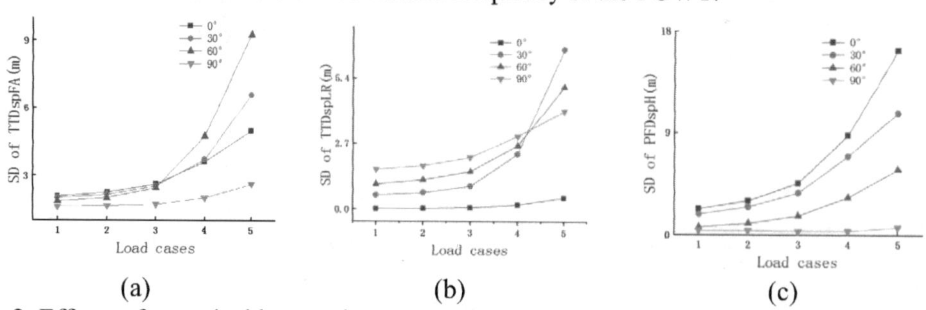

(a) (b) (c)

Figure 3. Effects of wave incident angles on spar FOWT vibration responses: (a) SD of TTDspFA (b) SD of TTDspLR and (c) SD of PFDspH.

Figure 3 gives the influences of different incident angles on FOWT vibration response under five load cases. Figure 3 shows that the vibration responses of FOWT will increase as the working conditions become more and more severe, the incident angles of 60° and 0° have great effects on TTDspFA and PFDspH, respectively. Under normal load cases, 90° incident angle has a greater impact on TTDspLR, while TTRspLR increases significantly at 30° incident angle as the working conditions become worse.

4. Similarity theory

In marine engineering, the similarity theory refers to a certain similarity relationship that the experimental model and the physical model should satisfy. We usually choose appropriate similarity criteria according to specific experimental research objects.

The ratio of the linear scales of the experimental model and the physical model is constant, and the corresponding angles are equal, as shown in Eq. (1).

$$\frac{L_s}{L_m} = \frac{B_s}{B_m} = \frac{D_s}{D_m} = \lambda \qquad \frac{\alpha_s}{\alpha_m} = 1 \tag{1}$$

where λ is the scale ratio, L_s, B_s, D_s, α_s and L_m, B_m, D_m, α_m are the length, width, draught and included angle of physical model and experimental model respectively.

The marine engineering model experiment is mainly to study the structure's movement under the wind and wave conditions, where gravity and inertial force are the main factors that determine its force. Therefore, the most common method of scaling the model is to use Froude's scaling law, the scaling factor is defined by the Froude-number (Fr). I In addition, the structure has periodic motion on the waves. The physical model and the experimental model must keep the Strouhal-number (St) equal, as shown in Eq. (2).

$$\frac{V_s}{\sqrt{gL_s}} = \frac{V_m}{\sqrt{gL_m}} \qquad \frac{V_s T_s}{L_s} = \frac{V_m T_m}{L_m} \tag{2}$$

where V_s, L_s, T_s and V_m, L_m, T_m are the characteristic velocity, characteristic line scale and period of the physical model and the experimental model, respectively.

Marine engineering model experiment is usually conducted in freshwater tank, but the actual working environment is the marine environment. Therefore, in the water tank experiment, the density of the fluid needs to be corrected, and the ratio of seawater to freshwater is γ (γ=1.025). For future experiment preparation, the scale ratio is selected as 150, and the experimental model is calculated by the physical model and scale ratio. The main experimental model design parameters are shown in Table 3.

Table 3. Experimental model design parameters.

Parameter	Scaled ratios	Experimental model
Rotor, Hub Diameter	λ	0.8m, 0.02m
Hub Height	λ	0.6m
Rotor Mass	$\gamma\lambda^3$	0.032kg
Nacelle Mass	$\gamma\lambda^3$	0.069kg
Tower Mass	$\gamma\lambda^3$	0.101kg
Platform Mass (Including Ballast)	$\gamma\lambda^3$	2.158kg
Coordinate Location of Overall CM	λ	(-0.00167m, 0, -0.525m)

According to the conversion relationships given above, the relationships between various physical quantities such as stiffness, weight per unit length, elastic coefficient, restoring force coefficient, and wind coefficient can be further calculated.

5. Comparisons of physical model and experimental model

By comparing the relative errors of the maximum and SD of TTDspFA, the accuracy of the scale model is verified, as shown in Eq. (3).

$$\sigma_x = \frac{X_b - X_a \times \lambda}{X_b} \times 100\%$$
$$\sigma_s = \frac{S_b - S_a \times \lambda}{S_b} \times 100\% \quad (3)$$

where X_b, S_b are the maximum and SD of the TTDspFA of the physical model respectively, X_a, S_a are the maximum and SD of TTDspFA of the experimental model, and λ is the scaled ratio.

Table 4. Relative errors of the maximum of TTDspFA.

Load case number	1	2	3	4	5
Physical model displacements(m)	10.16	11.36	13.89	20.61	29.00
Similar conversion values(m)	9.75	10.95	13.2	19.35	25.95
Relative errors	4.00%	3.60%	4.97%	6.11%	10.50%

Table 5. Relative errors of the SD of TTDspFA.

Load case number	1	2	3	4	5
Physical model displacements(m)	3.09	3.43	4.15	6.02	8.19
Similar conversion values(m)	2.98	3.29	3.95	5.66	7.42
Relative errors	3.56%	4.08%	4.82%	5.98%	9.40%

Table 4-5 show that the relative errors between the physical model and the experimental model basically increases with increasing wind and wave loads. While the overall relative errors are small and can basically meet the experimental requirements.

6. Conclusions

The influences of wave spectrum peak frequency and incident angle on the vibration response of spar FOWT are studied in this investigation, and the accuracy of the next experimental model is verified. The main conclusions are as follows:

1) The peak frequency of the wave spectrum has a significant influence on the vibration responses of FOWT. The FOWT will have a large vibration response when the peak frequency of the spectrum is close to critical natural frequency of the wind turbine.

2) The incident angle has a certain influence on the dynamic response of the wind turbine. The incident angles of 60° and 0° have great effects on TTDspFA and PFDspH, respectively.

3) By comparing the maximum and SD of TTDspFA with the physical model and the experimental model, the minimum errors are 3.60% and 3.56%, respectively.

References

[1] Cumali L, Hüseyin A, Rasim B. The Current Status of Wind Energy in Turcritical and in the World[J]. Energy Policy, 2011, 39(2): 961-967.

[2] Breton SP, MoeG. Status, Plans and Technologies for offshore Wind Turbines in Europe and North America[J]. Renewable Energy,2009,34(3): 646-654.

[3] Snyder B, Kaiser MJ. Ecological and Economic Cost-Benefit Analysis of offshore Wind Energy[J]. Renewable Energy,2009,34(6): 1567-1578.

[4] A. Sultania, L. Manuel. Extreme Loads on a Spar Buoy-Supported Floating Offshore Wind Turbine. AIAA 2010-2738.

[5] Jonkman J M, Buhljrml. Loads analysis of a floating offshore wind turbine using fully coupled simulation[R]. National Renewable Energy Lab. Golden, CO, 2007.

[6] F. G. Nielsen, T. D. Hanson and B. Skaare. Integrated Dynamic Analysis of Floating Offshore Wind Turbines. Proceedings of OMAE2006 25th International Conference on Offshore Mechanics and Arctic Engineering, Hamburg, Germany, 2006.

[7] T. Utsunomiya, Eitaro Nishida. Wave Response Experiment on SPAR-type Floating Bodies for Offshore Wind Turbine. Proceedings of the Nineteenth International Offshore and Polar Engineering Conference Osaka, Japan, 2009.

[8] T. Utsunomiya, T. Sato, et al. Experimental validation for motion of a SPAR-type floating offshore wind turbine using 1-22.5 scale model. International Conference on Ocean, Offshore and Arctic Engineering, Honolulu, Hawaii, USA, May 31-June 5, 2009.

[9] T. Ishihara, P. V. Phuc, et al. A study on the dynamic response of a semi-submersible floating offshore wind turbine system Part 1: A water tank test. The Twelfth International Conference on Wind Engineering, Cairns, Australia. 2007.

[10] Karimirad M. Floating Offshore Wind Turbines[M]. Offshore Energy Structures. Springer International Publishing,2014:53-76.

[11] Dinh V N, Basu B. On the modeling of spar-type floating offshore wind turbines[J]. Key Engineering Materials, 2013: 636-643.

[12] Jonkman J. Definition of the Floating System for Phase IV of OC3[R]. Golden, National Renewable Energy Laboratory, 2010.

[13] Jonkman J. Definition of a 5-MW Reference Wind Turbine for Offshore System Development[R]. Golden, National Renewable Energy Laboratory, 2009.

[14] Jonkman M J, Dynamics Modeling and Loads Analysis of an Offshore Floating Wind Turbine[R]. Golden, CO, USA: National Renewable Energy Laboratory, 2007: NREL/TP-500-41958.

Research on the Simplified Calculation Model of Large Phased Array Radar Reflector Mechanics Based on NASTRAN

YeQing Gu, JinWei Wang and BaoFu Tang

Nanjing Research Institute of Electronics Technology, Nanjing 210039, China

13913365261@139.com

Abstract. there are many plates and beams on the radar array. In the process of finite element calculation, choosing the appropriate shell mesh finite element model to replace the solid finite element model can reduce the calculation amount. In this paper, the finite element solid mesh model is established based on the reflection panel model of a radar antenna array, and the simulation results are obtained. Secondly, the shell grid is used to perform finite element modeling and correction calculations on the reflective panel. The results show that the modified shell grid calculation results are very close to the solid grid model results, and can replace the solid grid to simplify the calculation of the array.

1. Introduction

The radar antenna array is one of the most important and most complicated main structures of the radar. Its mechanical properties play an extremely important role in the overall electrical performance of the radar [1]. Therefore, in the design process of radar antenna array, it is often necessary to carry out simulation calculation [2]. However, because the radar antenna array is mainly composed of antenna beam, folding beam, antenna element blade, reflection panel and other plate parts, the structure is very complex, and the workload of finite element modeling and calculation is very large, so it is necessary to simplify the finite element model, and then complete the calculation and analysis of the whole array quickly and reasonably [3] [4]. It is pointed out that for the plate and shell parts, the work of generating the plate and shell finite element model can be greatly reduced and the calculation efficiency can be improved by extracting the mid plane model to establish the shell mesh[5].Therefore, in the process of simulation calculation of antenna array, the shell grid method can be used for some plate parts to reduce the workload of simulation calculation.

In this paper, taking a key component on a radar antenna array-reflecting panel as an example, the geometric model of the finite element solid mesh model and the finite element shell mesh model are established and simulated, and the calculation results are obtained and compared and analyzed. It is found that the displacement of the shell mesh and the solid mesh finite element model in the X direction is quite different。 Then select the mass, centroid and displacement of the relevant nodes in the X direction as the target value, establish the response target, and modify and calculate the thickness of different areas of the shell grid model as the modified parameters. The results show that the calculation results of the modified shell grid finite element model are the same as those of the solid grid under the same constraint conditions and uniform load The calculation results of the lattice finite element model are very close. The corresponding displacement error, centroid error and mass error are

IOP Conf. Series: Materials Science and Engineering **926** (2020) 012016 doi:10.1088/1757-899X/926/1/012016

all within the acceptable range, which can effectively reduce the finite element calculation model of the reflection panel, and can provide reference for the simulation and simplified calculation of the antenna array of similar structure.

2. Simulation calculation of finite element model of solid mesh of reflection panel

The antenna array of the large-format element array has excellent electrical properties, but a large number of antenna elements and reflection nets (as shown in Figures 1 and 2) bring great challenges to the simulation calculation. In this paper, the key component of the radar antenna array, the reflector, is selected for finite element calculation. The shell mesh finite element model is used instead of the solid mesh finite element model to reduce the calculation amount.

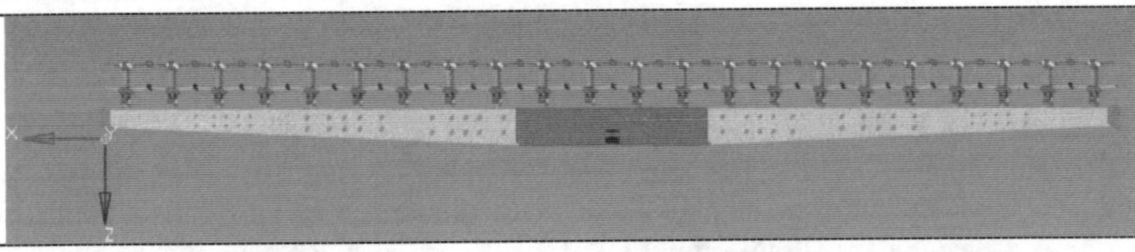

Figure 1. antenna array model

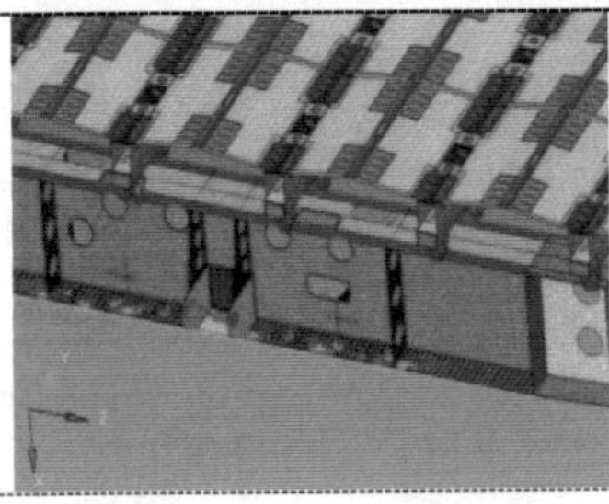

Figure 2. local schematic diagram of antenna array

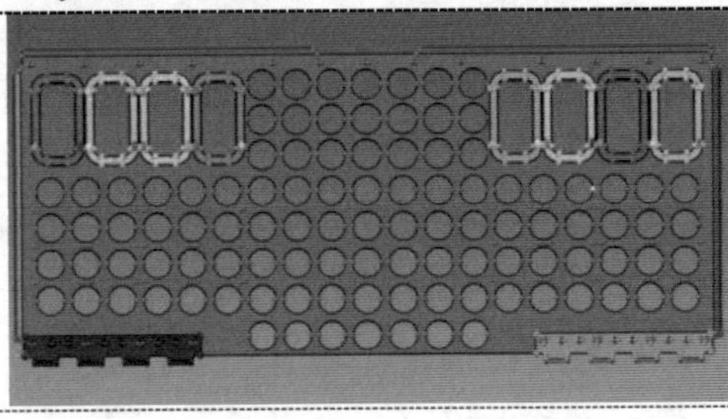

Figure 3. structural diagram of reflection panel.

The reflective panel on the antenna array is made of stainless steel, with the structure as shown in figure 3. In the middle area, there is a weight reducing hole with a thickness of 3mm. The thickness of the peripheral reinforcement of the hole on the upper side of the panel is 6.5mm, and the thickness of the hinge installed on the lower side is 1.2mm. The solid element model of the reflection panel is established. The reinforcement around the hinge and hole is made of tetrahedral mesh, and the solid mesh element size is 2 mm. The material property of the solid element is set as stainless steel, the elastic modulus is 210000Mpa, Poisson's ratio is 0.3, and the density is 7.9e-9 (t / mm^3). The uniform load is applied on the panel, and RB2 connection is adopted for the connection between the reflection panel and the hinge. At the same time, the stiffener around the hole on the upper left and right sides of the reflection panel is connected through RB2, and 6 degrees of freedom and Z-direction translational freedom of one end of the hinge on the reflection panel are constrained as shown in figure 4.

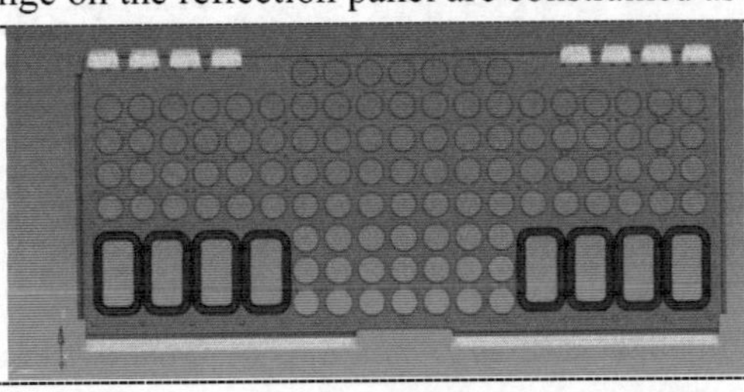

Figure 4. grid model constraints of reflection panel

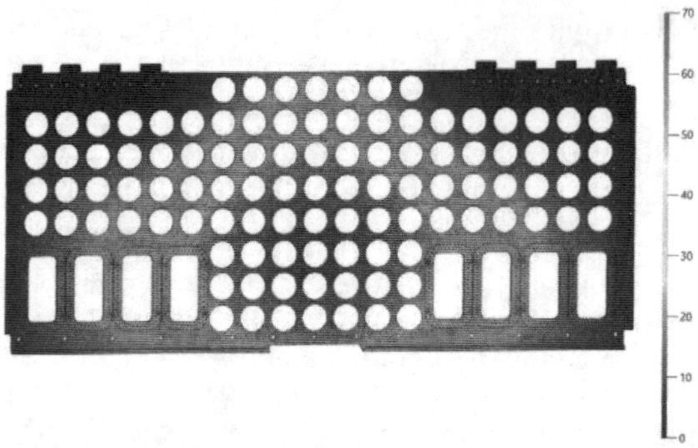

Figure 5. calculation results of solid grid model of reflection panel

After calculation, the displacement nephogram of the reflection panel in X direction is as shown in figure 5. After calculation, the displacement nephogram of the reflection panel in X direction is as shown in figure 5. As can be seen from figure 5, the maximum displacement of the reflection panel is 62.98mm, and the maximum displacement in X direction is 62.58mm.

3. Finite element model simulation of reflecting panel shell

According to the thickness and shape of the reflection panel structure shown in figure 3, it is divided into 8 areas first, then the weight reduction holes on the reflection panel are removed, the middle surface is extracted and the mesh is divided, and the corresponding mesh is given initial thickness according to different areas as shown in figure 6. Set the material property of the shell element as stainless steel, elastic modulus as 210000Mpa, Poisson's ratio as 0.3, density as 7.9e-9 (t / mm^3). Then the corresponding uniform load is applied to the corresponding area of the shell grid, and the total load applied to the reflection panel is the same as the load applied to the solid grid model, which constrains the six degrees of freedom of the hinge on the reflection panel and the z-direction translational degrees of freedom at one end. Finally, the simulation calculation is carried out, and the displacement nephogram of the reflection panel in the X direction is shown in figure 7.

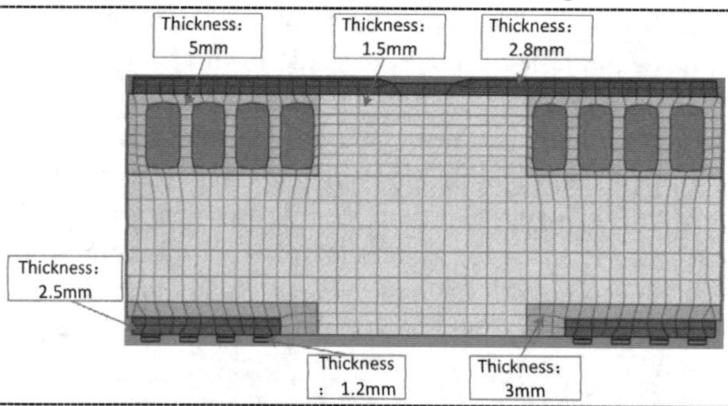

Figure 6. Schematic diagram of the partition thickness of the reflective panel

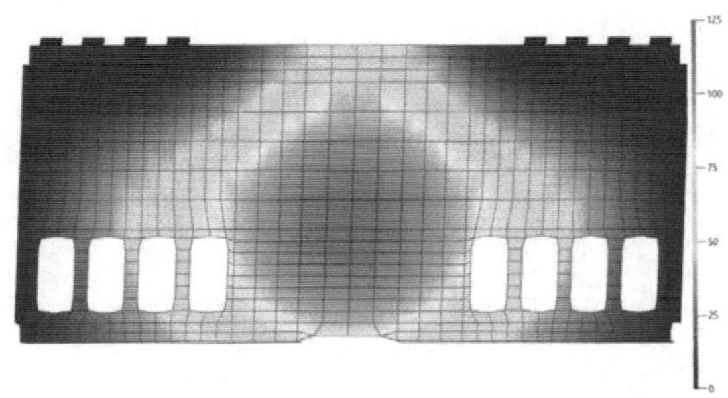

Figure 7. Calculation results of the reflective panel shell mesh model

From figure 7, it can be seen that the maximum displacement of the panel is 115.3mm, and the maximum displacement in the X direction is 114.6mm, which is quite different from the result of the solid mesh finite element model, so it is necessary to modify the shell mesh finite element model.

4. Reflective panel shell mesh finite element model correction calculation

4.1. Modified calculation of the node X displacement on the solid grid as the target value

The thickness of different areas of the shell grid is corrected according to the node displacement of the grid model of the reflective panel, and the finite element model of the reflective grid is reduced. Select the nodes 105687, 106226,106656, 107421, 107828, 132092, 100170, 94477, 88649, 88820, 89098, 89297, and 89524 in the area shown in figure. 8 as the target values in the X direction to establish the response target, and at the same time, use the reflection shell The thickness of different regions of the mesh model is the correction parameter.

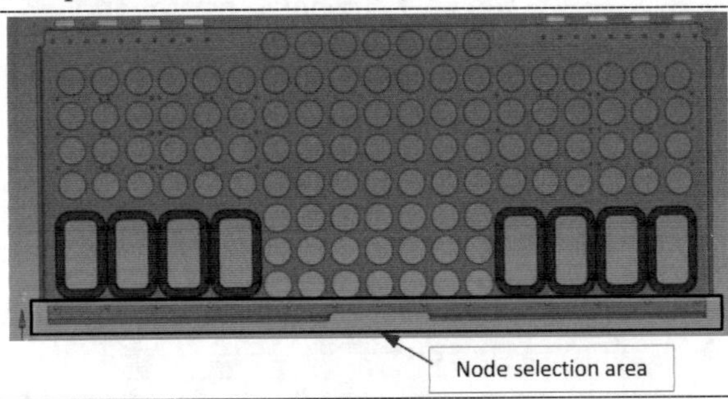

Figure 8. Location of the reflection panel

Set the number of model iterations and start to modify the model. After the final correction, the displacement comparison of each node and the reflective panel body mesh model at the corresponding node and the thickness change in the corresponding area are shown in table 1 and table 2, respectively. Similarly, the material properties of the shell element are set to stainless steel, the elastic modulus is 210000MPa, the Poisson's ratio is 0.3, the density is 7.9e-9 (t/mm^3), the uniformly distributed load is applied and the finite element calculation is performed. The calculation results are shown in figure 9.

Table 1. Comparison of displacement of the reflective panel shell mesh model after correction and displacement of the volume mesh model

ID number of volume grid point	Volume grid displacement in X direction	ID number of shell grid point	Corrected displacement in X direction	Difference(%)
105687	10.4217	21714	10.3035	-1.134

106226	17.4804	21718	17.4215	-0.337
106656	24.9909	21722	24.9098	-0.325
107421	34.8041	21712	34.6476	-0.45
107828	40.5274	21747	40.3591	-0.415
132092	46.523	22030	46.2652	-0.554
100170	48.2979	22031	48.1991	-0.205
94477	46.523	22424	46.2652	-0.554
88649	40.5274	22079	40.3591	-0.415
88820	34.8041	22231	34.6476	-0.45
89098	24.9909	22241	24.9098	-0.325
89297	17.4804	22237	17.4215	-0.337
89524	10.4217	22233	10.3035	-1.134

Table 2. Thickness variation of corresponding area

	Level	Type	Elem/Set	Old	Actual	Difference(%)
1	GLOBAL	H	2	2.50000E+00	2.67339E+00	6.93545E+00
2	GLOBAL	H	3	1.50000E+00	2.20093E+00	4.67286E+01
3	GLOBAL	H	4	5.00000E+00	4.48724E+00	-1.02551E+01
4	GLOBAL	H	5	2.80000E+00	2.90939E+00	3.90668E+00
5	GLOBAL	H	6	1.20000E+00	1.18192E+00	-1.50632E+00

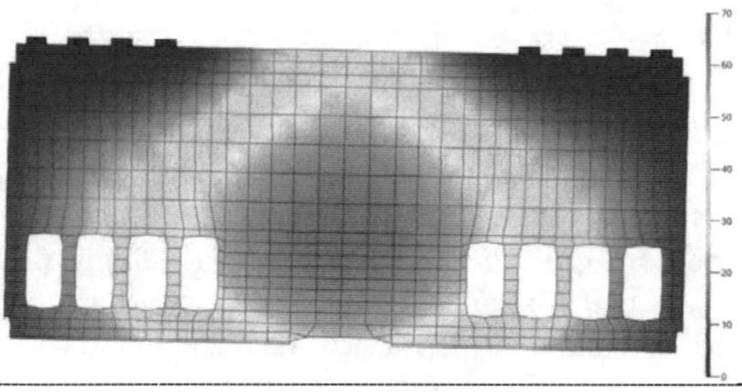

Figure 9. The calculation results of the reflection panel shell mesh model after modification

According to table 1, table 2 and figure 9, when a fixed constraint is applied to the hinge and a symmetrical constraint is applied to the reflection network, and the x-direction displacement of the node of the reflection panel mesh model is the response target, the maximum X direction displacement after the correction of the reflection panel mesh model is 63.86mm, while the maximum X-direction displacement of the reflection panel mesh model is 62.58mm, with an error of 2%, which can be ignored. In addition, the error of node X direction displacement and node X direction displacement is within 1.2% after the mesh model of reflecting panel shell is modified, so the modified mesh model of reflecting panel shell can replace the mesh model of reflecting panel shell and reduce the finite element calculation model of reflecting panel

4.2. The mass, center of mass and X-direction displacement of nodes of solid mesh model are taken as the target to modify the calculation

In the actual simulation calculation process, if only the x-direction displacement of the nodes on the solid grid of the reflection panel is taken as the target for correction calculation is not enough, the mass and the center of mass of the model need to be corrected, so as to reasonably reflect the accuracy of simulation calculation.Therefore, the response target is established by taking the mass, centroid and X-direction displacement of nodes on the solid mesh model of reflection network as the target value, and the node105687, 106226,106656, 107421, 107828, 132092, 100170, 94477, 88649, 88820, 89098, 89297, and 89524 in the area shown in figure 8 as the target value, and then the response target is established by taking the x-direction displacement of nodes in different areas of reflection network shell mesh model as the target value The thickness of the domain is the modified parameter to modify the shell mesh finite element model. See Table 3 for the comparison of the weight and centroid of the modified mesh model of the reflective shell and the mesh model of the reflective mesh.

Table 3. Comparison of the weight and centroid of the modified reflection panel shell mesh model and the reflection panel body mesh model

	Reflection mesh model	Modified shell mesh model
Weight (kg)	0.9695	1.1
Centroid coordinates	(-4978.32, 3991, -480.23)	(-4977.41, 3991, -471.43)
X maximum displacement(mm)	62.58	63.91

Similarly, the material properties of the shell element are set to stainless steel, the elastic modulus is 210000MPa, the Poisson's ratio is 0.3, the density is 7.9e-9 (t/mm^3), the uniformly distributed load is applied and the finite element calculation is performed. The calculation results are shown in figure 10.

Figure 10. The calculation results of the reflection panel shell mesh model after modification

Combining table 3 and figure 10, it can be seen that when a fixed constraint is imposed on the hinge, a symmetric constraint is imposed on the hinge, and the X direction displacement, weight, and center of mass of the node of the reflector mesh model are used as response targets, the reflector mesh model The maximum displacement in the X direction is 62.58mm, and the maximum displacement in the X direction is 63.91mm after the reflection grid model is modified, with an error of 2%, which can be ignored. After the correction of the reflective mesh shell model, the node X direction displacement and the reflective mesh body model node X direction displacement error are within 1.2%, and the mass increases by 0.13Kg, which is 0.67% different from the quality of the reflective mesh body model, and the reflection The modified centroid of the reticulated mesh model is very close to the center of mass of the reflective mesh model, so the revised reflective mesh model can completely replace the reflective mesh model for simulation calculations, which can reduce the limitation of the reflective mesh. The error of node X direction displacement of the modified reflection panel shell mesh model and the node X direction displacement of the reflection panel body mesh model is within 1.2%, the mass increase is 0.13kg, and the difference between the mass of the modified reflection panel shell mesh model and that of the reflection panel body mesh model is 0.67%, and the centroid of the modified reflection panel shell mesh model is very close to the centroid of the reflection panel body mesh model, so the modified reflection panel shell mesh model It can completely replace the body mesh model of the reflection panel for simulation calculation, and can reduce the finite element calculation model of the reflection panel.

5. Conclusion

In this paper, the key component of a radar antenna array, the reflection panel, is taken as an example to establish the finite element solid mesh model and the finite element shell mesh model respectively for simulation calculation under the same load and constraint conditions. The calculated results show that the displacement of the shell mesh and the solid mesh finite element model in the X direction is quite different.

Taking the mass, centroid of the solid mesh model of the reflecting panel and the displacement of the relevant nodes on the model in the X direction as the target value, the response target is established, and the thickness of different areas of the shell mesh model is modified and calculated. The results show that the calculation results of the modified shell mesh finite element model under the same constraint conditions and uniform load are respectively the same as those of the solid mesh model The calculation results of the mesh finite element model are very close, and their centroid positions are very close. The corresponding displacement errors are 2% and 0.67% respectively, which are within the acceptable range and the centroid positions are very close. Therefore, the modified mesh model of the reflection shell can completely replace the mesh model of the reflection network for simulation calculation, which can reduce the finite element model of the reflection network, It can provide a reference for the simulation of similar structure of antenna array.

References

[1] Zhao Yaosheng. Structural design and analysis of radar array antenna [D]. University of national defense science and technology, 2008
[2] Zhang Genji, Cha Jinshui, Hu Jinsong. Structural design of large phased array radar [J]. Radar science and technology, 2016, 14 (3): 337-342
[3] Lou Huawei, Du Chunjiang. Design and simulation analysis of radar antenna inverted mechanism [J]. Electronic mechanical engineering, 2016, 32 (5): 61-64
[4] Zhang Jicheng. Research on the antenna array structure technology of a phased array radar [D]
[5] Yang Xiaodong, Shen Changyu, Zhao Zhenfeng, et al. Automatic transformation from 3D solid model to finite element plate and shell model [J]. Journal of mechanical engineering, 2006, 042 (010): 219-223

IOP Conf. Series: Materials Science and Engineering **926** (2020) 012017 doi:10.1088/1757-899X/926/1/012017

Aerodynamic Performance and Flow Structure Investigation of Contra-rotating Wind Turbines by CFD and Experimental Methods

M Faisal, X Zhao*, M-H Kang and K You

School of Aeronautics, Northwestern Polytechnical University, Xi'an, Shaanxi, 710072, China

*Corresponding author: xuzhao@nwpu.edu.cn

Abstract. Wind tunnel experiment was carried out to measure mechanical power of the contra-rotating wind turbine (CRWT) and single rotor wind turbine (SRWT) under various working conditions. Particle Image Velocimetry（PIV）device was employed to measure the transient flow field on the meridional plane of the rotors. The azimuthal velocity vector, magnitude and streamline were obtained by the average PIV image of ten intervals in a rotation cycle. CFD model of steady flow was developed to predict aerodynamic performance and flow structure of the CRWTs using S-A turbulent model. The single channel model was employed with periodic boundary condition. CFD results were validated with wind tunnel test in literature and experiment in this paper. Results show that CFD method has a satisfied accuracy in power prediction, with a relative error less than 34%. The azimuthal average velocity distributions predicted by experimental and CFD methods are close and the maximum value are approximate to each other. Results demonstrate a high velocity zone at the outside of the front rotor tip and a low velocity zone near the middle of the rear rotor. CRWT has a complicated flow structure. Particularly at the rear rotor, flow becomes uneven and streamline curvatures are more ununiformed.

1.Introduction

Contra-rotating wind turbines (CRWTs) are investigated for high energy efficiency of power plant and high wind energy utilization rate of wind farm. According to theory, the maximum energy conversion efficiencies are 0.59 and 0.64 correspondingly for single and two turbines having identical radius[1]. Besides, CRWT has other attractive features such as near-zero reactive torque on the tower, near-zero swirl in the wake, relative high inter-rotor rotational speed and high extraction of energy at low wind speed[2]. It enhances turbine efficiency[2-5], stability, densification, annual energy production[6] and therefore improves wind farm efficiency, reduces cut-in wind speed[2] and generation cost. It is particularly suitable for the fixed cross-sectional area requirement or small turbine spacing situation[7].

The flow structure of the CRWT hasn't been fully understood, leaving the flow interaction between the front and rear rotor unrevealed. Although researchers have carried out some investigation of flow field structure of the CRWT based on free wake vortex lattice method[8] or PIV measurement of the flow field[9]. The tip vortex in CRWT were found to expand wider than in SRWT. The results are limited on theory assumptions or based on the mini scale of the turbine. To obtain a real flow structure of the 1.1m diameter turbines, both the wind tunnel experiment[10] and CFD simulation were carried out.

2.Validation of CFD result with predecessors experimental data

Usui et al. [11] carried an experiment of CRWT in Japan in 2012. It is composed of a 3-bladed front rotor and a 5-bladed rear rotor. The radius of the front and rear rotors are 0.25m and 0.21m. The blades were configured by airfoil MEL002. The wind tunnel experiment were carried out to obtain power and velocity component in the flow field. The velocity component were measured by the five holes pitot tube at different section perpendicular to the axis away from the downstream of the flow (M1-M3). During the experiment, the wind speed, rotor speed and pitch angle were verified.

CFD software NUMECA was employed to validate the test data. Navier-Stokes equation was discretized with S-A turbulent model employed. The computational domain is a combined region $[0.1R_f,5R_f]\times[0,2\pi/3]\times[-3R_f,0.1R_f]$ with $[0.1R_f,5R_f]\times[0,2\pi/5]\times[0.1R_f,7R_f]$, where R_f is the front rotor radius, see Figure 1. Mesh division is $87\times51\times107$ and $87\times63\times99$ respectively. Total grid number is 1.54 million in 18 blocks. Similarly, 17 layers of grids are arranged in the boundary layer, thickness for the first is 3×10^{-5}m, expansion ratio is 1.4, resulting in Y^+ on the blades less than 10. The interface between the front and rear domain is set conservation coupling by pitching row type to guarantee the conservation of mass, moment and energy. The boundary conditions of the model are set as: far field for the circumferential and bottom surfaces of the model, periodic boundary for two pairs of shaft sections, Navier-Stokes wall for the rotating blades and Euler wall for the hub.

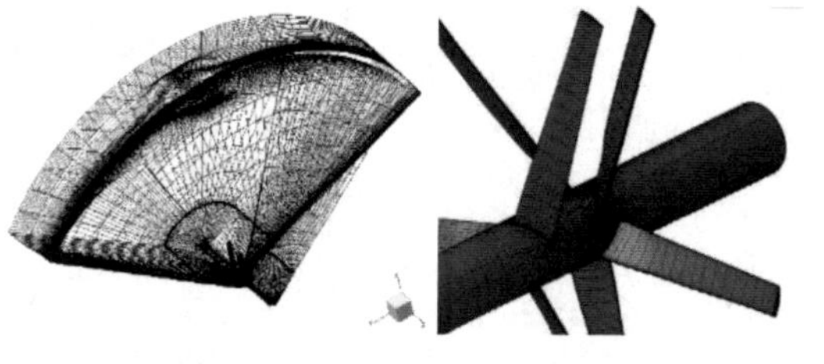

Figure 1. CFD model and surface mesh on blades

Table 1. Energy efficiency comparison for the CRWT

	C_{1p}/%	C_{2p}/%	C_p/%
EXP[11]	18	13	31
CFD	19.14	13.42	32.56

A typical working condition was selected for benchmark. The free stream velocity is 12m/s, the front rotor tip installation angle is 5.7 °and tip speed ratio is 3.4. The rear rotor tip installation angle is 15° and tip speed ratio is 2.5. The axial distance between the front and the rear wind rotors is 0.04m. Turbine power efficiency was calculated and given in table 1. It can be found that CFD result is greater than that of the experiment[11]. The energy efficiency difference for the front, rear and CRWT are 1.14%, 0.42% and 1.56% respectively. The errors are quite small which demonstrated that CFD prediction is accurate.

To validate the flow details across the CRWT, average tangential velocity (V_t) and axial velocity (V_m) at different cross sections perpendicular to the axis were compared, see Figure 2. Section M1 is in the upstream of the front turbine, M2 in the middle of the front and rear turbine, and M3 in the downstream of the rear turbine. It is clearly that the CFD result is close to that of the experiment, except small difference near the hub. This is due to the reason that in CFD, hub is treated as inviscid Euler wall. But in experiment, it was viscous and rotating wall. However, the velocity comparison shows that CFD simulation is accurate in flow field prediction.

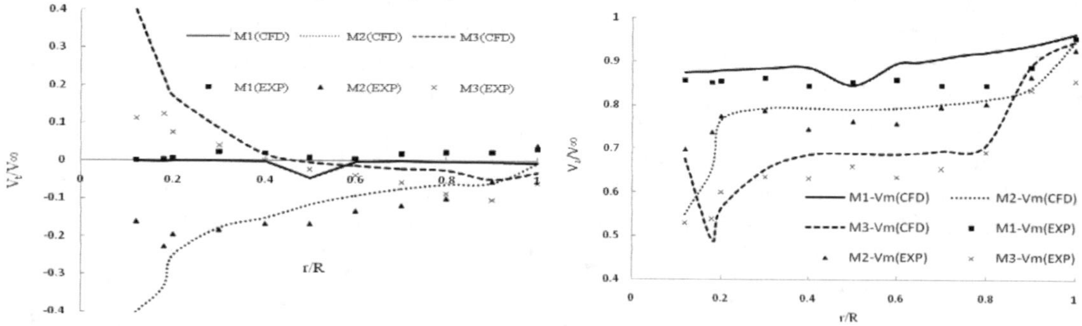

Figure 2. Average tangential velocity and average axial velocity at different cross sections

3.CFD result comparison with authors' experimental data

3.1. Wind tunnel experiment

The CRWT experiment was carried out in the low speed acoustic wind tunnel in Northwest Polytechnic University. It is a recirculation tunnel with an open test section in plenum chamber. The free stream velocity at the test section is 5~45 m/s with accuracy 0.5%, while the flow turbulence no more than 0.2%. The layout of the power measurement is shown in figure 3. Both turbines are composed of 3 blades and drive their own permanent magnetic generator via a torque meter. They have identical radius of 0.55m. The blades were made by wood and configured by NACA4412 airfoil. The front blades have a smaller chord length than the rear blades. They are different twist angles. The geometry of the blades were designed by aerodynamic coupling design and more details can be found in literature[10]. With the connection of a chain wheel, the rear rotor drived a parallel shaft. The output of each generators were linked with rectifier bridges. Therefore, the alternating currents were rectified to direct currents. As variable loads, two slide rheostats were adjusted to change the direct currents, resulting in turbine rotation speed change. The rotation of the front and rear rotor were at range 200~900rpm at wind velocity range 6~14m/s. For SRWT measurement, the rear rotor was removed and the single rotor was fixed at the front rotor location.

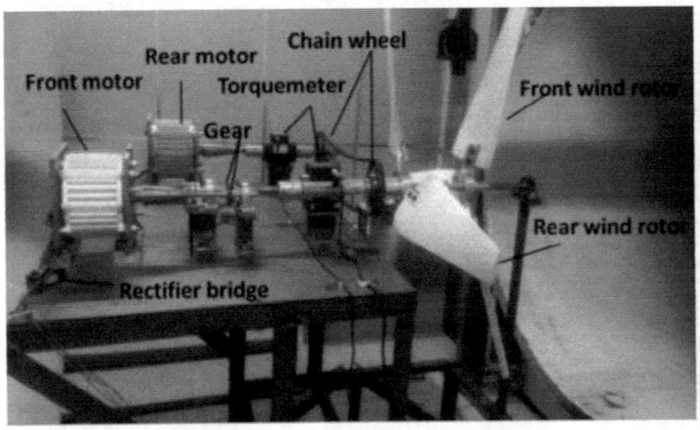

Figure 3. Test model of CRWT

Figure 4. PIV measurement of flow field

Figure 4 shows the flow field measurement layout. The meridional plane was considered to be the key zone for flow field investigation. Therefore, transient flow measurement was carried on. The blade tip zone was focused at range 293.5 mm × 293.5mm, see more details in section 3.3. The three-dimensional PIV system was made by Dantec company in Denmark, which is mainly composed of laser, CCD camera, synchronous controller and dynamic studio, a professional image processing software. Among them, the laser model is class 4 laser product of Dantec company. The pulse duration is 4ns. The laser wavelengths are 1064nm and 532nm. The CCD camera pixel is 2048 × 2048. The lens is Nikon 85mm f/1.4 lens. The maximum sampling frequency is 10Hz. In the experiment, 400W smoke generator was used as smoke generation. The azimuth locking technology adopts self-developed synchronous controller, which mainly includes Omron E6B2-CWZ6C encoder, signal acquisition and trigger circuit board and LCD. The trigger frequency and phase can be set through the touch of screen display window.

3.2.Power comparison

Several working conditions were selected for power comparison, see table 2. As for the single turbine, when the free stream velocity is 14m/s, the blade root installation angle is 45.54° and the rotating speed is 730rpm. For free stream velocity 10m/s, the blade root installation angle is 40° and the rotating speed is 519rpm. For the CRWT, when the free stream are 14m/s and 10m/s respectively, the blade root installation angles are both 35° for front blade and 42° for rear blades. However, the rotating speeds are 848rpm for the front rotor and 469rpm for the rear rotor in free stream 14m/s case. The rotating speeds are 200rpm for the front rotor and 180 for the rear rotor in free stream 10m/s case. The distance between front rotor and rear rotor is 0.22m.

In the experiment, the thickness of the blade trailing edge were manufactured in 3mm for strength consideration. While the CFD model applied the full size of airfoil, leaving the trailing edge smaller than 1mm. Considering the mean chord length ratio of the test model to CFD model is 0.932. Take the assumption that the power is proportional to cord length of the blade. The original mechanical power obtained from the experiment was modified, derived by ratio 0.932. The modified value were shown in table 2 as the experimental power.

The numerical models of single and CRWT were created separately. The mechanical power obtained by experiment were used as a reference. In table 2, the power ratio of CFD method to experimental data were listed in cracket.

Table 2. Power comparison of SRWT and CRWTs

Free stream(m/s)	SRWT(W)		CRWT					
			Front(W)		Rear (W)		Total	
	EXP	CFD (ratio)	EXP	CFD (ratio)	EXP	CFD(ratio)	EXP	CFD(ratio)
14	456	611(1.34)	511.8	570(1.11)	157.73	205.57(1.30)	669.53	727.73(1.09)
10	187.78	202.68(1.08)	22.53	23.64(1.05)	39.57	36.21(0.92)	62.10	59.85(0.96)

As can be seen from table 2, power predicted by CFD simulation is 0.92~1.34 of the experimental data. Generally, simulation powers are a little bit greater than the experiment. This is because in the experiment, the friction of gear, bearing, torque meter and other mechanical structures consume a certain amount of power, while they are ignored in CFD model. Meanwhile, the test rig has a certain obstruction effect on the free stream, which is also ignored in CFD model. Never the less, CFD model has a satisfying accuracy in power prediction.

3.3. Flow field comparison

The PIV measurement were taken near blade tip. Figure 5 shows the shooting zone with size 293.5 mm × 293.5mm near blade tip. The captured region is 0.25m~0.61m on radial direction, which is from middle of the blade to the outside of the tip. Once the blade started to rotate, CCD camera caught the picture at 0.1s interval. And the 10 intervals results were averaged. For example, at free stream velocity 14m/s, the averaged velocity vector was shown on the left of figure 6.

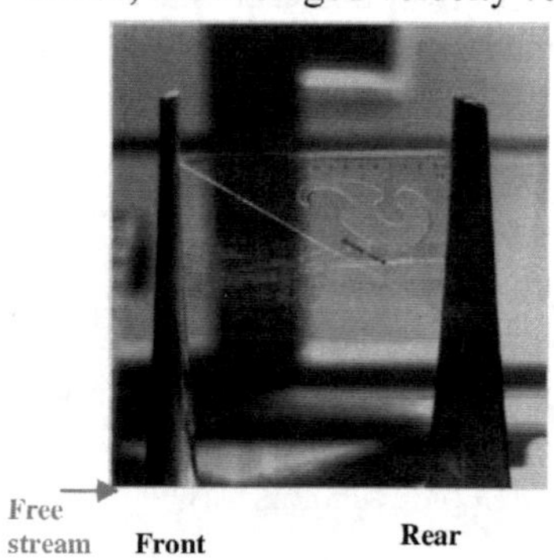

Figure 5. Exposure rate adjustment via image acquisition software for static blades

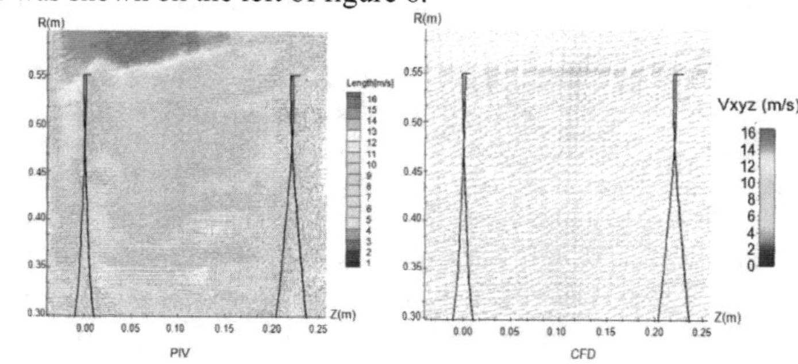

Figure 6. Azimuthal averaged velocity vector distribution for V=14m/s

The free stream is in right direction. The front rotor is on the left while rear rotor is on the right. Take the free stream velocity 14m/s for an example, figures 6~8 show the comparison of the velocity vector, velocity magnitude and streamline between two methods.

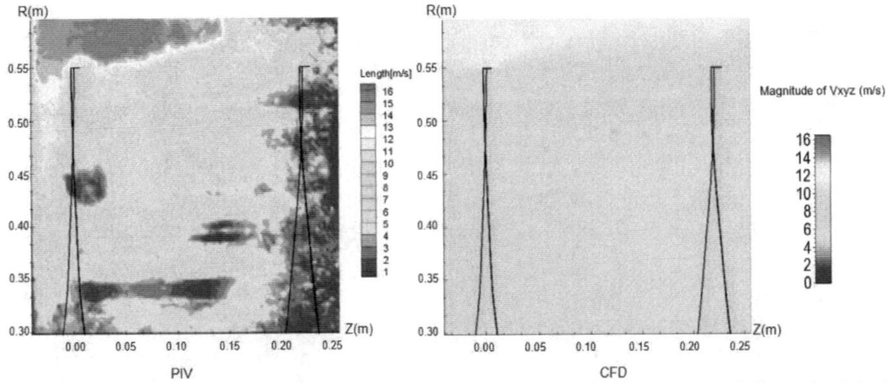

Figure 7. Azimuthal averaged velocity magnitude distribution in EXP and CFD for V=14m/s

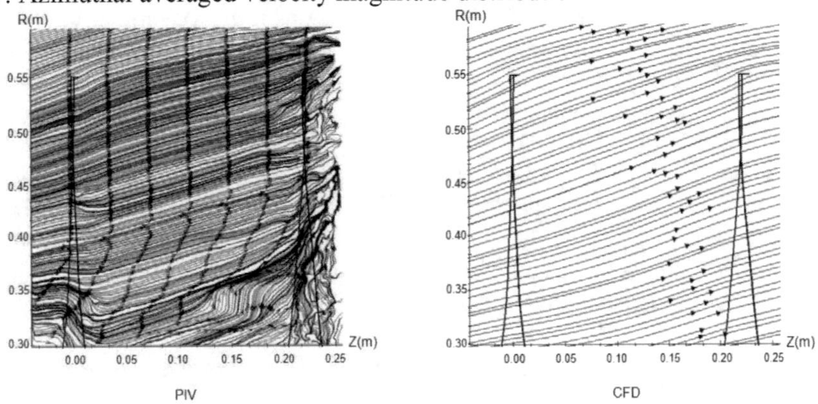

Figure 8. Azimuthal averaged streamline distribution in EXP and CFD for V=14m/s

It is obviously that both methods reveal similar velocity distribution trend. In figure 6, there is a high velocity range near the front rotor tip. The vector is along radial outward and downstream direction, which can be demonstrated in figure 8. It represents the streamline curvature and demonstrates the expansion of flow tube. Although figure 7 reveals that the maximum velocity magnitude (16m/s) in experiment in is greater than that of CFD result (13m/s), but the distribution pattern is quite similar. Both demonstrate the low velocity domain near the middle of the rear rotor. Figure 7 shows that the CFD result has a more smooth velocity distribution than the experiment. Figure 8 shows that in experiment, the streamline is more uniformed near blade tip than middle of the blade. The curve change in experiment is more complicated than that in CFD result. This is because flow was treated as steady flow in CFD model with periodic boundary, which brings the even distributed azimuthal averaged peformance.

4.Conclusions

The power and flow field prediction of the CRWT were predicted by CFD and experimental methods. Results show that the steady flow CFD models have a satisfied accuracy in power prediction. Referring to experimental power, for Usui's test [11], the relative error of CFD is less than 1.56%; for authors' experiment, the relative error less than 34%. CFD method predicts a quite close average tangential and axial velocity distribution on the cross sections to the measured data by Usui et al. For the authors' experiment, both methods predict the similar velocity distribution near blade tip. The maximum velocity magnitude of PIV result is a bit bigger than CFD result. It is regarded as test rig obstruction effect. It shows that the CRWT has a complicated velocity distribution. A relative high velocity zone existed at the external of the front blade tip and a low velocity zone in the middle of the rear blade. Particularly, the velocity stream line curvature at the rear rotor is more uneven than the front rotor.

IOP Conf. Series: Materials Science and Engineering **926** (2020) 012017 doi:10.1088/1757-899X/926/1/012017

References

[1] Newman BG. Actuator-disc theory for vertical-axis wind turbines. *Journal of Wind Engineering and Industrial Aerodynamics* 1983; 15: 347-355.

[2] Riadh WH, Yoicu G, Yeu Y et al. Performance of a contra-rotating small wind energy converter, *ISRN Mechanical Engineering* 2011; Article ID 828739.

[3] Appa K. Energy innovation small grant (EISG) program (counter rotating wind turbine system) *final report,* California, US: EISG; 2002.

[4] S Lee, H. Kim, S Lee. Analysis of aerodynamic characteristics on a counter-rotating wind turbine. *Current Applied Physics* 2010; (10): 339-342.

[5] Jung S, No T, Ryu K. Aerodynamic performance prediction of a 30kW counter-rotating wind turbine system. *Renewable Energy* 2005; 30(5): 631-644.

[6] Shen WZ, Zakkam VAK, Sørensen JN et al. Analysis of counter-rotating wind turbines. *Journal of Physics: Conference Series* 2007; IOP Publishing, 75(1): 012003.

[7] Yuan W, Tian W, Ozbay A et al. An experimental study on the effects of relative rotation direction on the wake interferences among tandem wind turbines. *Science China Physics Mechanics & Astronomy* 2014; 57(5):935-949.

[8] Lee S, Son E, Lee S. Velocity interference in the rear rotor of a counter-rotating wind turbine. *Renewable Energy*, 54:235-240, 2013.

[9] Ozbay A, Tian W, Hu H. Experimental investigation on the wake characteristics and aeromechanics of dual-rotor wind turbines, *Journal of Engineering for Gas Turbines and Power,* 138, 2016

[10] Zhao X, Zhou P, Liang X. The aerodynamic coupling design and wind tunnel test of contra-rotating wind turbines, *Renewable Energy,* 146,1-8, 2020.

[11] Usui Y, Kubo K, Kanemoto T. Intelligent wind power unit with tandem wind rotors and armatures (optimization of front blade profile). *Journal of Energy and Power Engineering*, 6, 1791-1799, 2012.

IOP Conf. Series: Materials Science and Engineering **926** (2020) 012018 doi:10.1088/1757-899X/926/1/012018

Design and Optimization of a High-altitude Long Endurance UAV Propeller

K You, X Zhao*, S-Z Zhao and M Faisal

School of Aeronautics, Northwestern Polytechnical University, Xi'an, Shaanxi, 710072, China

*Corresponding author: xuzhao@nwpu.edu.cn

Abstract. This paper presents a propeller designed for a solar unmanned aerial vehicle (UAV) at 22km altitude. The design objective for the 3-blade propeller is to achieve 72% efficiency and 7 N thrust, with 0.5588m diameter propeller rotating at 5500rpm under freestream velocity 50m/s. For such High-Altitude Long Endurance (HALE) flight vehicle, firstly, three low Reynolds number airfoils were assessed by airfoil aerodynamic performance analysis software Profili. As a result, FX 63-137 was selected for the best lift to drag ratio at given altitude. Secondly, a MATLAB program based on Joukowski propeller vortex theory was employed to determine three dimensional configuration of the blades. Then, CFD simulation was carried out on the preliminary design. Furthermore, optimal design method was employed to improve propeller efficiency of the preliminary design using software ISIGHT. Ten variables including chord lengths and blade pitch angles distribution at five spanwise sections were employed. Finally, the optimized propeller with power 467W was obtained, performance satisfying the requirement.

1. Introduction

Over the last 20 years, there are growing interest in the field of unmanned high-altitude aerial vehicles. Especially the high-altitude long endurance aircraft become popular. The propeller drive still remains the best propulsion system for airplanes powered by electricity or by fuel[1,2]. High altitude planes powered by propellers are able to fly over the wide range of altitude from 0 to 25000 meters. However, lots of difficulties have been found in propeller design in research of high-altitude aircraft propulsion system. Such wide range of altitude is connected with variability of propeller work conditions (for example the density of the air is changing 10 times) where the air density is much lower (low Reynolds number below 10^5) and the advance velocity of aircraft is extremely slow (about 10–30m/s). Therefore, the airfoils generate smaller lift and higher drag. This is connected with laminar flow separation phenomenon[3], which makes it a big challenge to design efficient propeller propulsions for stratospheric aircrafts[4]. Hence, the focus on improving the propulsive efficiency is a top priority.

There are two main propeller geometrical adjustments that could be made to improve the propeller performance[5]. Firstly, different airfoil cross-sectional shapes will affect the lift coefficients at different altitudes, affecting the propeller efficiencies under different conditions[6,7]. So, need to select some good functioning airfoil at high altitude. Secondly, the variation of blade pitch angles and chord lengths will also alter the propulsive efficiency[8]. Hence a theory must be chosen for accurate measurement.

This paper has focus on only designing the propeller efficiency for cruising condition at high altitude (Low Reynolds number) of UAV, ignoring the takeoff and landing conditions which work at low altitude (high Reynolds number). Therefore, particularly important is to find the airfoils, which are able to achieve good performance at low altitude. Airfoils for such aircraft typically operate in the Reynolds number range $2 \times 10^5 \sim 5 \times 10^5$. For computational tests, three of airfoils were selected. All of them represent airfoils dedicated for low Reynolds number flow regime: S1223, FX 63-137 and LOCKHEED L-188[9]. The Wortmann FX 63-137 is one of the most desirable airfoils for high-lift low Re models. The high-lift capability and mild-stall characteristics are among its key attributes[9]. the design philosophy employed in the design of the S1223 airfoil involves combining the favorable effects of both a concave pressure recovery and aft loading to achieve maximum lift at a design Reynolds number of 200k[9].

In this study, it is significant to evaluate the aerodynamic characteristics of the full-scale propeller for the purpose of validation. There are several methods, such as the momentum theory, blade element theory (BET), and vortex method. This papers is based on the vortex theory. Joukowski propeller vortex theory employed to determine the blade section pitch distribution and angle of flow.

To complete the proposal, the algorithm for the propeller optimizer is described, followed by the analysis of the aerodynamic characteristics of the best airfoil. Then parameters were applied established the 3-blade CATIA model. Afterwards, the validation of the design were carried by CFD method. Finally, the analysis of the optimum efficiency and design parameters of propellers are presented.

2. Primary design of the propeller

2.1. Problem description

The design objective for the 3-blade propeller is to achieve 72% efficiency and 7 N thrust, for the lowest power, with 0.5588m diameter rotating at 5500rpm under freestream velocity 50m/s.

At 22km altitude, air temperature is 218.6K, the density is 0.0645kg/m3, the speed of sound is 296.4m/s. The static pressure is 4048Pa, viscosity is 1.4×10^{-5}kg/(m·s). At cruise status, the relative velocity at 0.7R of the propeller is 127m/s. The local Ma number is 0.43 and Reynolds number is around 1.47×10^4. The relative blades chord length c/D and blade thickness distribution were selected, shown in figure 1. More details can be found in table 2.

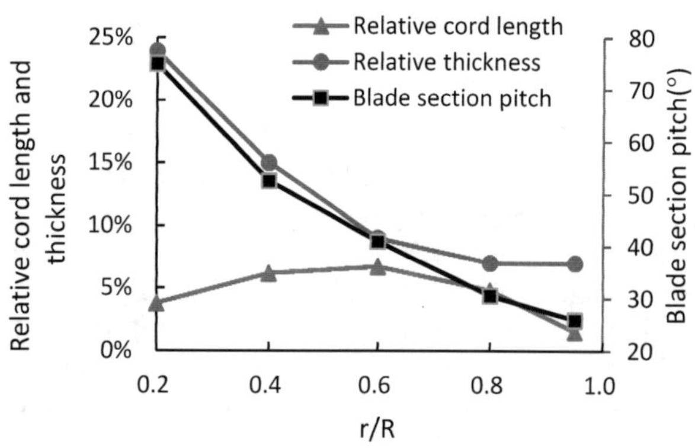

Figure 1. Blade cord length, thickness and section pitch distribution

Figure 2. Geometry of the primary designed propeller and blade

2.2. Low Reynolds airfoil selection

For high-altitude long endurance UAVs, the chord length of the propeller is small, and the air density is quite low, therefore resulted in low Reynolds number flow across the propeller. Generally, low Reynolds number airfoils has a poor performance of lift to drag ratio, which makes the high performance propeller design to be difficult.

Low Reynolds number airfoils were selected, such as S1223, FX 63-137 and LOCKHEED L-188. The aerodynamic performance under $Re=2\times10^4$ and Ma=0.43 were compared according to the same thickness design. Considering a higher lift to drag ratio and a relative steady pitching moment coefficient, airfoil FX 63-137 (thickness 13.7%) and its series thickness airfoils were selected. According to figure 1, the thickness distribution of the airfoil is related to radial location. At each section, the typical relative velocity of the blade element has typical Reynolds number. Therefore, corresponding lift to drag ratio of each airfoils were estimated via airfoil aerodynamic performance analysis software Profili. The best angles of attack were obtained under the highest Cl/Cd by of those airfoils, shown in table 1.

Table 1. Airfoil performance data

r/R	Airfoil	Thickness %	Re	Max Cl/Cd	Cl	Angle of attack (°)
0.2	FX 63-239	23.93	14000	2.65	0.5	13
0.4	FX 63-150	15.04	19000	6.36	0.69	7
0.6	FX 63-101	10.12	25000	16.34	1.4	11
0.8~0.95	FX 63-081	8.07	37000	35.53	1.38	7.5

3. Primary propeller designed by Joukowski propeller vortex theory

To obtain the chord length and blade section pitch distribution, Joukowski propeller vortex theory was employed and developed into a MATLAB program, where data in table 1 is involved. The thickness of the airfoils were modified a little bit. The preliminary design results is in table 2. The geometry of the propeller is shown in figure 2.

Table 2. Primary design result of propeller blade at each sections

r/R	Relative chord length c/D(%)	Chord length c(mm)	Blade section pitch (°)	Relative thickness(t/c) %	Airfoil
0.036	2.21	12.27	72.4	30	Ellipse
0.1	3.20	17.78	79.2	24.1	FX 63-241
0.2	3.78	21.00	75	24.1	FX 63-241
0.4	6.18	34.33	52.5	15.2	FX 63-152
0.6	6.71	37.30	41	10.3	FX 63-103
0.8	4.81	26.72	30.5	8.2	FX 63-082
0.95	1.52	8.43	25.9	8.2	FX 63-082

4. CFD simulation of the primary propeller

To evaluate the aerodynamic performance of the propeller, CFD method was carried out based on a steady flow simulation. Software NUMECA is employed to create structural grid and execute the solution.

In the numerical model, propeller is fixed in radial direction, with a radius R. Free stream flow is arranged in z direction. The computational domain covers a domain$[0,5R]\times[0,2\pi/3]\times[-10R,25R]$ in radial, azimuth and axial directions, respectively where R is the rotor radius, see figure 3. The propeller is placed at z=0. The boundary conditions include: (1) far field boundary for front, rear and circumferential faces of the cylinder, where the free flow is 50m/s in right direction; (2) periodic boundaries for the longitudinal sections; (3) Navier-Stokes wall for blade surface; (4) Euler wall for hub. The whole domain is rotation at 5500RPM around negative z direction, and the discrete precision is second order.

Structured mesh were created, the division is $129\times73\times215$ in radial, azimuth and axial directions, the total grid is 2 million in 33 blocks. Near wall boundary, the thickness for the first layer is 1×10-5c (c is the cord length of the blade) and around 15 layers in boundary layer zone. Steady flow Shear Stress

Transport (SST) turbulent flow model and Turbulent Navier-Stokes mathematical model were built on the model using Fine Turbo. The RANS equation is applied to simulate the steady flow across the propeller. Central difference method and multigrid approach are applied. The flow is considered to be converged under the condition that mass residual is less than 10^{-3}, thrust and torque of the rotor are constant.

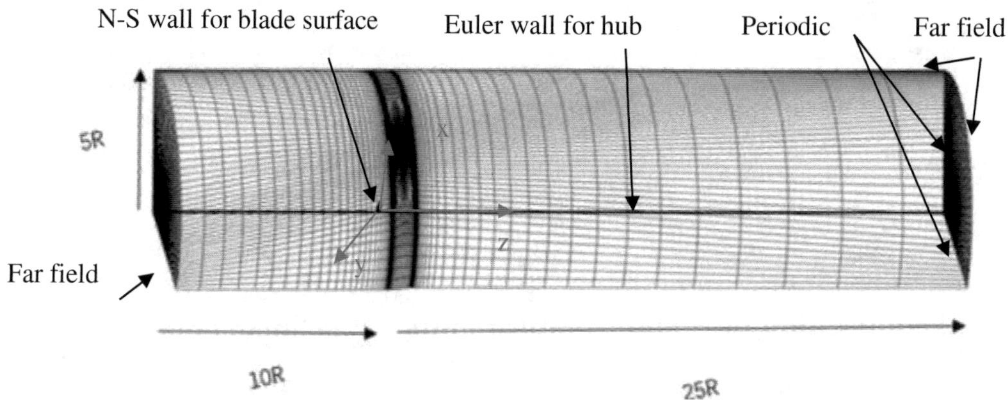

Figure 3. CFD model of propeller

The thrust and torque of the propeller were shown in table 3. And efficiency is calculated according to equation (1). In equation 1, T represents thrust, V is the cruse velocity and P is the power of the propeller, it is the product of torque with rotating speed (M·Ω).

$$\eta = \frac{TV}{P} \tag{1}$$

Compared to the design objectives, the thrust of the primary propeller is a little smaller. Therefore, the chord lengths of each sections were increased 15%, and optimal design were applied to adjust the chord length of each sections.

Table 3. CFD results of the propeller

	Thrust (N)	Torque (N·m)	Efficiency %
Design objectives	7	0.844	72
Primary design	6.1	0.759	69.76
Optimal design	7.47	0.92	70.49

5. Geometry optimization of the primary propeller

To improve the performance of preliminary design, the optimization problem was proposed to maximize the propeller efficiency at the design point with a constraint on thrust. The controlled variables are chord length at different sections. Altogether, there are 5 variables for sections 0.2R to 0.95R. The purpose of the approximate model is to produce a rapid method to evaluate propeller aerodynamic performance. This method requires the existence of a database containing several blade geometries and their associated aerodynamic performances. For better results, the NLPQL method is used. The NLPQL algorithm expands the objective function with a second-order Taylor series, and linearizes the constraints to obtain the next design point by solving the quadratic programming. Then perform a linear search based on two alternative optimization functions, where the Hessian matrix is updated by the BFGS formula, the algorithm is stable.

$$\textit{search}\ \textbf{chord length}\ x = (x_1, x_2, \cdots x_5) \qquad \textit{Design variables}$$

$$\textit{make the}\ \textbf{efficiency}\ \textit{largest} \qquad \textit{Objective function}$$

IOP Conf. Series: Materials Science and Engineering **926** (2020) 012018 doi:10.1088/1757-899X/926/1/012018

$$\textbf{thrust } F \leq -7N \qquad \textit{Inequality constraint}$$

$$0.8x_i \leq x_i^* \leq 1.2x_i \quad i = 1,2,\cdots 5 \qquad \textit{Boundary constraint}$$

The geometrical parameters and CFD results of this sample are stored into the database and the process goes on. As the number of samples in the database grows after each design cycle, the approximate model will become more and more accurate and better blade shapes will be found rapidly. Insight software were applied in optimal approach. The chord length distribution were given in figure 4. CFD results were shown in table 3. The power of the optimal result is 530W. It meets the thrust request but a little smaller than the efficiency request.

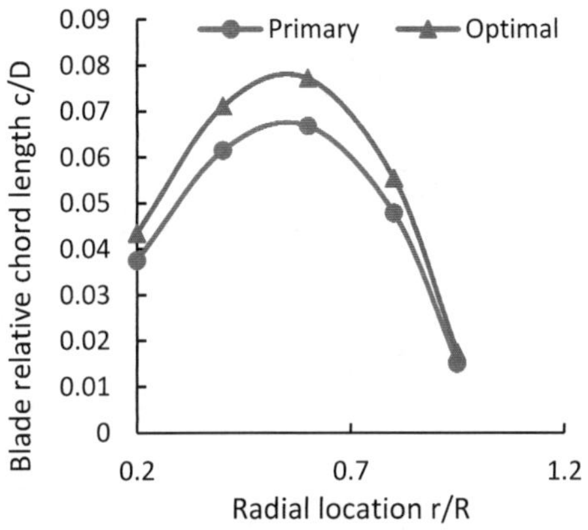

Figure 4. Chord length comparison of the primary and optimal design

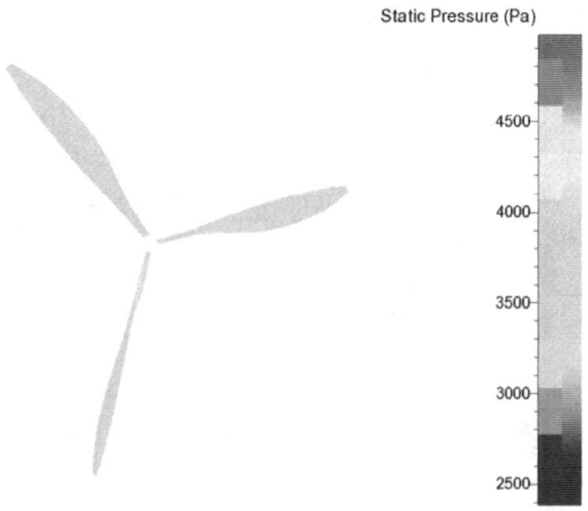

Figure 5. Static pressure distribution on the propeller

The CFD result show that the static pressure distribution on the blade, there is no flow separation on the blade wall, see figure 5. The streamline of the flow across the propeller is shown in figure 6.

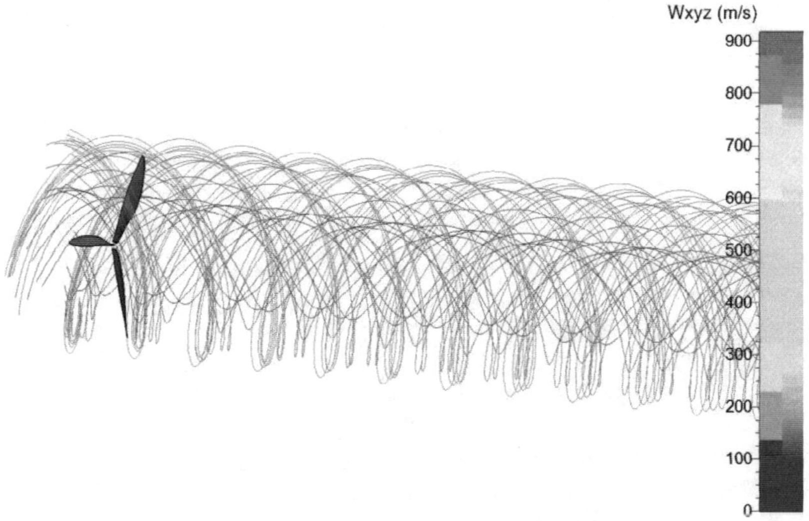

Figure 6. Streamline across the propeller

6. Conclusions

The low Reynold number airfoils FX 63-137 was selected and transformed for 4 thickness series. The primary design a HALE UAV propeller was completed by Joukowski propeller vortex theory and validated by CFD simulation. Results shows that Joukowski theory underestimates the chord length.

Finally, an optimal design approach was applied using software INSIGHT to obtain the best blade section pitch distribution. CFD result of the optimal design shows that the thrust meet the design requirement, the efficiency is 70.49%, which is a little bit smaller than the requirement 72%.

References

[1] Ilieva G, Páscoa J, Dumas A and Trancossi M. A critical review of propulsion concepts for modern airships. *Open Engineering,* Vol. 2, No. 2, pp 189-200, 2012.
[2] Liu P, Duan Z and Ma L. Aerodynamics properties and design method of high efficiency-light propeller of stratospheric airships. *International Conference on Remote Sensing,* Environment and Transportation Engineering (RSETE), pp 8041-8044, 24-26 June 2011.
[3] Hu H and Yang Z. An experimental study of the laminar flow separation on a low- Reynolds-number airfoil, *Journal of Fluids Engineering,* Vol. 130 / 051101-1, 2008.
[4] Tang Z, Liu P, Chen Y, and Guo H. Experimental study of counter-rotating propellers for high-altitude airships. *J Propulsion and Power*, 31, (5), pp 1491-1496, 2015.
[5] Nie Y. The design and experiments study on high efficient propeller used for stratospheric airship. *MS thesis,* Center for Space Science and Applied Research, Graduate University of Chinese Academy of Sciences, Beijing, China, 2008(in Chinese).
[6] Koch LD. Design and performance calculations of a propeller for very high altitude flight. *MS thesis,* Case Western Reserve University, Cleveland, Ohio, US, 1998.
[7] Ma R and Liu P. Numerical simulation of low-Reynolds-number and high-lift airfoil S1223. *Proceedings of the World Congress on Engineering,* Vol. 2. 2009.
[8] D'Angelo S, Berardi F and Minisci E. Aerodynamic performances of propellers with parametric considerations on the optimal design. *Calculus of Variations,* Vol. 4, No. 5, p 6, 2002.
[9] Selig M S, Guglielmo J J, Broeren A P and Giguere P. *Summary of low-speed airfoil data,* Volume 1. 1995.

The Simulation of Aircraft Ditching Based on ALE Method

Sun Feng[1,2,*], Wang Mingzhen[1,2], Wu Bin[1,2] and He Chao[1,2]

[1] Key Aviation Scientific and Technological Laboratory of High-Speed Hydrodynamic, Hubei, Jingmen, 448035, China
[2] China Special Vehicle Research Institute, Hubei,Jingmen, 448035, China

[*]E-mail: sunfeng686@126.com

Abstract: Ditching is an emergency measure for landing on the water when the aircraft cannot continue to fly in an accident. In order to verify the performance of the aircraft ditching, this paper carry out a simulation study on ditching of fixed-wing aircraft on the water. First determine the simulation method, material model and constitutive equation, and then analyze the influence of the sinking speed and attitude changes on the ditching of the aircraft. The simulation method adopts arbitrary Lagrange-Euler method (ALE). The results show that as the sinking speed increases, the load will increase accordingly. As the initial attitude angle increases, the vertical acceleration gradually decreases, and the longitudinal acceleration is basically the same, the pressure acting area gradually moves backward, and the pressure area gradually decreases. It is necessary to take appropriate local strengthening measures on the dangerous region.

1. Introduction

Ditching is an emergency measure for landing on the water when the aircraft cannot continue to fly in an accident. Ditching can be divided into planned ditching and unplanned ditching. Civil aviation regulations stipulate that to apply for certification with ditching, various practical design measures compatible with the general characteristics of the aircraft must be taken to minimize the probability that the occupant will be injured or unable to evacuate due to the movement and state of the aircraft during an emergency landing.

Ditching is a typical fluid-structure coupling problem. When the structure is impacted, it will also stimulate the movement of the fluid. The movement and deformation of the structure during the impact will affect the movement of the fluid [1,2]. Fluid-structure coupling have always existed throughout the process and is complicated. The duration of water impact load is short and the peak value is high, which is easy to cause structural damage, instrument failure, control failure, personnel casualties and other hazards. Therefore, carrying out research into the water load of structures has important engineering application value. With the development of numerical calculation methods and computer hardware, computer simulation has become more and more widely used in ditching. Olovsson and Souli [3-5], Aquelet and Souli[6], Bereznitzki[7], and Le Sourness[8] analyzed a series of fluid-structure coupling problems by this method. Chen Zhen[9,10] applied MSC. Dytran software to analyze the influence of air cushion on slamming during the process of flat-bottomed structure entering the water , and concluded that the slamming pressure peak was mainly caused by the compression of the air layer. Chen Xiaoping[11] applied experiments and numerical methods to analyze the influence of elastic effect on the slamming pressure and structural stress response of the wedge. Luo Hanbing[12] simulated the hydroelastic slamming of a three-dimensional stiffened wedge-shaped structure using the ALE method.

2.Calculation conditions

The maximum take-off weight state is selected for simulation to study the influence of the sinking speed and initial attitude on the water load and motion. The horizontal speed is 40m/s, the vertical speed is 1.0m/s and 1.5m/s. The initial attitude angle of the aircraft is controlled within the range of 6°～13°.

The arbitrary Lagrange-Euler (ALE) description method can well solve the problem of free boundary and motion interface tracking in the fluid-structure coupling. The ALE method makes the grid motion independent of material and space by introducing a reference coordinate system, which not only overcomes the large deformation problem described by Lagrange, but also solves the interface tracking problem described by Euler[13]. The basic equation is as follows:

$$\begin{aligned} &\frac{\partial \rho}{\partial t}+\vec{m}\cdot\nabla\rho=-\rho\nabla\cdot\vec{u} \\ &\rho\frac{\partial \vec{u}}{\partial t}+\rho\vec{m}\cdot\nabla\vec{u}=-\nabla p \\ &\rho\frac{\partial e}{\partial t}+\rho\vec{m}\cdot\nabla e=-\nabla(p\vec{u}) \end{aligned} \tag{1}$$

where: $\vec{m}$ is the convection speed; $\vec{u}$ is material velocity; ρ is the density of matter; e is the total energy per unit mass; p is the external load.

The penalty function method is selected to process fluid-solid interface coupling, which is equivalent to defining a series of spring-damping systems between the fluid and the structural nodes. Penalty coupling force is a function of relative displacement and relative velocity, as shown in equation (2).

$$F=kd+c\dot{d} \tag{2}$$

Where k and c represent the spring stiffness and damping coefficient, respectively.

In LS-DYNA, the constitutive model and the state equation are used to describe the fluid material at the same time. The material property is defined by the empty material model * MAT_NULL. The relationship between the pressure and volume of water and air is given by the linear polynomial state equation[14]

$$p=C_0+C_1\mu+C_2\mu^2+C_3\mu^3+(b_0+b_1\mu+b_2\mu^2)E \tag{3}$$

where, p is pressure; μ is the rate of change of density; E is the internal energy per unit volume; C_i is the material constant.

The fuselage element type is shell163, and the flow element type is solid164. Figure 1 shows the finite element model.

Figure 1 Finite element calculation model

3. Calculation result analysis

Based on the analysis of the effects of different sinking speeds, different initial attitude angles and other factors on the ditching load and motion, the change law of the load is summarized. The impact occurs in a short period of time and then slides.

3.1 Effect of different sinking speeds on acceleration

Selecting the initial state of attitude 11°and simulating the water landing at 1m/s and 1.5m/s sinking speed. Figure 2 shows the comparison results of the vertical acceleration and heave at the center of gravity of the aircraft at different sinking speeds. The greater the sinking speed, the greater the peak value of vertical acceleration, the shorter the time to reach the peak value of acceleration. The depth of water immersion is zero at the initial moment. The aircraft begins to sink after landing. At the same time, the landing of the stern causes the aircraft to bow down and further intensify the vertical movement of the center of gravity. The greater the vertical velocity, the shorter the time to reach the maximum water depth.

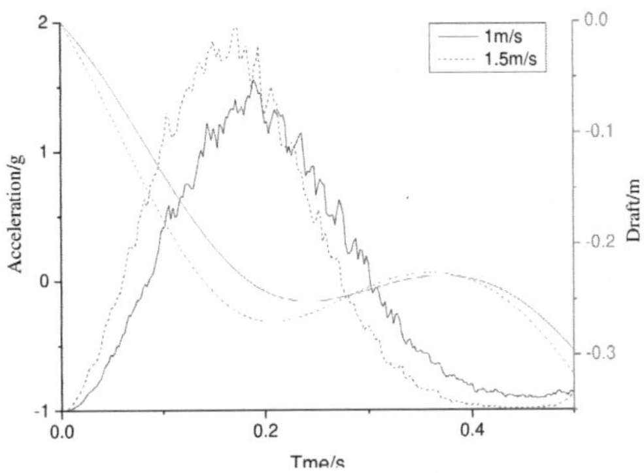

Figure 2 Acceleration at different sinking speeds

3.2 Effect of different pitch angles on acceleration

Figure 3 and Figure 4 respectively show the vertical acceleration and heading acceleration curves at the center of gravity at the initial pitch angles in the range of 6-13°. Comparison shows that horizontal acceleration is much smaller than vertical acceleration, and the maximum value of horizontal acceleration is 0.49g. The maximum value of vertical acceleration is 3.99g, and the time of occurrence of the both is basically within 0.4s, which reflects the characteristic of short impact load time.

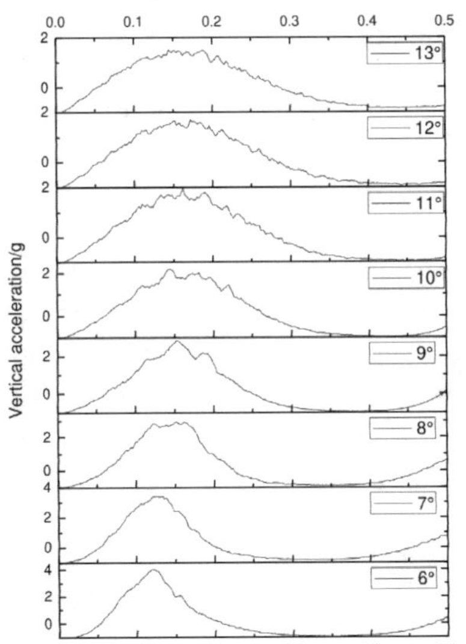

Figure 3 Vertical acceleration

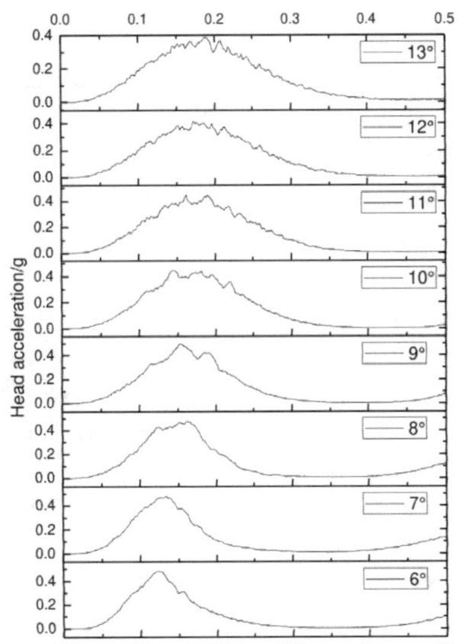

Figure 4 Heading acceleration

The landing position gradually moves back as the initial pitch angle increases. The farther the landing position is from the center of gravity, the greater the influence of the rotation of the fuselage on the acceleration at the center of gravity. Changes in the form of the fuselage under different landing positions will also cause changes in external loads, so the acceleration at the center of gravity is affected by the location and the configuration of landing point. It can be seen from Figure 5 that the horizontal acceleration peak at all angles has almost no change and the vertical acceleration peak decreases with the increase of the initial pitch angle, The main reason for this phenomenon is that as the initial pitch angle increases, the water landing part moves toward the tail of the fuselage, the farther the force action point is from the center of gravity. The greater the bowing moment that the aircraft receives, the greater the turning angular velocity, which reduces the acceleration at the center of gravity.

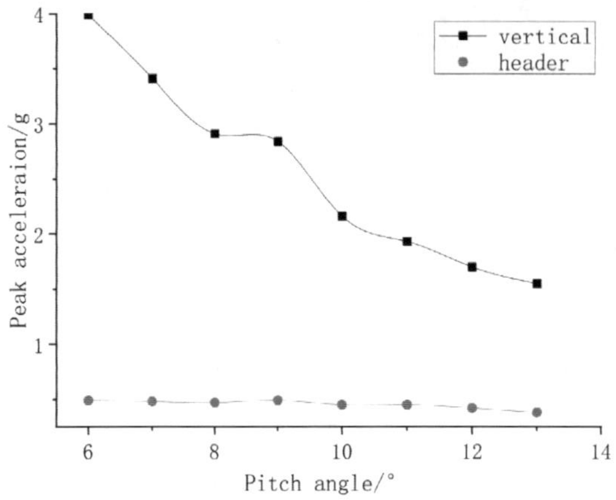

Figure 5 Peak acceleration comparison

3.3 Pressure distribution analysis

Arrange the pressure measuring points at the bottom of the fuselage in the horizontal and vertical directions. As shown in Figure 6, 25 pressure measuring points are arranged at the intersection of 8 transverse sections and 4 longitudinal sections on the one side of the fuselage to analyze the pressure distribution characteristics. In Table 1, the peak pressure of a certain working condition is counted and dimensionless processing is performed. It can be seen that with the increase of the initial attitude angle, the maximum pressure and the area of pressure distribution moves to the rear of the fuselage as a whole, and the pressure area gradually became smaller. The two areas circled in Figure 6 are the arcuate transition area at the belly of the fuselage and the rear of the fuselage, respectively. The arc-shaped transition zone will not directly collide with water, so the water pressure in such areas will be small or zero. As can be seen from Table 1, the cross-sections ④ and ⑧ have no pressure measured under all working conditions. , The cross section ⑦ is near the starting point of the arc-shaped transition of the tail. When the pitch angle of the fuselage is less than 11°, the position is always facing away from the direction of the incoming flow, so no pressure fluctuation is measured. When the pitch angle of the fuselage is greater than or equal to 11°, the location gradually changes from the backflow state to the up-flow state, and the pressure value and the pressure action range gradually increase.

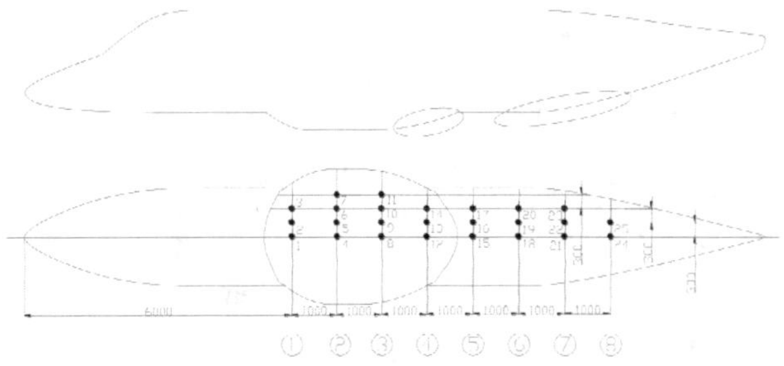

Figure 6 Arrangement of pressure measuring points

Table 1 Peak pressure statistics

Pitch		Peak pressure							
		h1	h2	h3	h4	h5	h6	h7	h8
6°	z1	0.000	0.230	0.297	0.000	0.089	0.294	0.000	0.000
	z2	0.000	0.225	0.070	0.000	0.115	0.278	0.000	0.000
	z3	0.000	0.192	0.000	0.000	0.064	0.211	0.000	-
	z4	-	0.205	0.000	-	-	-	-	-
7°	z1	0.000	0.126	0.317	0.000	0.140	0.366	0.000	0.000
	z2	0.000	0.115	0.120	0.000	0.174	0.304	0.000	0.000
	z3	0.000	0.106	0.000	0.000	0.090	0.270	0.000	-
	z4	-	0.096	0.000	-	-	-	-	-
8°	z1	0.000	0.004	0.295	0.000	0.177	0.372	0.000	0.000
	z2	0.000	0.004	0.168	0.000	0.187	0.356	0.000	0.000
	z3	0.000	0.004	0.000	0.000	0.117	0.314	0.000	-
	z4	-	0.001	0.000	-	-	-	-	-
9°	z1	0.000	0.000	0.227	0.000	0.173	0.391	0.000	0.000
	z2	0.000	0.000	0.135	0.000	0.152	0.410	0.000	0.000
	z3	0.000	0.000	0.000	0.000	0.109	0.365	0.000	-
	z4	-	0.000	0.000		-	-	-	-
10°	z1	0.000	0.000	0.079	0.000	0.125	0.555	0.000	0.000
	z2	0.000	0.000	0.041	0.000	0.122	0.439	0.000	0.000
	z3	0.000	0.000	0.000	0.000	0.082	0.351	0.000	-
	z4	-	0.000	0.000	-	-	-	-	-
11°	z1	0.000	0.000	0.000	0.000	0.055	0.611	0.002	0.000
	z2	0.000	0.000	0.000	0.000	0.052	0.476	0.000	0.000
	z3	0.000	0.000	0.000	0.000	0.023	0.439	0.000	-
	z4	-	0.000	0.000	-	-	-	-	-
12°	z1	0.000	0.000	0.000	0.000	0.008	0.539	0.025	0.000
	z2	0.000	0.000	0.000	0.000	0.008	0.498	0.015	0.000
	z3	0.000	0.000	0.000	0.000	0.000	0.480	0.000	-
	z4	-	0.000	0.000	-	-	-	-	-
13°	z1	0.000	0.000	0.000	0.000	0.000	0.419	0.079	0.000
	z2	0.000	0.000	0.000	0.000	0.000	0.494	0.050	0.000
	z3	0.000	0.000	0.000	0.000	0.000	0.419	0.000	-
	z4	-	0.000	0.000	-	-	-	-	-
Note: h is a horizontal section and z is a vertical section.									

4. Conclusion

Through the simulation and analysis of the aircraft ditching, the main conclusions obtained are as follows:

1) The greater the sinking speed, the greater the peak vertical acceleration, the shorter the time to reach the peak acceleration, and the shorter the time to reach the maximum depth;

2) The peak value of vertical acceleration at the center of gravity is gradually decreasing with the increase of the initial pitch angle, and the heading acceleration is less affected by the attitude;

3) As the initial attitude angle increases, the pressure action range gradually moves back and decreases;

4) The arc-shaped transition streamline makes the area at the bottom of the fuselage facing away from the incoming flow, and such areas are hardly affected by the impact pressure;

IOP Conf. Series: Materials Science and Engineering **926** (2020) 012019 doi:10.1088/1757-899X/926/1/012019

5) The local pressure near the cross section ②, ③, ⑤, ⑥ is relatively large. In order to better ensure the safety of the aircraft's during ditching, appropriate local strengthening measures may be taken in the corresponding area.

References

[1] Yonghu Wang, Xiuhua Shi. Review on research and development of water2entry impact problem[J]. Explosion and shock wave, 28(2008): 276-282.

[2] SUN Qi, ZHOU Jun, LIN Peng. Dynamic Analysis of Fluid-Structure Interaction for Water Impact of Projectile Using LS-DYNA[J]. Journal of System Simulation, 2010,22(6):1498-1501

[3] L. Olovsson. ALE and Fluid-Structure Interaction Capabilities in LS-DYNA. Proceedings of the 6th International LS-DYNA Users Conference[J], 2000.

[4] L. Olovsson, M. Souli. Improved fluid-structure capabilities in LS-DYNA[C]. Proceedings of the 3rd European LS-DYNA Users Conference, 2001.

[5] L. Olovsson, M. Souli. Multi-material Eulerian capabilities in LS-DYNA[C]. Proceedings of the 2nd European LS-DYNA Users Conference, 1999.

[6] N.Aquelet, M.Souli. DAMPING EFFECT IN FLUID-STRUCTURE INTERACTION: APPLICATION TO SLAMMING PROBLEM[C]. Proceedings of ASME Pressure Vessel and Piping Conference, 2003.

[7] A. Bereznitski. SLAMMING: THE ROLE OF HYDROELASTICITY[J]. International Shipbuilding Progress, 2001, 48(4):333-351.

[8] N.C. H. Le Sourne, F. Besnier, C. Kammerer and H. Legavre. LS-DYNA applications in shipbuilding[C]. 4th European LS-DYNA Users Conference, 2003.

[9] CHEN Zhen, XIAO Xi. Simulation Analysis on the Role of Air Cushion in the Slamming of a Flat-Bottom Structure[J]. JOURNAL OF SHANGHAI JIAOTONG UNIVERSITY, 2005,39(5):670-673.

[10] CHEN Zhen, XIAO Xi. The Simulation Study on Water Entry of 2D Wedge Bodies [J]. JOURNAL OF SHANGHAI JIAOTONG UNIVERSITY, 2007,41(9)： 1425-1428.

[11] CHEN Xiao-ping, LI Jun-wei, WANG Hui1, QI En-rong. Experiments and numerical investigation of water entry of large-scale steel wedge models [J]. Journal of Ship Mechanics, 2012, 16(10): 1152-1163.

[12] LUO Han-bing, ZHANG Zhi, WANG Hui. Hydroelastic simulation of one 3D free-drop wedge with stiffened panels using ALE method[J]. Journal of Ship Mechanics, , 2013, 17 (1):84-91.

[13] Sun Feng, Zhang Jiaxu, Gu Biao. Analysis of water entry performance of amphibious aircraft hull based on ALE method[J]. Aeronautical Computing Technique, 2016,46(3):54-57.

[14] K. Hughes, J. Campbell, R. Vignjevic. Application of the finite element method to predict the crashworthy response of a metallic helicopter under floor structure onto water[J]. International Journal of Impact Engineering, 2008 (35) 347–362.

Aerodynamic Performance Prediction of XH-59A Helicopter in Pop-up Flight

K You[1], X Zhao[1,*], Y-H Zhang[1], S-Z Zhao[1] and Y-F Lin[2]

[1]School of Aeronautics, Northwestern Polytechnical University, Xi'an, Shaanxi, 710072, China
[2]China Helicopter Research and Development Institute, Jingdezhen, Jiangxi, 333001, China

[*]Corresponding author: xuzhao@nwpu.edu.cn

Abstract. The pop-up progress of the coaxial rigid rotor helicopter was decomposed into 5 stages. Quasi-steady flow assumption was adopted to analyze the helicopter performance. XH-59A helicopter was regarded as a solid body. Flow interactions between fuselage and rotor, rotor and tail were ignored. The aerodynamic performance of coaxial rotor, fuselage and horizontal tail were calculated based on Blade element moment theory, wind tunnel test data and theoretical calculation correspondingly. The Pitt/Peters inflow model was used to represent the inflow velocity on the rotor and flow interaction between upper and lower rotor were considered using interference factor. The net force and moment on the helicopter was derived and formulated into flight mechanics equations. To validate the calculation, a wind tunnel test of a coaxial rigid rotor helicopter was employed. The rotor lift coefficient, helicopter pitching and rolling moment coefficients were close to the experimental data. The controlled variables of the helicopter was composed of helicopter pitch angle (shaft angle), collective pitch, cyclic pitch of both rotors, deflection angle of the horizontal tail during the pop-up progress. The mathematical model presented in this paper provides the basic system structure to analyze the pop-up progress. The required power of the helicopter were formulated with altitude, rotor thrust, flight velocity, which help to obtain a rapid and primary evaluation of the integral performance of the progress.

Nomenclature

A_1	Longitudinal cyclic blade pitch	R	Rotor radius
a_0, a_1	Taper angle, Back chamfer angle of the rotor	S	Rotor side force
B_1	Lateral cyclic blade pitch	T	Rotor thrust
b_1	Roll angle of rotor	V_1	Induced velocity at rotor disk
CLR	Rotor lift coefficient	$\overline{V_0}$	Dimensionless mean induced velocity of the rotor
CXR	Net force coefficient on wind direction of the rotor	V_x, V_y, V_z	Flight speed
CYR	Rotor side force coefficient		Greek symbols
CMZ	Helicopter yawing moment coefficient	α	Attack angle of helicopter
CMY	Helicopter pitching moment coefficient	α_s	Shaft angle of attack
CMX	Helicopter rolling moment coefficient	α_0	Attack angle of rotor
H	Rotor backward force	β	Sideslip angle of helicopter
k	Number of blades per rotor	δ	Interference factor
θ, γ, φ	Pitch angle, Roll angle, Yaw angle of helicopter	τ	Swing angle
θ^*	Rotor blade collective pitch	φ^*	Blade azimuth

θ_p	Differential total interval	Ω	Rotor speed
λ	Inflow ratio	$\omega_x, \omega_y, \omega$	Angular velocity of the helicopter
μ	Advance ratio		

1. Introduction

Rigid coaxial rotor helicopter is an important development direction of armed helicopter due to high speed and high mobility. However, there are limited data regarding its performance in experimental test or flight test. In order to capture the helicopter aerodynamic performance with controlled variables, theoretical method based on Blade Element Moment (BEM) theory of the rotor and rigid-body kinematics were carried out.

The XH-59A helicopter created by Sikorsky Company applies the Advancing Blade Concept (ABC) [1] rotor and achieved symmetrical lift force from the contra rotating rotors. Two auxiliary jet engine are located at both side of the fuselage to supply extra thrust in flight direction. In 1970s, the demonstration helicopters reached speeds of 296 km/h and 356 km/h in level flight and dive without auxiliary thrust [2]. Although it was replaced by X-2 and S-97 for better performance later on. But it has the most adequate parameters and performance published in wind tunnel test and flight test. Figure 1 shows the picture of XH-59A helicopter and table 1 shows the detailed parameters, the parameters are directly or calculated from literature [3].

To analyze pop-up flight, the process is assumed to be decomposed into 5 stages: (1) A steady level flight at original altitude, (2) a circular movement in vertical direction, (3) vertical climb with constant pitch angle, (4) a circular movement in vertical direction, and (5) a steady level flight at increased altitude, shown in figure 2. The quasi-steady forward flight model based on BEM [4-5] will be applied for the maneuver process.

Figure 1. XH-59A helicopter

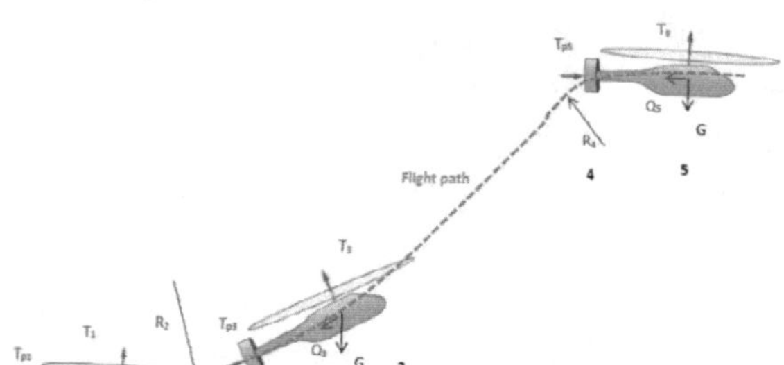

Figure 2. Helicopter in pop-up process

Table 1. XH-59A helicopter specifications

Gross weight, lb	13000 lb
Rotor radius, cutout radius	5.486 m, 1.1 m
Number of rotors, blades per rotor	2, 3
Rotor separation	2.5 ft
Blade tip Chord	0.938 ft
Blade taper Ratio	2:1
Blade twist (nonlinear)	-10°
Total rotor solidity	0.1267
Blade precone angle, prelag angle	3°, 1.4°
Shaft Tilt	0°
Design Rotor Tip Speed	650 ft/sec
Rotor drive system power	1500 HP
Airfoil section	NACA series
Horizontal/vertical surface area of tail	60 ft^2, 30 ft^2
Horizontal Tail Incidence	-5°
Single blade mass	133.60 kg
Single blade mass moment to flapping hinge	187.26 kg·m
Single blade polar moment of inertia	615.51 kg·m^2

2. Motion equations of the helicopter

To carry out the helicopter performance prediction, assumptions are proposed. (1) Helicopter is a rigid body; (2) The flow interaction among rotor, fuselage and tail is ignored; (3) Forces and moments on the helicopter are composed of that on rotor, fuselage and tail. (4) The inflow model is suitable for the coaxial rigid rotor with flow interaction factor derived from other coaxial rotor helicopter experiment.

(5)Then the forces and moments on each components of the helicopter were calculated and transformed from different coordinate systems.

2.1. Force and moment equations in body axis system

In order to describe conveniently, the external forces and moments are unified in the body axis system. Terrestrial coordinate system can coincide with body axis system after rolling three times $(\varphi, \theta, \gamma)$. Suppose the velocity of the incoming flow is V, then the component of velocity along the body axis should be $V_x = V\cos\alpha\cos\beta$, $V_y = V\sin\beta$, $V_z = V\sin\alpha\cos\beta$. Thus the force and moment equations in body axis system can be written as (1) and (2).

$$\begin{aligned} m(dv_x/dt + V_z\omega_y - V_y\omega_z) + mg\sin\theta &= F_x \\ m(dv_y/dt + V_x\omega_z - V_z\omega_x) + mg\sin\theta\cos\gamma &= F_y \\ m(dv_z/dt + V_y\omega_x - V_x\omega_y) + mg\sin\theta\sin\gamma &= F_z \end{aligned} \tag{1}$$

$$\begin{aligned} I_x d\omega_x/dt + \omega_y\omega_z\left(I_z - I_y\right) + \left(\omega_x\omega_z - d\omega_y/dt\right)I_{xy} &= \Sigma M_x \\ I_y d\omega_y/dt + \omega_x\omega_z\left(I_x - I_z\right) + \left(\omega_y\omega_z - d\omega_x/dt\right)I_{xy} &= \Sigma M_y \\ I_z d\omega_z/dt + \omega_x\omega_y\left(I_y - I_x\right) + \left(\omega_y^2 - \omega_x^2\right)I_{xy} &= \Sigma M_z \end{aligned} \tag{2}$$

2.2. Aerodynamic performance of fuselage

The aerodynamic characteristic matrix of the XH-59A fuselage can be achieved which is based on the wind tunnel experiments [3].

2.3. Aerodynamic performance of horizontal tail

The velocity of air flowing through the horizontal tail includes the relative flow of helicopter forward flight, the relative flow velocity caused by fuselage angular motion and the velocity caused by rotor downwash, side wash and fuselage downwash. Therefore, the velocity of the horizontal tail can write as (3). Then the force and moment equations can also be given in equation (4) and (5). Three variables x_H, y_H, z_H mean the longitudinal, vertical, and lateral positions of the horizontal tail relative to the center of gravity of the helicopter.

$$\begin{aligned} V_{xH} &= V_x(K_{QH})^{0.5} + \omega_y z_H - \omega_z y_H \\ V_{yH} &= V_y(K_{QH})^{0.5} + \omega_z x_H - \omega_x z_H + K_{MH}v_1 + \varepsilon v_x(K_{QH})^{0.5} \\ V_{zH} &= V_z(K_{QH})^{0.5} + \omega_x y_H - \omega_y x_H \end{aligned} \tag{3}$$

$$\begin{vmatrix} F_{x,H} \\ F_{y,H} \\ F_{z,H} \end{vmatrix} = \begin{bmatrix} -\cos(\alpha_H - \varphi_H) & \sin(\alpha_H - \varphi_H) & 0 \\ -\sin(\alpha_H - \varphi_H) & -\cos(\alpha_H - \varphi_H) & 0 \\ 0 & 0 & 1 \end{bmatrix} \begin{bmatrix} D_H \\ L_H \\ 0 \end{bmatrix} \tag{4}$$

$$\begin{aligned} M_{x,H} &= F_{y,H} Z_H \\ M_{y,H} &= -F_{x,H} Z_H \\ M_{z,H} &= F_{y,H} x_H - F_{x,H} y_H \end{aligned} \tag{5}$$

2.4. Aerodynamic performance of rotors

Firstly, the airfoils performance on each section of the blades were calculated using CFD software under various working conditions. The predicted lift coefficient, drag coefficient varying with angle of attck, Mach number, Reynolds number were formulated into a database for performance prediction of the rotating blades. According to the BEM theory, the helicopter is controlled by the forces and moments. And seven manipulated variables $(A_{1l}, A_{1u}, B_{1l}, B_{1u}, \theta_{pl}, \theta_{pu}, \alpha_s)$ can determine the forces and moments on the rotors.

$$T=k\frac{1}{2\pi}\int_0^{2\pi}d\varphi^*\int_0^R\frac{1}{2}a_\infty\rho c(\Omega R)^2$$

$$\left\{\left(\bar{r}\cos\tau+\mu^*\sin\varphi^*\right)^2\theta^*-\left(\bar{r}\cos\tau+\mu^*\sin\varphi^*\right)\left[\left(\bar{v}_1-\lambda_0\right)\cos\tau+\bar{r}\left(a_1\sin\varphi^*-b_1\cos\varphi^*\right)+\mu^*\cos\varphi^*\sin\tau\right]\right\}dr \tag{6}$$

$$H=k\frac{1}{2\pi}\int_0^{2\pi}d\varphi^*\int_0^R\frac{1}{2}\rho C_{xav}c\left(\bar{r}\cos\tau+\mu^*\sin\varphi^*\right)^2\sin\varphi^*+\frac{1}{2}a_\infty\rho c(\Omega R)^2$$

$$\left\{\left(\bar{r}\cos\tau+\mu^*\sin\varphi^*\right)^2\theta^*-\left(\bar{r}\cos\tau+\mu^*\sin\varphi^*\right)\left[\left(\bar{v}_1-\lambda_0\right)\cos\tau+\bar{r}\left(a_1\sin\varphi^*-b_1\cos\varphi^*\right)+\mu^*\cos\varphi^*\sin\tau\right]\right\}\beta^*\sin\varphi^* \tag{7}$$

$$-\frac{1}{2}a_\infty\rho c(\Omega R)^2\left\{\left(\bar{r}\cos\tau+\mu^*\sin\varphi^*\right)^2\theta^*-\left(\bar{r}\cos\tau+\mu^*\sin\varphi^*\right)\left[\left(\bar{v}_1-\lambda_0\right)\cos\tau+\bar{r}\left(a_1\sin\varphi^*-b_1\cos\varphi^*\right)+\mu^*\cos\varphi^*\sin\tau\right]\right\}\frac{\partial y}{\partial r}\cos\varphi^*\}dr$$

$$S=k\frac{1}{2\pi}\int_0^{2\pi}d\varphi^*\int_0^R\frac{1}{2}\rho C_{xav}c\left(\bar{r}\cos\tau+\mu^*\sin\varphi^*\right)^2\cos\varphi^*+\frac{1}{2}a_\infty\rho c(\Omega R)^2$$

$$\left\{\left(\bar{r}\cos\tau+\mu^*\sin\varphi^*\right)^2\theta^*-\left(\bar{r}\cos\tau+\mu^*\sin\varphi^*\right)\left[\left(\bar{v}_1-\lambda_0\right)\cos\tau+\bar{r}\left(a_1\sin\varphi^*-b_1\cos\varphi^*\right)+\mu^*\cos\varphi^*\sin\tau\right]\right\}\beta^*\cos\varphi^* \tag{8}$$

$$-\frac{1}{2}a_\infty\rho c(\Omega R)^2\left\{\left(\bar{r}\cos\tau+\mu^*\sin\varphi^*\right)^2\theta^*-\left(\bar{r}\cos\tau+\mu^*\sin\varphi^*\right)\left[\left(\bar{v}_1-\lambda_0\right)\cos\tau+\bar{r}\left(a_1\sin\varphi^*-b_1\cos\varphi^*\right)+\mu^*\cos\varphi^*\sin\tau\right]\right\}\frac{\partial y}{\partial r}\sin\varphi^*\}dr$$

$$M_{yhu}=\frac{k}{2\pi}\int_0^{2\pi}\int_0^R\left[-\left(\frac{dX}{dr}+\frac{dY}{dr}\beta_*\right)-ma_{px1}\right]rdrd\varphi^* \tag{9}$$

$$a_{px1}=-2\dot{y}\left(-\omega_x\cos\varphi^*+\omega_z\sin\varphi^*\right)+\omega_y r-\left(\omega_z\sin\varphi^*-\omega_x\cos\varphi^*\right)y+\omega_y\left(\omega_x\sin\varphi^*+\omega_z\cos\varphi^*\right)y+\left(\omega_x\sin\varphi^*-\omega_z\cos\varphi^*\right)\left(\omega_x\sin\varphi^*+\omega_z\cos\varphi^*\right)r$$

The above equations require the variable $\bar{v_1}$ to get the final result, however it is not a constant value. So equations (10-11) were added based on Pitt/Peters inflow model [6]. The equation (10) shows the induced velocity of upper rotor, while (11) shows that of lower one, where δ_u is the flow interaction factor of upper rotor to lower rotor, and δ_l is the verse versa.

$$v_{1u}=v_{0u}+\delta_u v_{0l}+(r/R)\left[\left(v_{cu}+\delta_u v_{cl}\right)\cos\varphi_u+\left(v_{su}-\delta_u v_{sl}\right)\sin\varphi_u\right] \tag{10}$$

$$v_{1l}=v_{0l}+\delta_l v_{0u}+(r/R)\left[\left(v_{cl}+\delta_l v_{cu}\right)\cos\varphi_l+\left(v_{sl}-\delta_l v_{su}\right)\sin\varphi_l\right] \tag{11}$$

Rotor blade pitches can also be expressed as the function of blade azimuth, lateral cyclic blade pitch, longitudinal cyclic blade pitch.

$$\theta_u^*=\left(\theta_u^*+\theta_p\right)-\left(A_1+A_{1p}\right)\cos\varphi_u-\left(B_1+B_{1p}\right)\sin\varphi_u \tag{12}$$

$$\theta_l^*=\left(\theta_u^*-\theta_p\right)-\left(A_1-A_{1p}\right)\cos\varphi_l-\left(B_1-B_{1p}\right)\sin\varphi_l \tag{13}$$

3. Performance validation by wind tunnel test at advancing ratio 0.7

Falarski [7] investigated the full-scale advancing blade concept rotor system at high advance ratios in 1971. The experiment covers an advance ratio range of 0.2 to 0.9, with all the force and moment trimmed. The experimental data of Run 21 was used to validate the BEM theory. The parameters of the experiment is shown in table 2. Figure 3 shows the ABC helicopter during the test.

The rotor performance predicted by Blade Element Moment (BEM) theory are compared with the experimental data in Figure 4. The x coordinate PT (point) represents the serial number.

Figure 4 shows that the prediction of CLR, CMX, and CMZ are close to experimental result. However, CXR, CYR and CMY have a certain disparity with experiment. The reasons are supposed to be: (1) flow interaction of the rotor with faired body in the test, while the body is ignored in the BEM method; (2) The moment coefficients on the helicopter in the experiment are compared with

IOP Conf. Series: Materials Science and Engineering **926** (2020) 012020 doi:10.1088/1757-899X/926/1/012020

moment coefficients on the rotor in BEM theory; (3) The trimmed result refers to variables with quite small value, which exaggerates the error of BEM results.

Figure 3. ABC helicopter in test

Table 2. Rotor parameters in Run 21 of ABC helicopter in test

Parameter	value
Advancing ratio	0.7
Free steam velocity	138m/s
Rotor radius, $M_{tip,}$	6.098m, 0.58
Rotating speed, Tip speed	309RPM, 197.2m/s
Shaft angle	-4°~ 4°
Precone angle	5°(upper), 0°(lower)
Collective angle	3°~10°
longitudinal cyclic pitch A_1	-7.5°~ 1°
lateral cyclic pitch B_1	-5.6°~ -0.1°
Airfoil section	NACA 0030- NACA 0006

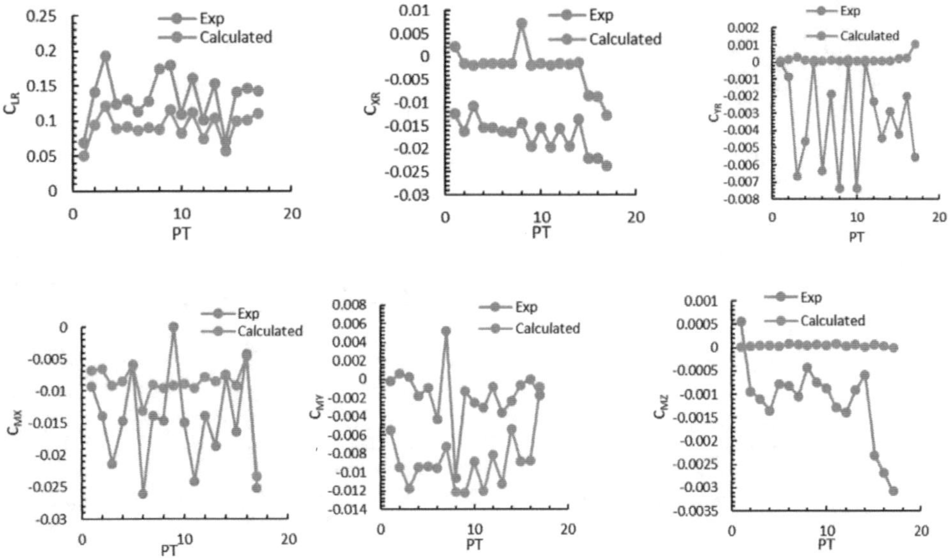

Figure 4. Result comparison of theoretical prediction and experiment

4. Simplified verification formula of forces and moments

The control variable of the helicopter is assumed to be eight, despite the previous variable mentioned in section 2.4, deflection angle of the horizontal tail were added. The forces and moments of the helicopter can be received from solving the integral equations (6)-(9). However the process is difficult and because of the lack of important data, those equations can't be solved for now. So simplified verifications formula of forces and moments are necessary. Here the simplified equation for proximate calculation were given in equation (14).

$$
\begin{aligned}
&T_u\left(\alpha_{su}-a_{1u}\right)+T_l\left(\alpha_{sl}-a_{1l}\right)=mg\theta \\
&T_u+T_l=\left(1+\frac{P}{1000}\right)mg \\
&T_u b_{1u}+T_l b_{1l}=-mg\gamma \\
&T_u b_{1u} y_{Mu}+T_l b_{1l} y_{Ml}=0 \\
&T_u b_{1u} x_{Mu}+T_l b_{1l} x_{Ml}=0 \\
&T_u\left(x_{Mu}-y_{Mu}\alpha_s\right)+T_l\left(x_{Ml}-y_{Ml}\alpha_s\right)+T_u a_{1u} y_{Mu}+T_l a_{1l} y_{Ml}=0
\end{aligned}
\tag{14}
$$

Simplified verification formula of power and torque of the helicopter were given in equation (15), where lift, backward force, side force of rotor can be obtained.

$$
\begin{aligned}
&P=m_K\frac{1}{2}\rho\pi R^2(\pi R)^3 \\
&M_K=m_K\frac{1}{2}\rho\pi R^2(\pi R)^2 R \\
&m_K=\frac{1}{4}\sigma C_{x7}K_{p0}\left(1+5\mu^2\right)+C_T\bar{v}_1 J_0\left(1+3\mu^2\right)+C_T\left(-\lambda_0\right)-C_H\mu
\end{aligned}
\tag{15}
$$

The parameters K_{p0}, J_0 mean type resistance power correction factor and induced power correction factor, and C_{x7} means lift coefficient of characteristic section.

5. Conclusions

The pop-up progress of the coaxial rigid rotor helicopter was decomposed into 5 stages. The mathematical equations for helicopter performance prediction was created based on assumptions. The BEM theory was employed for rotor performance prediction. The evaluation of the model was completed by the wind tunnel test of a rigid coaxial rotor helicopter. Results shows that method proposed in this paper has a satisfying accuracy for rotor lift coefficient, helicopter pitching and rolling moment coefficients prediction. It can be developed for future investigation of the pop-up process of the helicopter.

Acknowledgments

This work was funded by the Aeronautical Science Foundation of China 20175753030.

References

[1] Paglino V M. Forward flight performance of a coaxial rigid rotor. *Proceedings of the 27th American Helicopter Society Forum,* Washington, DC. 1971.

[2] Ruddell A J, Groth W, Mccutcheon R. Advancing blade concept (ABC) technology demonstrator, *book,* 1981.

[3] Felker F F. Performance and loads data from a wind tunnel test of a full-scale, coaxial, hingeless rotor helicopter. *NASA technical memorandum 81329,* National Aeronautics and Space Administration Moffett Field, Ames Research Center, 1981.

[4] Y-H. Zhang, X. Zhao, N-Z. Huang, A high efficiency trim method for the aerodynamic analysis of XH-59A rotor in forward flight, *AIAA 2019-4386*, AIAA Propulsion and Energy Forum, 19-22 August 2019, Indianapolis, IN, US.

[5] Zhou P, Zhao X, Deng L, Lin Y-F, Analysis of XH-59A rotor in forward flight with high efficiency trim method, *ICAS,* Belo Horizonte, Brazil, Sept 9-14, 2018.

[6] Chen R. A survey of nonuniform inflow models for rotorcraft flight dynamics and control applications, Vertica, 1990, 14(2): 147~184.

[7] Falarski M D, Mc Cloud J L, Soderman P T, et al. An investigation of a full-scale advancing blade concept rotor system at high advance ratio. *NASA technical memorandum.* Ames Research Center and U.S. Army Air Mobility R&D Laboratory Moffett Field, Calif. 1971.

IOP Conf. Series: Materials Science and Engineering **926** (2020) 012021 doi:10.1088/1757-899X/926/1/012021

Drilling and Migration Characteristics of Critical Size Lunar Soil Particles

Lijie Tong, Deming Zhao*, Ming Hu, Xingwen Gao and He Wang

Zhejiang Sci-Tech university, Hangzhou, 310018, China

*Corresponding e-mail: miancer@163.com

Abstract. Critical size lunar soil particles refer to particles with a particle size similar to the diameter of the drill bit, which is an important factor affecting lunar soil drilling sampling, and its influence is uncertain. In this paper, the discrete element method is used to simulate the drilling sampling process of the critical size lunar soil particles and the influence of the size, position and shape of the lunar particles with critical diameter on the drilling load is analysed. It is found that the critical size lunar soil particles have special migration trajectories and load characteristics during drilling sampling. The size and shape of lunar soil particles are the key factors affecting the drilling load. This study can support the design of drilling tools for lunar soil exploration and sampling.
Keywords: discrete element method; drilling sampling; lunar soil particles

1. Introduction

At present, the main task of the lunar exploration project is the sampling and return of lunar soil [1], in which drilling sampling is the preferred method to achieve lunar soil sample collection. Sampling drilling tools for lunar drilling are generally hollow spiral drilling tools, with a sampling port at the front of the drill bit [2]. In the early experiments, it was found that the lunar soil particles have a great influence on the sampling process, especially when the drilling tool encounters particles with a critical particle size close to the size of the drill bit, it will cause a large load change. In severe cases, it may cause the drilling load to exceed the driving capacity of the system. In addition, the large particles will block the sampling channel, making the lunar soil unable to enter the drilling tool, thus preventing sampling. In the former Soviet Union's Luna20 sampling activity [3], it also encountered stuck pipe conditions, which resulted in the drilling activity not reaching the expected target. It is judged that the failure is most likely caused by the blockage of the lunar soil particles of critical size.

There are two main types of simulation research on the drilling process of lunar soil, the finite element method and the DEM method. Quan et al. [4-5] studied the quasi-static stress distribution of the soil near the drill bit based on the finite element system. The finite element method cannot describe the rheological mechanical properties of lunar soil materials during drilling, so it is impossible to accurately simulate the characteristics of the actual lunar soil drilling force load. Wei et al. [6] used the DEM method to simulate the drilling process, and the obtained drilling pressure and torque increased with the increase of the drilling depth. The simulation did not consider the impact of the existence of critical particle size in the lunar soil. In this paper, the DEM method is used to study the impact of the critical particle size, shape and relative position on the drilling process, especially the drilling load.

2. Calibration of lunar soil mesoparameters

To establish a discrete element model [7] of lunar soil, in order to ensure that the model is close to lunar soil in mechanical properties, it is necessary to calibrate the meso parameters of the model. Use the bucket lifting method to carry out the stacking simulation test, and set the diameter range of the lunar soil particles to 0.1mm-1mm. As shown in Figure 1, the diameter of the drum is 50mm and the height is 100mm, and lunar soil particles are generated in the bucket. Lift the bucket at a speed of 0.1m/s, and measure the stacking angle after the lunar soil particles are stably accumulated on the bottom plate.

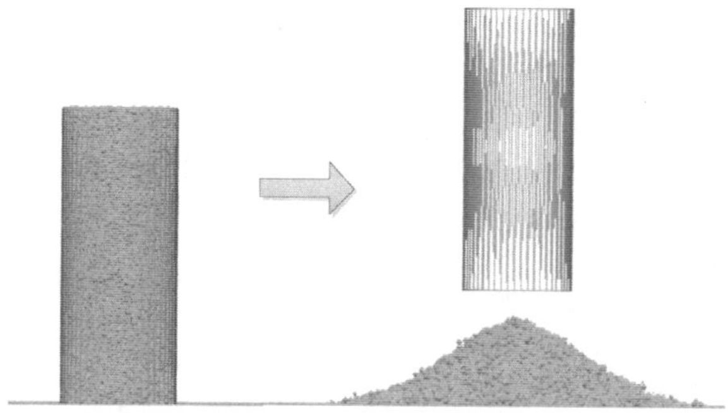

Figure 1. Simulation of stacking angle with bucket lifting method.

Since the main influencing factors of the static stacking angle of lunar soil particles are the inter-particle restitution coefficient, static friction coefficient and rolling friction coefficient, these three parameters are calibrated in this test, and other parameters remain unchanged during the test. According to Box-Behnken's central combination test, 17 sets of simulation tests are designed.

The test result was subjected to multiple regression fitting analysis to obtain the regression model of lunar soil accumulation angle:

$$\Phi = 17 - 0.625A + 2.63B + 6.25C - 0.75AB - 0.5AC + 4.5BC - 0.875A^2 - 1.38B^2 + 1.37C^2 \tag{1}$$

Where, A is the restitution coefficient between lunar soil particles (e); B is static friction coefficient between lunar soil particles (μ_s); C is rolling friction coefficient between lunar soil particles (μ_r).

The predicted value of the multiple linear regression equation (1) using the stacking angle of the lunar soil is compared with the experimental value, as shown in Figure 2. It can be seen that the predicted value and the actual value have a good fit. The measured stacking angle of lunar soil can be used to inversely determine the mesoscopic parameters of lunar soil according to the regression equation.

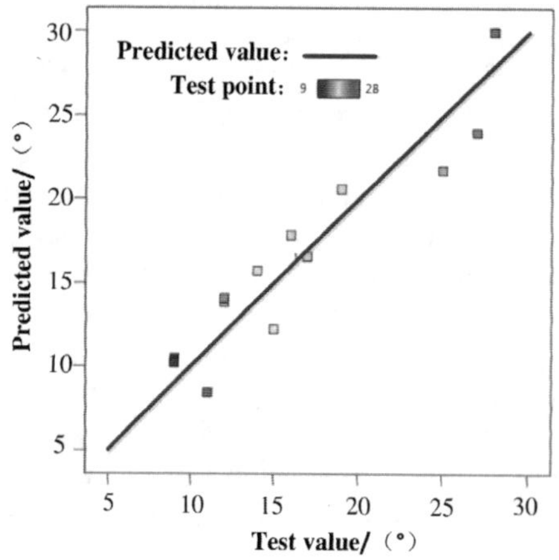

Figure 2. Comparison between predicted and test value of lunar soil stacking angle.

In order to obtain the meso-parameters of the lunar soil particle interaction model in the DEM simulation, this study takes the CUG-1A lunar soil simulant as the object, and uses the bucket lifting method mentioned above to conduct a stacking angle experiment on the simulated lunar soil after drying. The average stacking angle of lunar soil was measured 5 times and the average value was 29°. A set of optimal solutions can be obtained as: A=0.3, B=0.8, C=0.3, and the mesoparameters of the discrete elements of lunar soil are shown in Table 1.

Table. 1 Discrete element simulation parameters of lunar soil particles.

Young's modulus(Pa)	Poisson's ratio	Restitution coefficient (e)	Static-friction coefficient (μ_s)	Rolling-friction coefficient (μ_s)
10^8	0.5	0.3	0.8	0.3

3. Simulation and analysis

3.1. Drilling model

The critical particle size is defined as large particles with a diameter similar to the end size of the sampling tool. In this paper, the critical particle size is 10-20mm, which is similar to the drill bit size. Figure 3 shows three typical shapes of lunar rock samples: (a) spheroid, (b) angular and (c) elongated. In the simulation test, simulated particles of ellipsoid, tetrahedron and oblique body were established for the three lunar rocks.

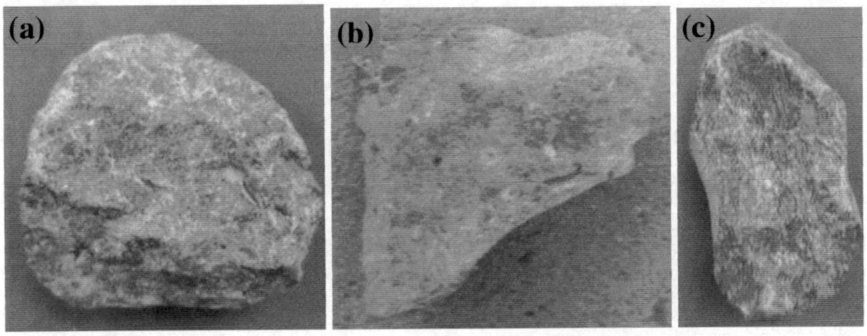

Figure 3. Typical critical size particle morphology of lunar soil.

Figure. 4 shows the drilling model of critical size lunar soil particles. Lunar soil drilling conditions are: simulated drilling depth is 40 mm, feed rate is 60 mm/s, rotation rate is 1000 rpm, lunar soil barrel radius is the same as the influence domain radius, lunar soil particle radius is set to 0.5-0.15 mm, and generate critical particle size particles in the subsurface layer. Nine sets of experiments were designed to explore the effect of critical particle size, relative position and shape on drilling load. The detailed design scheme is as follows: (a) non-critical size particles, (b) spherical particles with a diameter of 10 mm at the center, (c) spherical particles with a diameter of 15 mm at the center, (d) spherical particles with a diameter of 20 mm at the center , (e) Spherical particles with a diameter of 15mm at a center offset of 7.5mm, (f) Spherical particles with a diameter of 15mm at a center offset of 15mm, (g) Oval particles with a center offset, (h) Tetrahedral particles with a center offset , (i) Hexahedral particles with a center offset.

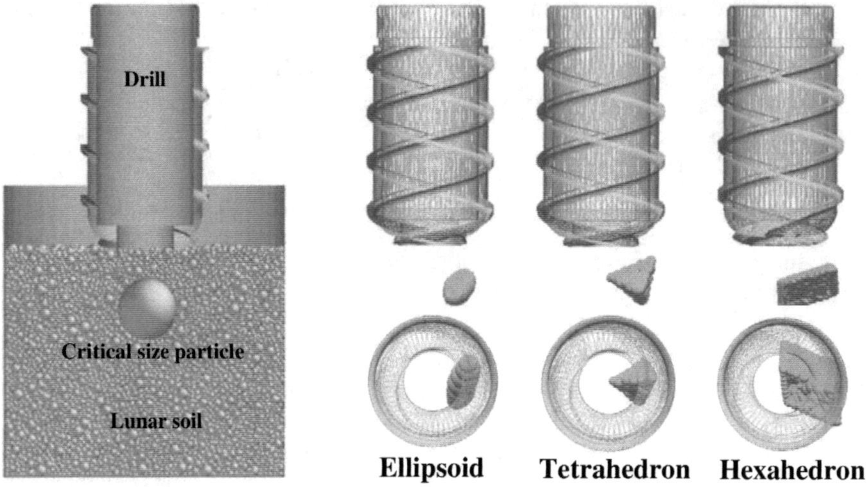

Figure 4. Sampling model for critical size lunar soil particle drilling.

3.2. The migration trajectory of critical size lunar soil particles

As shown in Figure. 5, the movement trajectory of the critical size particles in each working condition during the drilling process. It can be seen that the flow characteristics of critical size particles during drilling are divided into three cases. The first is that the critical size particles smaller than the cutting circle are located directly in front of the drill bit and enter the sampling tube to produce spiral movement. The second is that the critical size particles larger than the cutting circle are located directly in front of the drill bit, and they will be stuck at the drill bit and move down with the drill bit. The third type is the critical size particles located on the side of the drill bit, which will move upward spirally after moving outward under the screw action of the drill.

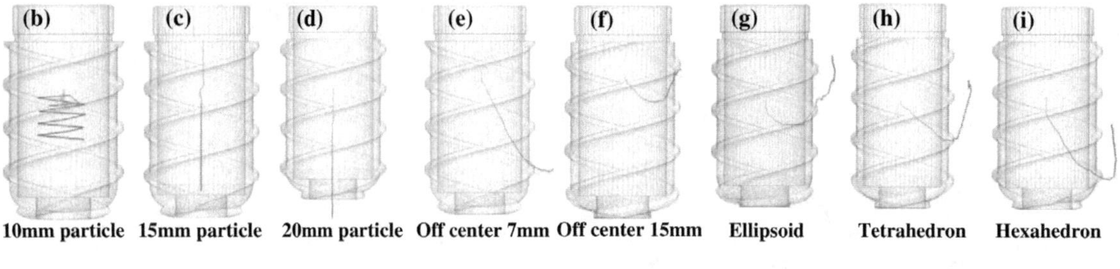

Figure 5. The trajectories of critical size lunar soil particles with different shapes.

3.3. Critical scale lunar soil particle velocity analysis

Lunar soil particles will squeeze and slip during the drilling process, and eventually enter the sampling tube or be transported around the drilling tool. Figure 6 shows the speed curve of the critical size lunar soil particles during the drilling process with time. Obviously, the velocity of particles entering the sampling tube and being squeezed to the outer surface of the drill bit is generally larger. Figure 7 shows the Z-axis velocities of critical size lunar soil particles with three shapes. During the drilling process, the particles will be spirally moved upward by the surrounding particles.

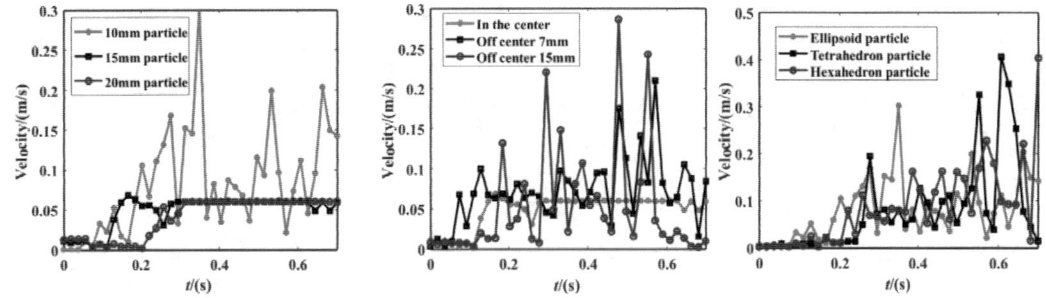

Figure 6. Velocity analysis of critical size lunar soil particles.

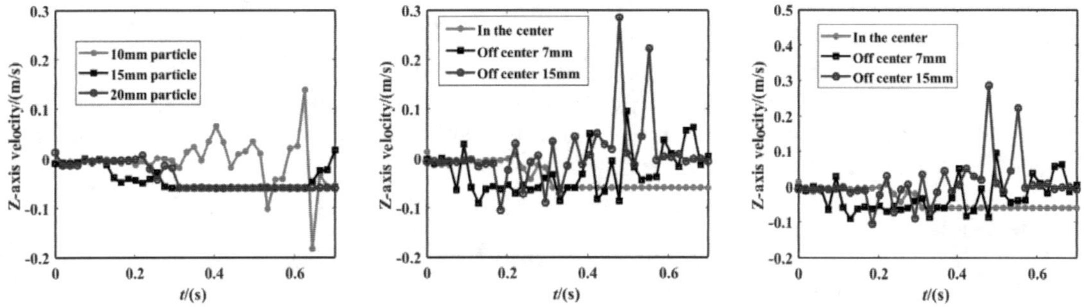

Figure 7. Z-axis velocity analysis of critical size lunar soil particles.

3.4. Drilling load

Drilling load is an important indicator of the drilling process, including bit load and bit torque. Figure 8 shows the bit load under different critical particle size conditions. The smaller particle has little influence on the drilling load, and the feed force increases with the drilling depth. When the particle size increases to 15mm, the load increases significantly. When the particle size reaches 20mm, the bit load increases sharply, and severe sticking occurred. The relative position and shape have little effect on the drilling load. However, angular particles will amplify the fluctuation of drilling load. This is due to the fact that the particles with critical diameters are turned by the surrounding small particles during the drilling process, and the irregular shape causes uneven turning resistance, which leads to fluctuations in the bit load. It shows that the shape of particles with critical particle size will not affect the drilling load and its tendency but will cause jitter of the drilling tool.

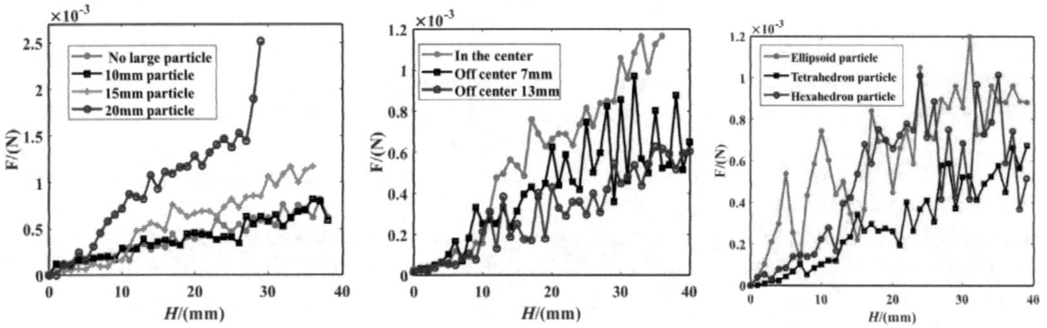

Figure 8. Trend of bit load.

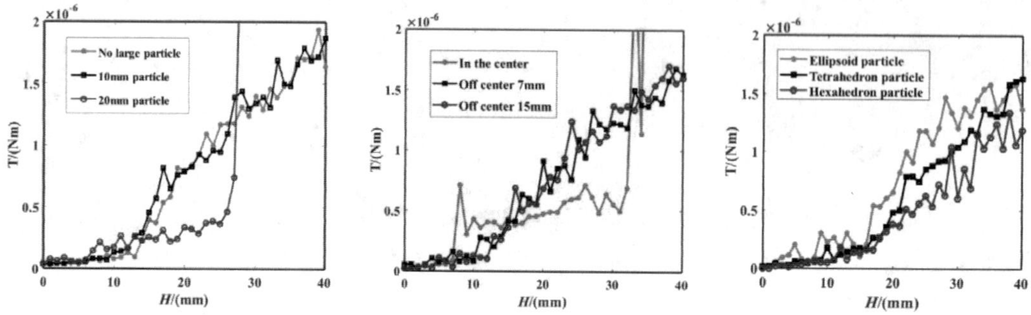

Figure 9. Trend of bit torque.

During lunar drilling, if the drill bit is stuck or blocked, the torque of the drill bit will rise sharply. As shown in Figure. 9, when 15mm and 20mm particles were placed in front of the drill bit, the drill bit blocked and the torque increased sharply. The size of the center-mounted particles is smaller than the cutting circle and offset particles of any size will not affect the bit torque. The shape of the critical size particles has no effect on the torque of the drill bit, but the more irregular the particles, the more obvious the fluctuation of the torque value.

4. Conclusion

In this paper, the discrete element method is used to simulate and analyse the drilling process of critical size lunar soil particles. According to the size, relative position and shape of the particles, 9 sets of simulation conditions were designed. By comparing the trajectory and load of particle velocity in each working condition, the conclusion is as follows.(1) The movement trajectory of critical size lunar soil particles during drilling is divided into two categories: one is that the particles move linearly with the drill bit during plugging, and the other is the spiral upward chip removal movement.(2) The critical size particles larger than the sampling port of the drill bit are easily stuck and blocked when the particle is directly in front of the drilling tool, and the load increases sharply. However, the particles smaller than the sampling port have no significant effect on the drilling load at any position. (3) Although the shape of the critical size particles is not affect the load and its trend, it will increase the fluctuation of the load curve and make the drill jitter.

Acknowledgements

The authors thank financial support from the National Natural Science Foundation of China (NSFC) under Grant No. 51805488, Science Foundation of Zhejiang Sci-Tech University(ZSTU) under Grant (No. 17022055-Y, No. 19022406-Y & No.19022104-Y).

References

[1] Anil B. Preface: Lunar science and exploration[J]. *Advances in Space Research*, 2014, 54(10).

[2] Tang J, Quan Q, Jiang S, et al. A soil flowing characteristic monitoring method in planetary drilling and coring verification experiments. *Advances in Space Research*, 2016, 59(5): 1341–1352.

[3] Zhao D, Tang D, Hou X , et al. Soil chip convey of lunar subsurface auger drill[J]. *Advances in Space Research*, 2016:S0273117716300060.

[4] Quan Q, Tang J, Yuan F, et al. Drilling load modelling and validation based on the filling rate of auger flute in planetary sampling. *Chinese Journal of Aeronautics*, 2017, 30(1): 434–446.

[5] Quan Q, Chen C, Deng Z, et al. Recovery rate prediction in lunar regolith simulant drilling. *Acta Astronautica*, 2017, 133: 121–127.

[6] Wei C, Hongliu W, Tianxi L. Mechanical model of hollow-external-screw drill rod for lunar soil particle vertical conveying // *IEEE International Conference on Control and Automation. Hangzhou*, 2013: 1240–1245.

[7] Cundall P A, Strack O D L . Discussion: A discrete numerical model for granular assemblies[J]. *Géotechnique*, 1980, 30(3):331-336.

IOP Conf. Series: Materials Science and Engineering **926** (2020) 012022 doi:10.1088/1757-899X/926/1/012022

Study of Tail Rudder Deflection Angles for Stabilizing the Twin Turboprop Small Passenger Aircraft in Critical Flight due to One Engine Failed Condition

Y. V. Agatha[1,a] and W. Nirbito[1,*,b]

[1)]Department of Mechanical Engineering, Universitas Indonesia, Depok, West Java, 16424, Indonesia

[*]Corresponding author: [a]yuke.vahira@ui.ac.id; [*b]bito@eng.ui.ac.id

Abstract. An aircraft must have durability, whether for normal flight condition and for a critical flight condition. One of the critical flight conditions of a twin-engines aircraft is the failure of one engine while the aircraft is cruising. The aircraft with only one live engine on will still have enough power to generate thrust. However, the aircraft will experience a moment couple due to the thrust on the remaining engine that makes the aircraft to yaw. This yaw effect must be compensated by the flight control in order to maintain a stable flight condition. The rudder as one of the flight control systems manages the aircraft yaw motion. So, therefore the rudder deflection angle must be set properly as a treatment to overcome the moment force of the live engine. Study to determine best approximated optimum rudder deflection angle setting were conducted to get the figures of how the counter side forces generated on the rudder can maintain a stable flight. The result of the study can be applied as important guidance for a pilot to control the aircraft in a critical flight condition due to one engine fails. Considerations on the strength and integrity of the rudder structure especially at the hinge pivot points between the dynamic and the static parts are taken account as well.

1. Introduction

Twin turboprop engines are more fuel-efficient than the one that is using jet engines in slow to medium flight [1]. The turboprop engines make it possible for the aircraft to operate and take off from shorter runways and perform take-off and landing on unpaved surfaces [2], which is suitable to reach the varied condition of terrain in Indonesia. The twin engines aircraft must have durability for normal flight condition as well as for critical flight condition, i.e. one engine is failed. On period 2001-2016, engine failure is number 3 ranked out 10 leading cause of fatal general aviation accidents. The published accident investigation reports revealed that during the last 25 years there are more than 300 propulsion system malfunctions of multi-engine airplanes accidents happened. Also, it took casualties of more than 3,100 people lost their lives [3]. In such a critical flight condition, the aerodynamic consequences from the one engine failure lead to a very significant flight performance penalty. While the failure of an engine represents a 50% loss of available power, it can result in as much as an 80% loss of performance [4]. In general, aircraft with twin engines already being prepared to handle this kind of situation. With only one engine left the aircraft will experience yaw effect because of asymmetric thrust produced by remaining engine. The rudder needs to be deflected properly to handle the drag force until it able to land safely, or else the aircraft will experience stall. The information of optimum rudder deflection to handle this situation is important as the pilot guidance. As a result, for the rudder deflection in certain time, the rudder will also experience forces that might affect the structure of the tail. The second deadliest plane crash that kills in total 265 passengers, American Airlines flight 587 to Santo Domingo (2001) happened because of the failure of the plane's tail structure due to overstress when excessive rudder deflection applied. This accident provides a strong example that the study to inspect the effect of deflection in tail rudder structure is important. For safety precautions and improvement, it is needed to evaluate whether the structure able to handle the stress due to the rudder deflection in a one engine failed condition.

2. Literature Review

2.1 Asymmetric Thrust Due to One Engine Failed

Since all engines in a multi-engine aircraft are not in a same centreline and if one engine fails, then the thrust of the remaining operative engine will cause a force moment. In order to maintain a straight flight path, a yawing trim must be adjusted by controlling the rudder. The appropriate rudder deflection angles must be set to generate side forces according to the speed for countering such yawing motion. The illustration of the force moment due to the right engine and its conservation are described in figure 1 and figure 2.

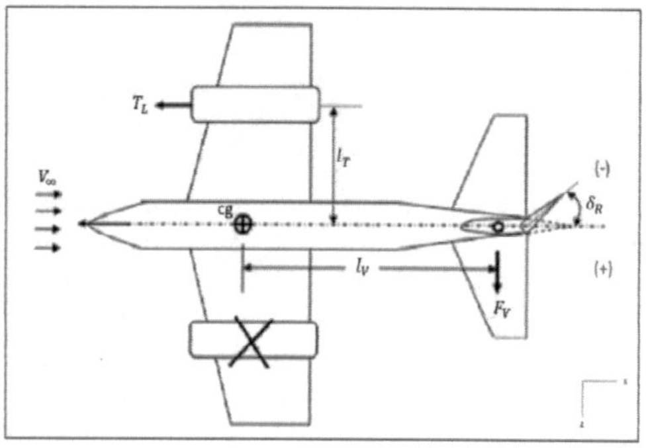

Figure 1. Force moment due to right engine fail [5].

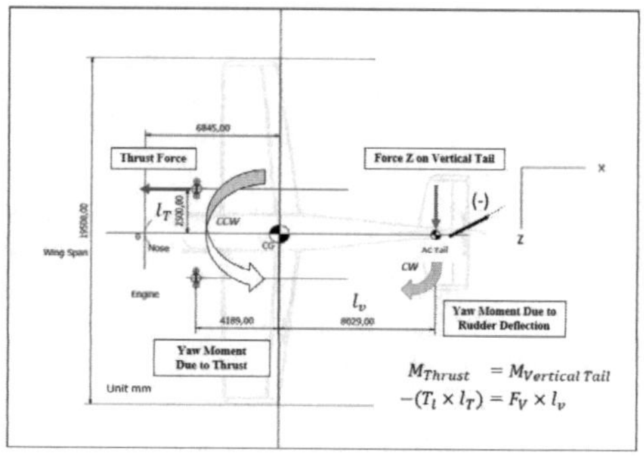

Figure 2. Conservation of Moment in one engine fail condition [5].

Each engine is at a distance of l_T to the fuselage centreline. The asymmetrical thrust in one engine fail situation is resulting a yawing motion due to a force moment (M_T) about the aircraft centre of gravity (CG), as described in the equation [5]:

$$M_T = T_L \times l_T \tag{1}$$

The convention for the negative rudder deflection is defined as the deflection to the right from pilot point of view. negative ruder deflection creates a positive side-force in the z direction, which generates a positive yawing moment (clockwise). In reverse, positive rudder deflection (to the left) resulted to negative side-force in z direction which generates a negative yawing moment (counter-clockwise). For a symmetrical aircraft with a zero-sideslip angle and zero-aileron deflection, a yawing moment (M_V) of the vertical tail is determined from multiplication the side lift force of the vertical tail air foil with the vertical tail arm length [5]:

$$M_V = L_V \times l_V \tag{2}$$

In the steady-state trimmed straight path of this critical flight, the total aircraft drag is overcome just by the single thrust (T_L) of the operative engine. In order to maintain a relatively straight flight path, the summation of yawing moments is equal to zero [5]. Therefore:

$$\sum M_{CG} = M_T + M_V = 0 \tag{3}$$

Aerodynamic yawing moment (M_V) of an aircraft can be calculated from the following equation [5]:

$$M_V = q\,S\,b\,C_n \tag{4}$$

Which (q) is the dynamic pressure, (S) is the wing area, and (b) is the wing span, as well as C_n is the yawing moment coefficient that is a function of aircraft configuration of sideslip angle, rudder deflection and aileron deflection. In order to get the possible maximum rudder deflection for balancing the couple moment of the remaining engine thrust, the aircraft is assumed to be symmetrical about x z plane, with no sideslip and no deflection at the aileron. So, therefore the considered yawing moment coefficient is the yaw moment coefficient due to the rudder deflection, $i.e. (C_{n_{\delta R}} \delta_R)$ [5].

$$M_V = q\,S\,b\,C_{n_{\delta R}} \delta_R \tag{5}$$

In an asymmetrical thrust condition, the necessary rudder deflection angle to trim the straight flight path is [5]:

$$\delta_R = (-)\,\frac{T_L l_T}{q\,S\,b\,C_{n_{\delta R}}} \tag{6}$$

The sign in equation represent the direction of the rudder, positive sign represent that the rudder is deflected into the left side of the pilot. In vise versa, minus sign represent that the rudder is deflected into the right side of the pilot [5] [6].

2.2. Minimum Control Speed (V_{mc})

The only aerodynamic control surface that could provide the required side force to counteract the asymmetrical thrust yawing moment is the vertical tail stabilizer with rudder. The rudder must be able compensate the yaw turning tendencies caused by the operating engine, the drag of the wind milling propeller, and the aileron drag caused from trying to hold the wings level. From **Table 1**, it can be seen that the speed directly affect the thrust generated by the engine which have inverse relationship, this affect the magnitude of the force needed to couple the thrust. In parallel, speed also affect the dynamic pressure which is one of the denominator parameters to determine the rudder deflection as used in **equation (6).** Which the Dynamic pressure (q) can be evaluated by the equation below:

$$q = \frac{1}{2}\, V^2 \times \rho \tag{7}$$

(V) in the equation above represent the speed. It can be understood when the air speed experienced by the aircraft is minimum, the thrust generated will become bigger and the deflection needed to be able to generate the force to handle it is also become greater. Thus, as airspeed decreases, the rudder becomes less effective until it reaches lowest possible minimum level when it loses its ability to control the yaw moment afterward. This speed level is called the minimum control speed (V_{mc}). Which is used for maintaining straight flight after one engine fails [6].

3. Methodology

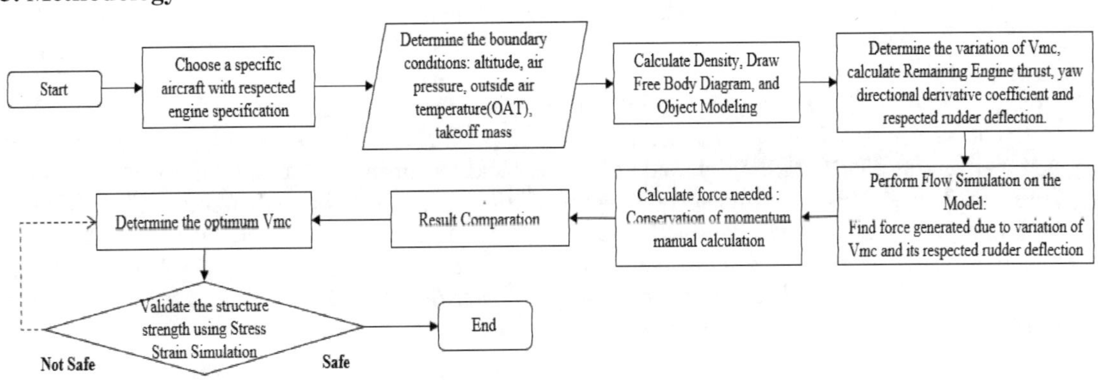

Figure 3. Flowchart of the Methodology

The whole process in this paper is illustrated in the figure 3. The object in this research is an original multipurpose aircraft made by Indonesian Aerospace categorized as part 23 FAR aircraft. It can carry loads up to 6,940 kg on landing, with V stall at take-off mass 6700 Kg is 69 knots. Powered by Pratt Whitney PT6-42, connected with 4 Hartzell blades. The engine performance while in one engine failure condition is shown in table 1. The aircraft has 41,5 m^2 wing area, 19.5m wingspan, thrust line for each engine is 2.5 m from cg (Center of Gravity) and distance from cg to ac (Aerodynamic Center) tail is 8 m. The calculation is performed under specific condition which are altitude flight of 6000 ft with respected pressure 81199.6 Pa, and outside temperature of ISA +15 Celsius. Most of the calculation is done on a spreadsheet in Microsoft Excel®. The modelling and simulation were conducted using SOLIDWORKS 2018® software. The accuracy of the simulation is really dependent on the assumption of the object modelling and the boundary condition, which is in research the model is the simplified version and the boundary condition assumed is the ideal condition at respected altitude. The flow simulation is done with resolution 5 out of 7 which already describe not the best but good accuracy of the result.

Table 1. Engine Performance [7]

Np	Speed		Thrust		Fuel Flow		Efficiency
rpm	knots	m/s	Kgf	N	lb/hr	g/s	-
2000	60	30.87	860.98	8443.3	485.56	61.1796	0.45
2000	80	41.16	815.34	7995.8	488.24	61.5173	0.562
2000	100	51.44	763.61	7488.5	491.52	61.9305	0.652

3.1. Calculation of Generated force Due to Rudder Deflection Rounding

For the realistic practical condition when the pilot adjusts the rudder degree in flight test and to eliminate redundancy, the degrees of deflection are being rounded to the nearest possible non-decimal number. The rounding process is performed with considering the structural limit of deflection is 25 degree for each side. The number of forces generated supposed to increase with the increment of the deflection. Thus, the rounding up method is endeavoured for safety flight reason to make the force is great enough to be able to counteract the asymmetric thrust and easier to reduce if the force is excessed. The expected generated force by rounding the rudder deflection, can be evaluated using the **equation (5)**, divided by the distance from cg to ac tail:

$$F_V = \frac{q\, S\, b\, C_{n_{\delta R}} \delta_R}{l_V} \tag{8}$$

3.2. Calculation of Needed force

Needed side force from the vertical tail can be evaluated by applying conservation of moment at equation (3) without considering the rudder deflection as shown in figure 2.

4. Results and Discussions

The flow trajectories for speed in x direction on its respected rudder deflection angle are illustrated in figure 4. The velocity was changing according to the air foil shape, where the higher velocity in the left side and lower velocity in the right side of the vertical tail of the deflection was heading to. The maximum velocity is shown on the surface of the vertical tail hinge for each figure. The vortexes occurred on the root of trailing edge of the vertical tail and on the horn feature. It increased as the increment of the deflection. It also worked accordingly for the horn feature. The velocity of the massive vortex generated by the horn feature became slower in both side, this would help to alleviate the overall force that experienced by the rudder.

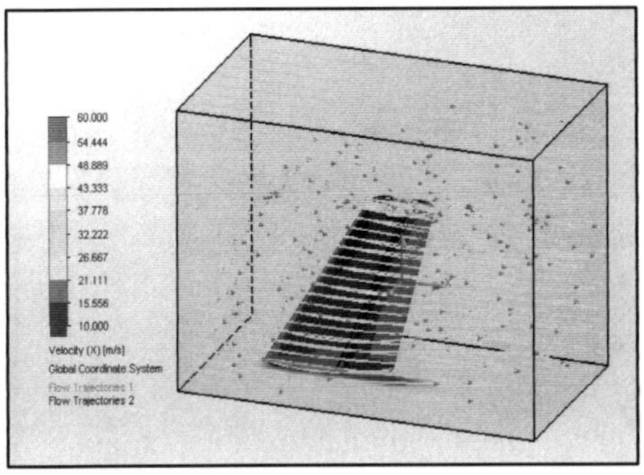

Figure 4. Flow Trajectories at 78 knots with 18 Degrees of Rudder Deflection

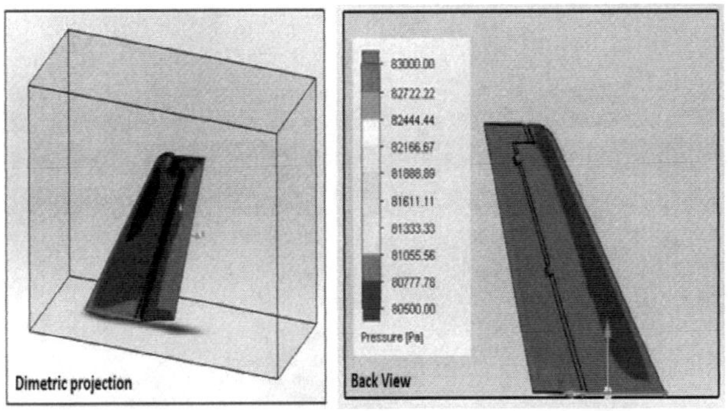

Figure 5. Surface Pressure Distribution at 78 Knots with 18 Degrees of Rudder Deflection

The load pressure on the surface of the tail is illustrated in figure 5. The pressure distribution can be observed from front and back view or the vertical tail. It is shown that the pressure increased when the air flew through the leading edge of the vertical stabilizer, and then the pressure decreased in the outside of the generated curvature of the deflecting rudder. After that, the pressure increased again when it passed the end of the rudder tip. From the back view it can be observed that the pressure was already in average magnitude between 81055.56 – 81611.11 Pa, then it reached its maximum at the hinge area with a value of 82166.67 Pa approximately. The pressure then slightly decreased as the flow travelled to the end of the rudder tip. By evaluating table 1, it was known that the fuel consumption decreased as the speed decreased, thus by operating aircraft at the lowest V_{mc}, the fuel consumption would be minimum. To determine the lowest possible optimum V_{mc}, the forces generated from simulation as well as due to rounding rudder deflection angle were compared with the forces needed from manual calculation. The comparation among those three results are shown in the table 2 and in the figure 6 and 7. From figures 6 and 7, it can be observed that the rudder worked properly with the decreasing airspeed from 82.8 knots. Then it lost its effectiveness after 22 degree of rudder deflection at 72 knots airspeed. Decreasing airspeed of 70 to 68 knots were considered as stall speed since the rudder could not generate forces as needed even though already at its maximum deflection angle of 25 degree. The rudder capability limit was considered as the best approximated optimum minimum control speed at airspeed of 78 knots and rudder deflection angle of 18 degree, for long time operation and safety reasons. At those airspeed and degree of rudder deflection, the generated forces by the manual calculation and the simulation were nearly the same and adequate for handling the yawing moment force since had only small difference from the needed force.

Table 2. Needed side force (Manual Calculation) and Generated force (Simulation)

Speed	Rudder Deflection $\boldsymbol{\delta_R}$		Needed force (calculation)	Rounding $\boldsymbol{\delta_R}$ F. Generated (Calculation)	Generated force (Simulation)
knots	Calculated Degree	Rounding Degree	N	N	N
68	24.1	25	2544.28696	2640.98275	2514.43232
70	22.6	23	2531.22431	2574.72972	2531.04000
72	21.3	22	2518.16166	2605.52596	2589.40000
74	20.0	20	2505.09902	2502.07985	2497.92000
76	18.9	19	2492.03637	2507.19732	2526.53000
78	17.8	18	2478.97372	2501.89708	2512.93735
82.8	15.6	16	2447.62337	2506.04224	2471.53000

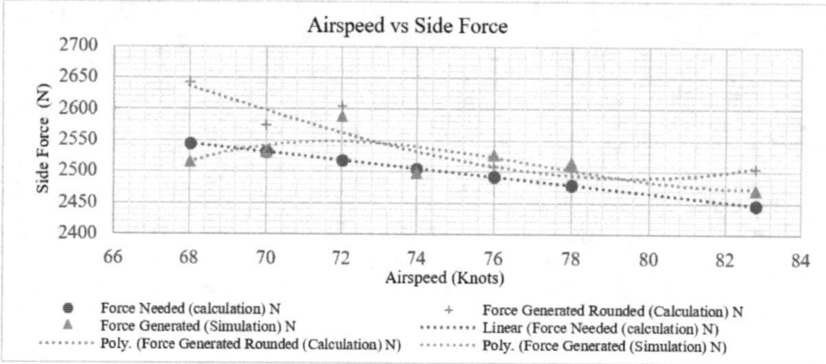

Figure 6. Airspeed vs Side Force Comparation

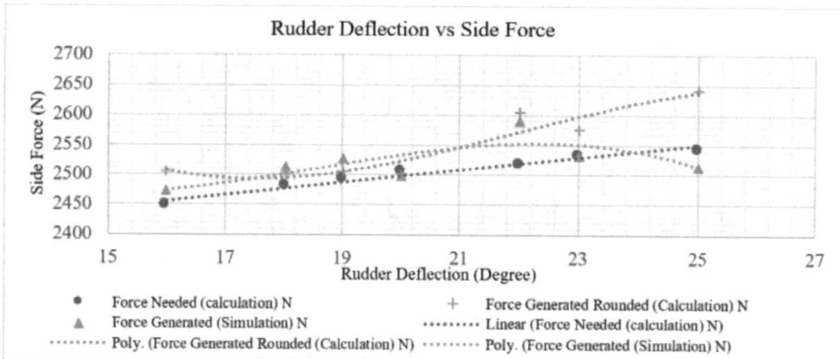

Figure 7. Rudder Deflection vs Side Force Comparation

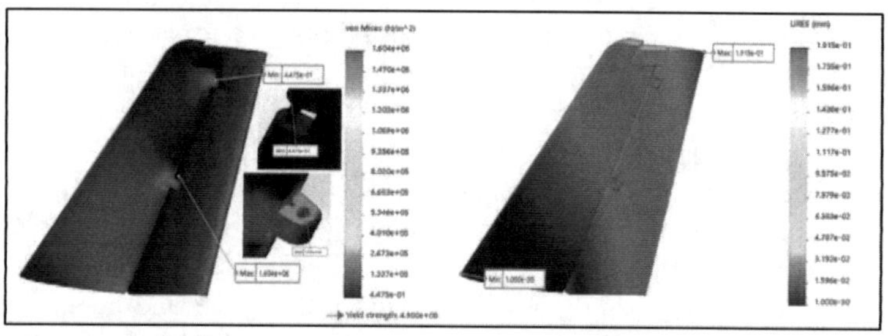

Figure 8. Stress and Displacement distribution at 18 Degrees of Rudder Deflection with Loaded Force 2512.9 N

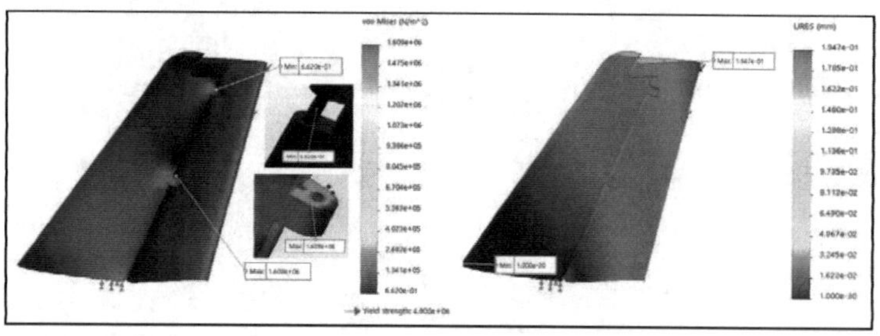

Figure 9. Stress and Displacement distribution at 22 Degrees of Rudder Deflection with Loaded Force 2589.4 N

Static stress simulation results for the rudder model with material of Aluminium Alloy 7050-T7651 was also conducted with consideration of the structure as one solid body to give approximation of the location of overall structure critical stress which were shown in figure 8 - 9 and in the table 3 as well. Observing from figures 8 and 9, the high stress for each deflection happened mainly on hinge area and the tip end of the rudder, but with the displacements were progressively decreasing to the root. At 18 degree of deflection angle, the maximum stress is located inside the pivot hinge of the tail. But at 22 degree, the stress inside the pivot hinge is also high but the maximum stress was at the end tip of the rudder. However, the maximum stress at each deflection angle was not over the material's yield strength

Table 3. Static Stress Simulation Result

Simulation Result		Rudder Deflection		Units
		18 Degree	22 Degree	
Side Force	Max	2512.937353	2589.4000	N
Von Mises Stress	Max	1.604 e+06	1.609 e+06	N/m2
	Min	4.475 e+01	6.620 e-01	N/m2
Yield Strength [8]		4.900 e+08	4.900 e+08	N/m2
Strain	Max	1.725 e-05	1.706 e-05	-
	Min	1.127 e-11	1.379 e-11	-
Displacement	Max	1.915 e-01	1.947 e-01	mm
	Min	1.000 e-30	1.000 e-30	mm
Safety Factor	Max	1.095 e+09	7.402 e+08	-
	Min	3.055 e+02	3.045 e+02	-

5. Conclusion

The rudder system of the twin engines aircraft model has an best approximated optimum capability to maintain the stable flight in a critical condition due to one engine failed, which are at 18 degree of rudder deflection angle with airspeed best approximated optimum V_{mc} of 78 knots. The best approximated lowest possible V_{mc} that this Aircraft could achieved is 72 knots with the rudder deflection of 22 degree. The maximum generated force is 2512.9 N at the optimum V_{mc} and 2589.4 N at the minimum possible V_{mc}. The tail structure of the twin engines aircraft model can stand the load due to the generated force by the rudder systems at the airspeed optimum V_{mc} and minimum possible V_{mc}. For safety precaution, the critical parts which experienced the high stress, i.e. the pivot hinge area and the rudder end tip, need to be designed well and maintained regularly.

Acknowledgment

The authors would like to thank The Directorate of Research and Development of UNIVERSITAS INDONESIA with PUTI grant programme for funding this study with contract number: NKB-1058/UN2.RST/HKP.05.00/2020. And high appreciation for PT. Dirgantara Indonesia (Indonesian Aerospace) regarding the opportunity, guidance, and facility for the development of this study.

References

[1] Kundu A K, Price M A, and Riordan D. 2019. *Conceptual Aircraft Design: An Industrial Approach.* United State of America, John Willet & Sons Ltd.
[2] FAA (Federal Aviation Administration). 2011. *Airplane Flying Handbook.* FAA-H-8083-3A. New York, Skyhorse Publishing.
[3] Dorr L. 2018. *Fact Sheet – General Aviation Safety.* FAA (Federal Aviation Administration).
[4] Admin. 2020. *Multi Engine Airplane.* Accessed on Monday, 20 April 2020 12.07 am from mycfibook.com.
[5] Sadraey M. 2012. *Aircraft Design: A System Engineering Approach.* Wiley Publication.
[6] Horlings H. 2012. *Control and Performance During Asymmetrical Powered Flight for Multi-engine.* Avio Consult.
[7] Commuter Aircraft with 2 Turboprop Engine and 19 Passengers. Engineering data. PT. DI.
[8] ASM Aerospace Specification Metal Inc. *Aluminium 7075-T651.* Accessed on Tuesday, 21 April 2020 at 9:23 pm from http://asm.matweb.com.

IOP Conf. Series: Materials Science and Engineering **926** (2020) 012023 doi:10.1088/1757-899X/926/1/012023

Design and Optimization of Multi-petal Damping Structure for Vibration Control of Satellite Antenna Component

Min Wang[1,2,a], Tingting Zhao[1,b], Qinda Liu[1,c], Jiaxing Liu[1,d]

[1] Institute of Intelligent Monitoring and Diagnosis, College of Mechanical Engineering and Applied Electronics Technology, Beijing University of Technology, Beijing 100124, China.
[2] Beijing Municipal Key Laboratory of EDM Technology, Beijing 100191, China.

[a] wangm@bjut.edu.cn, [b] Corresponding author: ztt9607@163.com, [c] liuqinda@hotmail.com, [d]1677486548@qq.com

Abstract. A preloaded multi-petal friction damping structure is proposed for deployment mechanism of large-scale flexible deployable satellite antenna. The damping structure is designed to suppress transverse vibration for promising stability and accuracy in the deployment process. In order to study the damping property of multi-petal damping structure and facilitate theoretical modelling, the model of deployment component with damping structure is simplified as double-layer cantilever beam model to study mechanical properties of friction and damping energy dissipation mechanism. The optimum parameters of damping structure are determined by numerical analysis of the simplified modal and the hysteresis curve can be obtained by finite element simulation to calculate loss factor. In the dynamic simulation, the loss factor can reach 0.2685 and the vibration amplitude can be reduced by more than 65%, which indicates the excellent effect for vibration suppression and the feasibility of the proposed structure.

1. Introduction

With the development of satellite technology, the application of satellites equipped with large-aperture antennas are increasing in many fields such as earth observation, telecommunications, scientific researches and remote sensing missions [1-2]. The deployment process stability and accuracy of large-scale flexible deployable satellite antenna are in high demand. Therefore, it is indispensable to develop a vibration control method with the characteristics of high working reliability, obvious vibration suppression effect and high vibration control stability to improve the precision and retention of antenna deployment process. In recent years, based on the development of artificial intelligence and intelligent materials including piezoelectric actuators with the feature of large rigidity, fast response and simple control structure, a large number of active control methods have been adopted to realize vibration control at home and abroad[3-5]. Due to the limitations that active control needs energy input and small noisy may lead to vibration aggravation, passive control is still the most extensive method of controlling vibration. Passive control restrains vibration mainly by increasing energy dissipation of structural vibration in damping system[6-7]. Xu has carried on the dynamic characteristic analysis of Additional Constrained Damping Layer. With the addition of constrained damping layer, the natural frequencies of the structure do not change much, but the vibration damping effect is nearly 20%[8]. At present, the aircraft mainly uses tuned mass dampers (TMD) to suppress vibration[9-10]. But for large flexible deployable antennas with low natural frequency, the design of mass and elastic components is difficult. The control effect of TMD is unstable and its frequency band is narrow, which cannot meet the requirements of reliability and stability for deployment mechanism vibration control. Therefore, the paper designs a new damping structure suitable for large flexible deployable antenna.

In this paper, taking advantage of the characteristics that nonlinear friction vibration has large energy dissipation and control bandwidth[11], it is considered to adopt nonlinear friction damping structure to improve effectiveness, stability and reliability of vibration control in deployment component. Firstly, the model of deployment component with damping structure is simplified. Then, in terms of the simplified model to study the mechanical properties of friction and damping energy dissipation mechanism. The optimum parameters of damping structure are obtained by theoretical analysis. Finally, the hysteresis curve is determined by finite element simulation. Loss factor is achieved to prove rationality of the proposed structure.

2. Structure design and preliminary optimization of friction damper

Under the vibration characteristics of large-scale flexible satellite antenna and the stability requirements of deployment process, a preloaded thin-walled multi-petal friction damping structure is designed. The energy dissipation mechanism of the damping structure is that microscopic relative sliding friction of the contact surfaces between the outer wall of the petals and the inner wall of the deployment component produces energy loss. The relative sliding state of the contact surface at different positions along the deployment component axis changes with different payload of the deployment component, including micro-slip zone and bonding zone. When effective bending load of the deployment component exceeds a certain threshold, the contact area between the deployment component and the petals of damping structure is in a micro-slip state. The mathematical model of micro-slip energy dissipation on the contact surface is established to obtain the maximum damping capacity of the preloaded thin-walled multi-petal structure. Through the numerical analysis of the model, the relationship among the loss energy, the preloading force and the deployment component structure is determined, and the optimal design of the preloading multi-petal damping structure is realized.

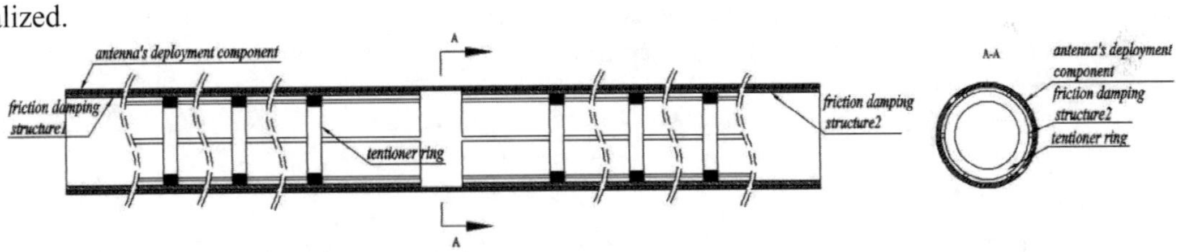

Figure 1. Schematic diagram of satellite antenna deployment component with damping structure.

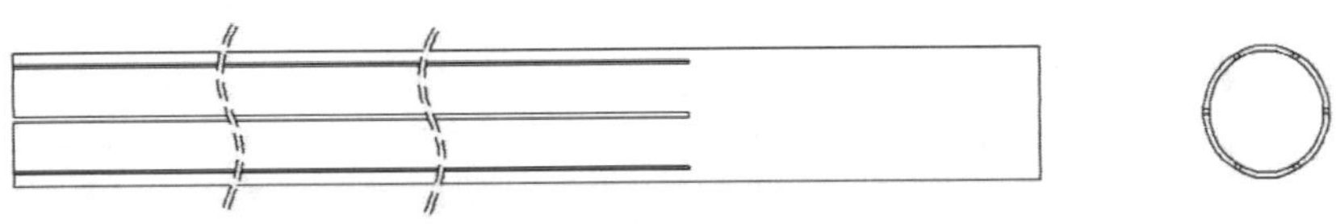

Figure 2. Thin-walled multi-petal damping structure.

In general, the structure of deployment component is a thin-walled circular ring cross-section beam structure with two ends hinged as shown in Figure 1, and the material is light and high rigid, such as carbon fiber. As shown in Figure 2, the preloaded thin-walled multi-petal friction damping structure is a multi-petal thin-walled ring cross-section structure. one end of the damping structure fixed with the deployment component has no petal and the other end preloaded by the tensioner ring on the inner wall of deployment component has petals. The friction loss energy is related to the bending stiffness and the structure of the petals, the preloading force and the external load of the expanded component. In order to verify the damping effect of damping structure, the model of deployment component with damping structure is simplified as double-layer cantilever beam model for theoretical analysis. Then an example is given to receive the appropriate parameters of the friction damping structure.

2.1. Calculation for loss factor of cantilever beam model with rectangular section

The contact surface between the deployment component and the damping structure is cylindrical, which can be simplified as plane contact in a tiny region. Therefore, it is reduced to double-layer cantilever beam model with rectangular section for numerical analysis firstly. Figure 3 shows the simplified double-layer cantilever model. The subscript 1 represents the upper beam and the subscript 2 represents the lower beam. The distance between the slip interface and the neutral plane of the beams is h. For rectangular section beams, the beam thickness is twice the neutral surface distance. The length of two beams is defined as l. The load consists of total preloaded distribution force p between two beams and concentrated load P at the free end. The friction at the connection interface is defined as F_f. Because of the external load, the double-layer cantilever beam will produce relative displacement $\Delta u(x)$ at the interface.

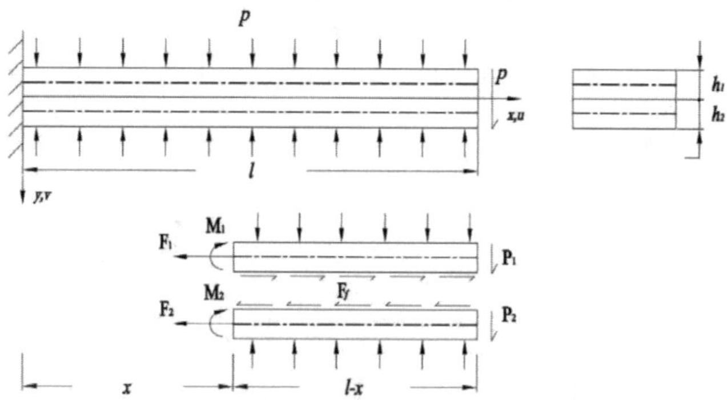

Figure 3. Double-layer cantilever beam (the simplified model).

The curvature at any position x of the upper and lower beams is obtained by force balance and the relationship between bending moment and curvature. Thus, the relationship between end load P and end deformation v is achieved:

$$v = \frac{Pl^3 - F_f l^2 (h_1 + h_2)}{3(E_1 I_1 + E_2 I_2)} \tag{1}$$

where E is elastic modulus of the beam, I is the cross-section moment of inertia of the beam. From the axial displacement $u_{1,2}$ of the upper and lower beams, the relative slip of the contact surface is obtained:

$$\begin{aligned} \Delta u(x) = u_2 - u_1 &= \left(\frac{1}{E_1} \int_0^x \sigma_{x_1} dx - h_1 \frac{dv_1}{dx} \right) - \left(\frac{1}{E_2} \int_0^x \sigma_{x_2} dx - h_2 \frac{dv_2}{dx} \right) \\ &= \frac{l}{2} \left[-\frac{F_f}{E_1 A_1} - \frac{F_f}{E_2 A_2} + \frac{(h_1 + h_2)(Pl - h_1 F_f - h_2 F_f)}{E_1 I_1 + E_2 I_2} \right] \left[2\left(\frac{x}{l}\right) - \left(\frac{x}{l}\right)^2 \right] \end{aligned} \tag{2}$$

where the σ_{x1} and σ_{x2} are the axial stresses at any position of the upper and lower beams. To simplify, some dimensionless parameters are defined as shown in table 1. So the simplified result is received:

$$\Delta u(\psi_x) = \frac{P}{2} \left[3\varepsilon\beta \left(1 - \frac{\beta}{\gamma} \right) \delta_{st} - \frac{1}{\sigma\gamma} \lambda_{st} \right] \left[2(\psi_x) - (\psi_x)^2 \right] \tag{3}$$

From the relative slip, the friction energy can be acquired by working integral on the friction of the whole contact surface:

$$E_{loss} = 4F_f \int_0^1 \Delta u(\psi_x) d\psi_x = \frac{4F_f^2}{3}\left[3\varepsilon\beta(\gamma-\beta)\delta_{st} - \frac{1}{\sigma}\lambda_{st}\right] \tag{4}$$

The energy storage of the system is composed of the energy storage E_{ne1} produced by bending moment and the energy storage E_{ne2} of interfacial friction force:

$$E_{nel} = \frac{1}{2}P\left(v - \frac{v_c}{P_c}P\right) \tag{5}$$

$$E_{ne2} = 4F_f\left(\frac{1}{E_1}\int_0^l\int_0^x \sigma_{x_1} dxdx + \frac{1}{E_2}\int_0^l\int_0^x \sigma_{x_2} dxdx\right) \tag{6}$$

Where v_c is the critical end deformation load, P_c is the minimum end concentrated load causing the initial relative slip. Therefore, loss factors of the system derived from friction energy dissipation and system energy storage:

$$\eta = \frac{E_{loss}}{2\pi E_{ne}} = \frac{4}{3\pi}\frac{3\sigma\varepsilon\beta\gamma - 3\sigma\varepsilon\beta^2 - \kappa}{\frac{3\sigma^2\varepsilon^2\beta^2}{\kappa + 3\sigma\varepsilon\beta^2}\gamma^2 - \sigma\varepsilon\beta\gamma + \frac{8}{3}\kappa} \tag{7}$$

Table 1. Dimensionless parameters and implications.

Dimensionless parameters	Implications
$\beta = (h_1 + h_2)/l$	Ratio of the sum of neutral surface distance of upper and lower beams to the whole length
$\gamma = P/F_f$	Ratio of end concentration force to interfacial friction force
$\sigma = (E_1A_1)/(E_1A_1 + E_2A_2)$	Ratio of tensile stiffness of upper beam to the sum of upper and lower beams
$\varepsilon = (E_1I_1)/(E_2I_2 + E_1I_1)$	Ratio of bending stiffness of upper beams to the sum of upper and lower beams
$\lambda_{st} = l/(E_1A_1)$	Static tensile deformation at end of upper beam under unit tensile load
$\delta_{st} = l^3/(3E_1I_1)$	Static bending deformation at the end of the upper beam under unit concentrated load
$\kappa = \lambda_{st}/\delta_{st}$	Ratio of static tensile deformation to static bending deformation
$\psi_x = x/l$	Ratio of any position of beam to the whole length

2.2. Numerical simulation of cantilever beam with rectangular section

To ensure consistency with subsequent finite element simulations, the double-layer cantilever beam interface is defined as a rectangular section, so that its length $l = 500mm$,the thickness of the upper layer beam are $1mm$(the distance from neutral surface to friction interface $h_1 = 0.5mm$). The material of the upper and lower beams is carbon fiber t700, and the elastic modulus $E_1 = E_2 = 210GPa$.

To simplify the formula, the neutral plane distance ratio of double-layer cantilever beams is defined as $\alpha = h_2/h_1$ .Thus, $\beta = (h_1 + h_2)/l = (\alpha+1)/500$, $\sigma = (E_1A_1)/(E_1A_1 + E_2A_2) = \alpha/(\alpha+1)$, $\varepsilon = (E_1I_1)/(E_2I_2 + E_1I_1) = 1/(\alpha^3+1)$, $\kappa = \lambda_{st}/\delta_{st} = 4\times10^{-6}$. The formula of loss factor is obtained:

$$\eta = \frac{1}{\pi} \frac{3000\alpha(\alpha+1)^3 \gamma - (\alpha+1)^6}{225\times10^4 \alpha^2\gamma^2 - 750\alpha(\alpha+1)^3 + 2(\alpha^3+1)(\alpha+1)^3} \tag{8}$$

The expression of the resulting loss factor η is biased about γ and makes it equal to zero. As a result, the maximum damping effect is obtained, that is, the optimum force ratio γ_{op} (the ratio of the end concentrated force to the interfacial friction force) when the loss factor is maximum:

$$\gamma_{op} = \frac{(\alpha+1)^3 + \sqrt{8(\alpha+1)^3(\alpha^3+1)}}{3000\alpha} \tag{9}$$

The expression of γ_{op} is brought into the expression of η (loss factor), and the optimal damping effect can be obtained in the case of α (neutral surface distance ratio) of different double-layer cantilever beams. The relationship between η and α is shown in Figure 4, and the relationship between γ_{op} and α is in Figure 5. When the thickness of the two beams is the same($\alpha=1$), the maximum loss factor($\eta=0.3326$) and the optimum force ratio($\gamma_{op}=0.006438$) can be obtained. Figure 6 shows the relationship between loss factor and force ratio when the neutral plane distance ratio is 1, 0.5 or 2 respectively. The relationship between loss factor and force ratio and neutral plane ratio is as shown in Figure 7.

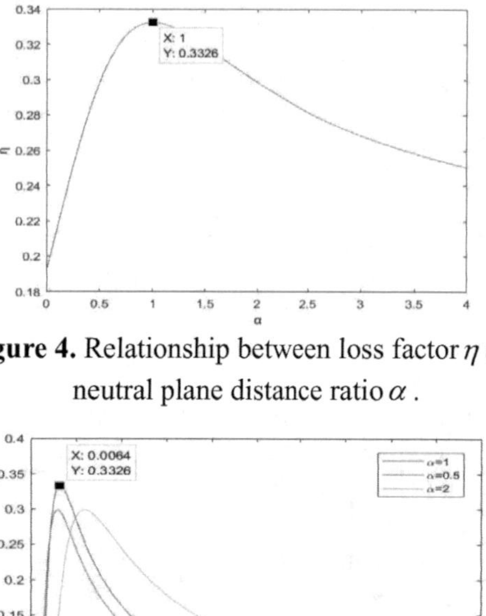

Figure 4. Relationship between loss factor η and neutral plane distance ratio α .

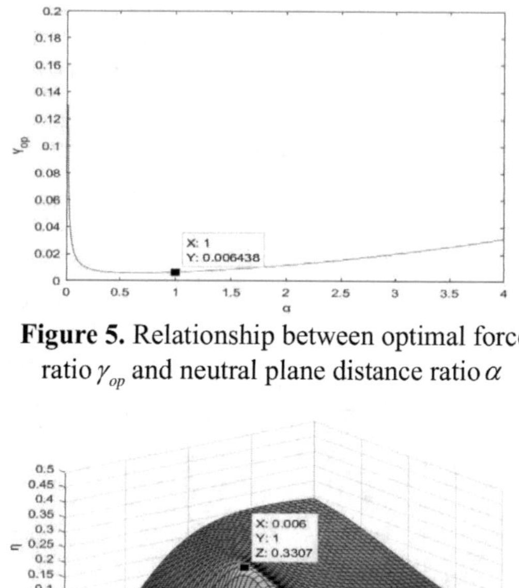

Figure 5. Relationship between optimal force ratio γ_{op} and neutral plane distance ratio α

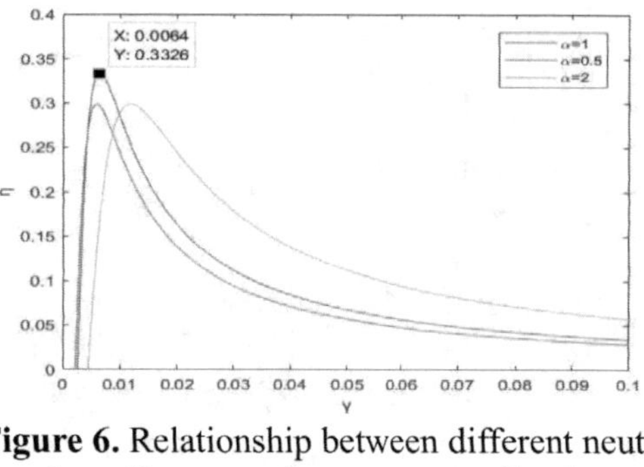

Figure 6. Relationship between different neutral plane distance ratio α corresponding loss. factor η and force ratio γ .

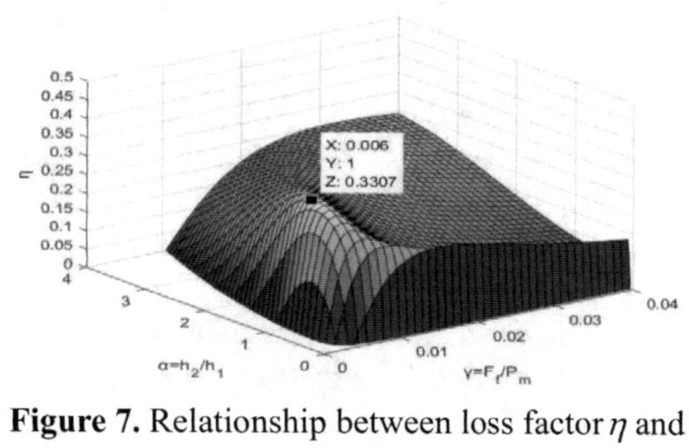

Figure 7. Relationship between loss factor η and ratio of force ratio γ and ratio of neutral surface α .

Through the theoretical analysis of the double-layer cantilever beam model, the parameters of multi-petal damping structure are preliminarily determined, which provides the foundation for finite element dynamic simulation.

IOP Conf. Series: Materials Science and Engineering **926** (2020) 012023 doi:10.1088/1757-899X/926/1/012023

3. Finite element simulation and analysis of multi-petal cantilever model

The multi-petal damping structure and the deployment component are hollow cylinders, and the length of them is l. The damping structure is inserted into the hollow cylindrical deployment component and is pressed on the inner wall of it. It is assumed that the gap of the contact surfaces between the damping structure and the deployment component is negligible ($r_2 = r_3$). The geometric parameters of the antenna-damping structure model are shown in table 2 and the cross section of the multi-petal cantilever model in figure 8.

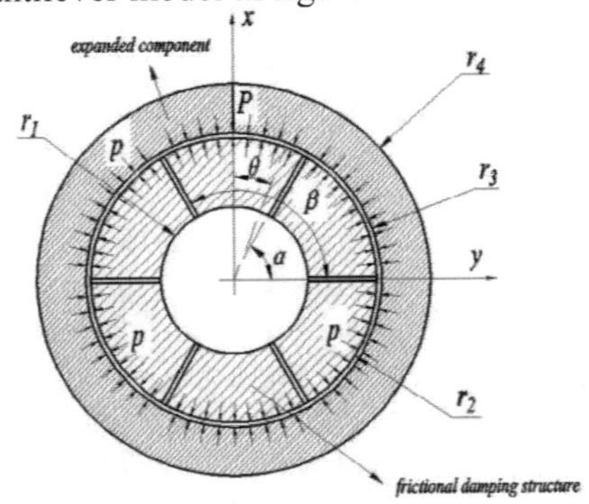

Figure 8. The cross section of the multi-petal cantilever model.

Table 2. The geometric parameters of deployment component and damping structure.

Implications of parameters		Measurement
Damping structure	Inner diameter	$r_1 = 10.5mm$
	Outer diameter	$r_2 = 11.5mm$
Deployment component	Inner diameter	$r_3 = 11.5mm$
	Outer diameter	$r_4 = 12.5mm$
Length of component		$l = 500mm$
Number of petals		$n = 6$
Friction factor		$\mu = 0.15$

In order to validate the validity of the theoretical analysis, undamped deployment component (the finite element models of the cantilever beam with ring section) and deployment component with damping structure are established, and the loss factor is obtained by hysteresis curve to verify the accuracy of the numerical analysis. Using ANSYS finite element analysis software, the material is carbon fiber t700, and the parameters are shown in table 3. During the process of ANSYS analysis, the two models are established and the grid is divided firstly. Then, the modal analysis is carried out. Finally, the harmonic response analysis is carried out and the two modals are analyzed to obtain their hysteresis curves respectively.

3.1. Finite element analysis of deployment component.

The model of deployment component is shown in Figure 9. By modal analysis, the first order natural vibration frequency is 197.17 Hz. When the load is 100N, through its harmonic response analysis, the hysteresis curve is obtained as shown in Figure 10. Its loss factor is 0.018 and the amplitude can reach 2.449 mm. And the first three modes are obtained in Figure 11.

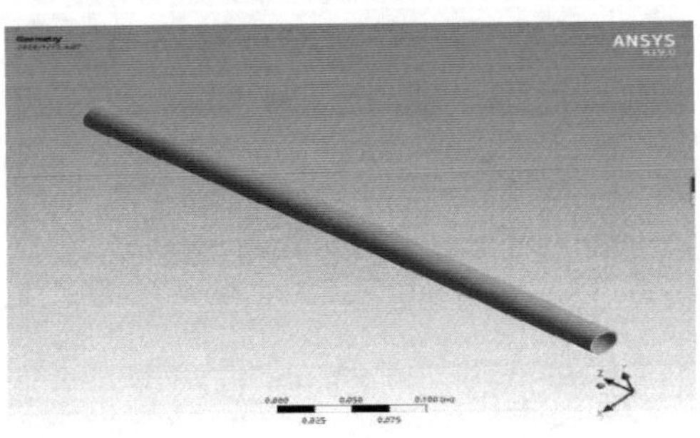

Figure 9. Deployment component without damping structure.

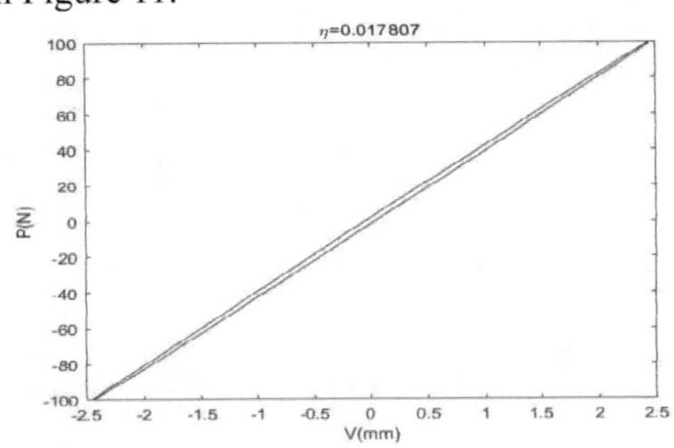

Figure 10. Hysteresis curve of deployment component.

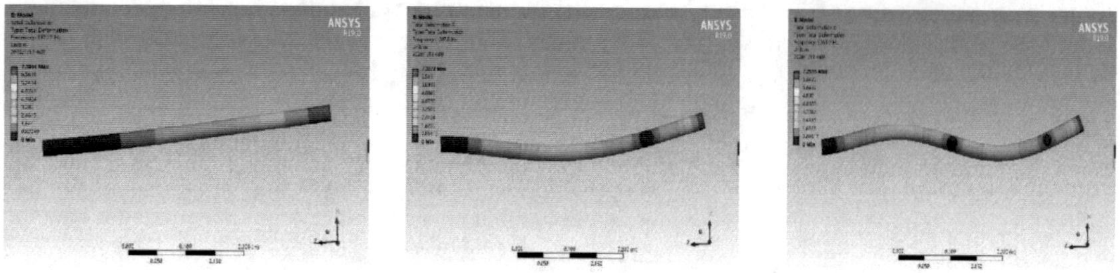

Figure 11. First three modes of deployment component.

3.2. Finite element analysis of deployment component with damping structure.

The model of deployment component with damping structure is shown in Figure 12. By modal analysis, the first three modes are obtained in Figure 13, and the first natural vibration frequency is 393.37Hz. When the load is 100N and the optimum force ratio is 0.0064, the corresponding optimal total preload force is 942 N. By harmonic response analysis, the hysteresis curves under different preloaded forces (when the forces are 906N, 942 N and 978N) are obtained respectively as shown in Figure 14, and the loss factor and amplitude are shown in table 3. When the preload force is 942 N, the loss factor and amplitude drop rate are maximum and the effect of vibration control is excellent.

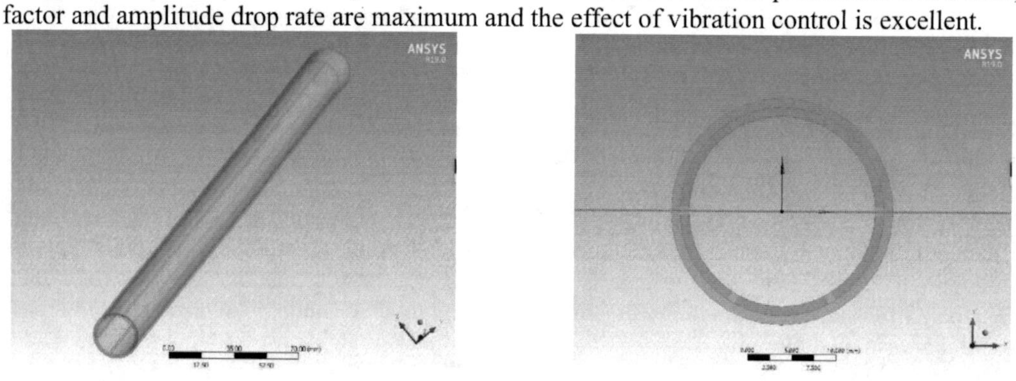

Figure 12. Model of deployment component with damping structure and its section diagram.

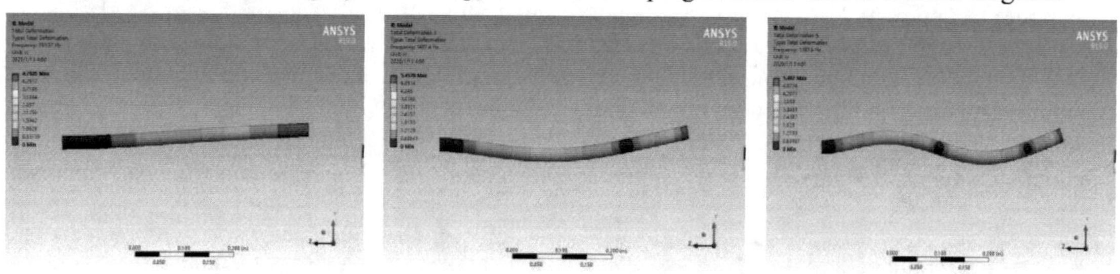

Figure 13. First three modes of deployment component with damping structure.

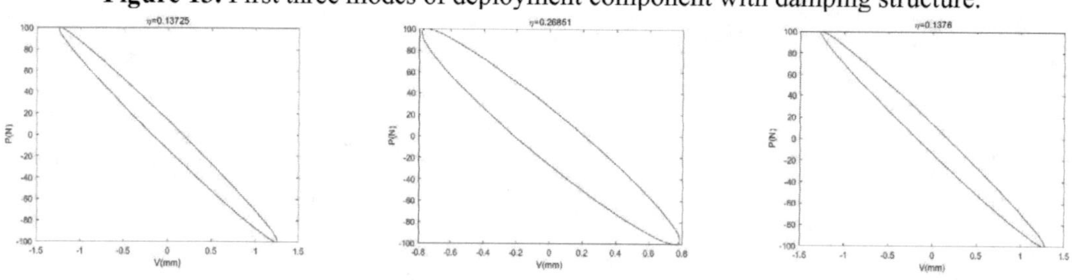

Figure 14. Hysteresis curves at 905.96N, 942.07N and 978.18N preload forces.

Table 3. Loss factors and amplitudes under different preload forces.

Preload (N)	906	942	978
Loss factor	0.13725	0.26851	0.1376
Amplitude (mm)	1.2486	0.78434	1.2693
Amplitude drop rate	49.02%	67.97%	48.17%

4. Conclusion

The paper introduces a preloaded multi-petal friction damping structure designed for deployment mechanism of large-scale flexible deployable satellite antenna. The friction damping structure is proposed to promise stability and accuracy of deployment process. Different from the complex structure of tuned mass dampers that most satellites use to restrain vibration, the proposed damping structure has eminent vibration suppression effect and simple structure. Theoretical analysis of the simplified model shows the loss factor can reach 0.3326. In finite element simulation of vibration dynamics, the loss factor of the deployment component with the damping structure can reach 0.26851 and the amplitude drop rate can reach 67.97%, which are obvious for the deployment mechanisms of large flexible deployable antenna with low-order natural frequency. Preloaded multi-petal friction damping structure provides a simple structure and obvious vibration suppression method for satellite antenna, which has a profound influence on the development and application of large flexible deployable antenna satellite in the future.

Acknowledgment

This work was supported by National Defense Equipment Pre-Research Foundation of China (Grant NO.61402100103).

References

[1] Qi X, Huang H , Li B and Deng Z 2016 A large ring deployable mechanism for space satellite antenna *Aerospace ence and technology* **58(NOV.)** 498-510

[2] Puig L, Barton A and Rando N 2010 A review on large deployable structures for astrophysics missions *Acta astronautica* **67(1-2)** 12-26

[3] Li W B, Li X R, Zhao Z G, Wang Y Y and Zhao Y 2012 Optimal piezoelectric sensors and actuators deployment for active vibration suppression of satellite antenna reflector *Advanced Materials Research* **479-481** 1490-94

[4] Qiu Z C, Zhang X M, Wu H X and Zhang H H 2007 Optimal placement and active vibration control for piezoelectric smart flexible cantilever plate *Journal of Sound & Vibration* **301(3-5)** 521-43

[5] Sharma A, Kumar R, Vaish R and Chauhan V S 2015 Active vibration control of space antenna reflector over wide temperature range *Composite Structures* **128** 291-304

[6] Wang P 1992 Passive control of a stiff flexible link *Int.j.robotics Research* **11(6)** 572-78

[7] Onoda J, Sano T and Minesugi K 1995 Passive damping of truss vibration using preloaded joint backlash *AIAA Journal* **33(7)** 1335-41

[8] Xu W, Wang J, Shao Y, Li L and Wang S 2015 Vibration damping design and analysis of additional constrained damping layer for satellite adapter *Missiles and Space Vehicles* **05** 95-99

[9] Bae J S, Hwang J H, Kim J H, Park J S, Pyeon B D and Roh J H 2018 Vibration suppression of a cantilever plate using magnetically multimode tuned mass dampers *Shock and vibration* **2018(PT.4)** 3463528.1-3463528.13

[10] Kienholz D A, Pendleton S C, Richards K E and Morgenthaler D R 1994 Demonstration of solar array vibration suppression *North American Conference on Smart Structures & Materials* International Society for Optics and Photonics.

[11] Lee J H, Berger E and Kim J H 2005 Feasibility study of a tunable friction damper *Journal of Sound & Vibration* **283(3-5)** 707-22

IOP Conf. Series: Materials Science and Engineering **926** (2020) 012024 doi:10.1088/1757-899X/926/1/012024

Elite Niche Particle Swarm Optimization for Energy Clustering in Aeronautical Wireless Sensor Network

Mengying Xu[1], Jie Zhou[1] and Rui Yang[1]

[1]College of Information Science and Technology, Shihezi University, Shihezi, 832000, China

jiezhou@shzu.edu.cn

Abstract. Recently, a wide variety of applications of aeronautical wireless sensor networks (AWSNs) make the clustering algorithm a paramount issue and bring profound changes to Internet of Things in AWSN-based applications. In AWSN, an energy-clustering algorithm is very significant to reduce the total energy cost of sensing data and solve the energy balance problem in AWSN. However, it is a challenging issue in AWSNs to have a low lifetime and energy utilization rate. In order to achieve a low energy consumption and further improve the lifetime of AWSNs, in this paper, we propose an elite niche particle swarm optimization (ENPSO), which combine the elite selection and niche sharing mechanism to develop the algorithm's convergence rate and robustness. In the simulation, we compare the energy cost optimized by ENPSO with that optimized by grey wolf optimization (GWO) and simulated annealing (SA). Simulation results demonstrate that the energy cost optimized by ENPSO is 9.63% and 19.54% smaller than GWO and SA when the number of nodes is 100 with 10% cluster heads. It obviously shows that the performance of proposed ENPSO is better than other two algorithms and have a faster convergent rate and better robustness.
Keywords: Aeronautical wireless sensor networks, Particle Swarm Optimization, Clustering, Energy Cost

1. Introduction

Wireless sensor networks (WSN) is composed of a large number of micro sensors distributed in the monitored area, which can send the collected information to external monitoring personnel [1]. WSN is widely used in many fields, such as military, agriculture, personal health management, transportation, smart home, etc. It has become a research hotspot in above field in the world [2-3].

Aeronautical wireless sensor network (AWSN) is a significant branch of WSNs [4]. The node is mainly composed of four modules, including power module, processor module, wireless communication module and sensor module. In AWSN, each node can collect data and transmit data at the same time. An AWSN is composed of nodes using a self-organized way. The energy consumption consists of transmitting energy and receiving energy [5]. However, most of the nodes in the actual monitoring area are distributed in the complex and harsh environment. There is a problem that the batteries cannot be replaced in AWSNs. Once the energy of nodes is exhausted, they cannot work and it is difficult to replace them.

Sensors assembled in Aeroengine usually work in a bad environment, which will undergo large gradient changes in temperature and a large number of loads in a very short time, resulting in large alternating stress and severe vibration fatigue damage. Therefore, it is necessary to study the durability of the sensor in the stage of design.

In order to guarantee the lifetime as well as other performance indexes of the aircraft, ensure the flight safety and meet the comprehensive requirements of long life, high reliability, high attendance and low maintenance cost of the aircraft structure, it is significant to study the cluster head allocation optimization algorithm with low energy consumption.

Clustering problem is the most important problem in the applications of AWSN. It is also a hot topic in recent years [6-8]. The quality of clustering strategy directly affects the quality of network lifetime. Clustering technology is also an important mean to reasonably allocate resources and improve the energy efficiency of nodes and extend the network lifetime. Therefore, low energy clustering is an important goal of lifetime optimization design in AWSNs. Therefore, the research on low energy clustering is very important for WSN.

In [9], authors combined the energy heterogeneity of nodes with the classic LEACH protocol and proposed distributed energy efficient clustering method, which dynamically adjusted the parameters of the threshold of cluster head to improve the performance of WSN and reduce the energy cost of WSN. However, the robustness of system is not high. In [10], authors consider the transmission energy, distance and number of hops to build a multi-hop routing protocol, which can effectively reduce the energy consumption in the transmission stage, but the cluster head is still overloaded. In [11], authors consider the distance to the base station and the rest energy of the nodes to improve the number of cluster heads. However, there is still energy consumption imbalance. Many heuristic algorithms are also proposed to reduce the energy cost of WSN [12]. In [13], authors proposed a grey wolf optimization (GWO) scheme to save the energy consumption for better clustering scheme in WSNs. GWO can reduce network communication costs than SLM. The GWO is flexible but suffers from the problem of high computational complexity. In [14], authors proposed a simulated annealing (SA) scheme to save energy consumption in WSNs. In their work they improved energy efficiency without considering distance constraint. However, its convergence speed is slow when the quantity of sensor nodes is large.

In this paper, we propose an elite niche particle swarm optimization (ENPSO) to reduce the energy cost of WSN based on the traditional particle swarm optimization, which simulate the process of communication and foraging of firefly. The proposed algorithm combines the elite selection and niche sharing mechanism to improve the convergence speed and quality of solution. In the experience, we compare the results and performance of proposed method with GWO and SA with different number of nodes. Simulation results demonstrate that the ENPSO effectively reduce the total energy cost of system and improve the convergent speed of algorithm when the number of nodes is huge.

2. Problem Description and System model

Low energy adaptive clustering hierarchy (LEACH) is one of the typical clustering algorithms. However, it is impossible to guarantee the amount of cluster heads in each round. In the total lifetime of network, the network may have no cluster heads in the round. The main goal of the clustering problem is to reduce the energy consumption of aeronautical nodes in the network during data transmission, so that the nodes can work longer and the effective working time of the network can increase. The lifetime of AWSN is usually measured by the number of rounds of network running when the energy of first node is exhausted or that of last node is exhausted.

In this section, clustering model is given to calculate the total communication energy in AWSN. The circuit energy consumption of transmitting data and receiving data in AWSN is expressed by $E_{elec} \cdot k$. The energy consumption is directly proportional to the length of bits of data. E_{elec} represents electronics energy parameter. In the transmission module, the energy consumption on the signal amplifier is also proportional to the data length and related to the distance of data transmission.

The energy consumption of transmitting k-bits data in the transmitting module is calculated in (1); energy consumption of receiving k-bits data in the receiving module is calculated in (2).

$$E_t(k,d) = E_{elec} \cdot k + \varepsilon_{amp} \cdot k \cdot d^2 \tag{1}$$

$$E_r(k) = E_{elec} \cdot k \tag{2}$$

where $E_t(k,d)$ represents the energy cost of transmitting k -bits data between two nodes with d meters. $E_r(k)$ represent the energy cost of receiving k -bits data between two nodes. E_{elec} in transmitting circuit is equal to that in receiving circuit. ε_{amp} is the parameter of signal amplification. d is the communication distance between two nodes. d^2 represents the channel transmission energy attenuation.

$$E_{sum} = E_t + E_r \tag{3}$$

Where E_{sum} represents the total energy cost of transmitting k -bits data and receiving k -bits data of a node. If there are N nodes in AWSN, the total energy cost of N nodes can be calculated in (4).

$$R = \sum_{n=1}^{N} E_{sum}(n) \tag{4}$$

3. Elite Niche Particle Swarm Optimization

Particle swarm optimization is an evolutionary algorithm based on swarm intelligence optimization. It was proposed by American social psychologist Kennedy in 1995 and inspired by the simulation of the migration and group behaviour of birds in the process of foraging [15]. The algorithm was found that birds can share the information about the location of food or target between birds, which can greatly improve the efficiency of finding food. Therefore, the cooperation between birds improves the search efficiency. At the same time, the fitness of population can be maximized. The search process of groups can be explained. When an individual searches for a certain target in the population, it usually adjusts the next search by referring to the individual currently in the optimal position in the population and the previous optimal position [16].

However, there are some defects in this algorithm, for example, it is easy to fall into local optimization and there are many optimized parameters, the coding scheme is not suitable for solving the energy consumption problem. This paper designs an elite niche particle swarm optimization, which combines elite selection and niche sharing mechanism, uses binary coding for the initial population, and improves the update strategy of particle swarm optimization, so that the algorithm can solve the problem more pertinently.

We assume that the search space is N -dimensional. there are M particles in the population. When the position of each particle is updated, the historical best position is taken into account. The steps of ENPSO are composed of population encoding, population initialization, fitness calculation, position updating of particles, termination condition.

3.1. Population encoding and Initialization

Assuming that the dimension of the population is N and the amount of the particles is M . N also represents the amount of sensor nodes. In (5), the population can be expressed as $Q=\{Q_1,Q_2,\cdots\cdots,Q_M\}$, where the position information of the m_{th} particle in the N -dimensional space can be expressed as $Q_m = \{q_{m,1}, q_{m,2}, \cdots\cdots, q_{m,N}\}$. The corresponding speed can be expressed as $V_m = \{v_{m,1}, v_{m,2}, \cdots\cdots, v_{m,N}\}$. q_{best}^{m} is used to represent the best position of the m_{th} particle, which records the best historical information of the particle during the whole searching period. The optimal location of the population is recorded as g_{best}.

In clustering problem, the values in population represent the mode of sensors. In (6), the number of cluster heads in each individual is fixed. The particle position information is encoded by binary code.

Each $q_{m,n}$ can only have two values, i.e. 0 or 1. When $q_{m,n} = 1$, the mode of sensors is cluster head. Otherwise, it represents a common node.

$$Q = \begin{bmatrix} q_{1,1} & q_{1,2} & \cdots & q_{1,N-1} & q_{1,N} \\ q_{2,1} & q_{2,2} & \cdots & q_{2,N-1} & q_{2,N} \\ \cdots & \cdots & q_{m,n} & \cdots & \cdots \\ q_{M-1,1} & q_{M-1,2} & \cdots & q_{M-1,N-1} & q_{M-1,N} \\ q_{M,1} & q_{M,2} & \cdots & q_{M,N-1} & q_{M,N} \end{bmatrix} \tag{5}$$

$$\sum_{n=1}^{N} q_{m,n} = K \tag{6}$$

3.2. Fitness calculation

The performance of individuals is evaluated by calculating fitness. Fitness represents the total communication cost of all cluster heads and common nodes in the process of transmitting and receiving data. Fitness can be calculated in (4) and we aim to find the individual with minimum energy cost. In (7), we aim to find the best individual in the population with the lowest energy cost.

$$f = \min(R) \tag{7}$$

3.3. Position updating of particles

In PSO, the velocity updating of particle is also composed of three parts: velocity inertia, self-learning and social learning. The updating of particle position in the $(T+1)_{th}$ generation is determined by the position of the T_{th} generation and the velocity vector of the $(T+1)_{th}$ generation. The details are calculated in (8) and (9):

$$v_{m,n}^{t+1} = \omega v_{m,n}^{t} + c_1 r_1 (p_{m,n} - q_{m,n}^{t}) + c_2 r_2 (g_{m,n} - q_{m,n}^{t}) \tag{8}$$

$$q_{m,n}^{t+1} = v_{m,n}^{t+1} + q_{m,n}^{t} \tag{9}$$

Where m represents the particle and n represents dimension. $v_{m,n}^{t+1}$ represents the velocity of particle in the $(t+1)_{th}$ generation. $q_{m,n}^{t+1}$ represents the position of particle in the $(t+1)_{th}$ generation. ω is inertia weight. t represents the iterations. c_1 and c_2 represent self-learning factor and social learning factor respectively. In general, $c_1 = c_2$. r_1 and r_2 are random number between 0 and 1. $p_{m,n}$ represent the best position of m_{th} particle. g_{best} represents the best position in the population.

In ENPSO algorithm, the state of particle motion is defined from the perspective of probability. In the population, the $q_{m,n}$ population is 0 or 1, $v_{m,n}$ represents the probability. The formula of position updating is given in (10) and (11):

$$q_{m,n} = \begin{cases} 1 & rand \le Sig(v_{m,n}) \\ 0 & rand > Sig(v_{m,n}) \end{cases} \tag{10}$$

$$Sig(v_{m,n}) = \frac{1}{1 + e^{-v_{m,n}}} \tag{11}$$

where $rand$ is a random number, $rand \in [0,1]$, Sig is a sigmoid function .

3.4. Termination condition

If the current number of iterations $T < T_{\max}$, recalculate energy cost and update the fitness of individuals. Otherwise, output the energy cost of the optimal particle and its cluster head allocation scheme.

3.5. Steps of ENPSO

This section gives the steps of ENPSO to optimize the allocation scheme of cluster heads.

Step1: Initialize parameters including number of individuals, maximum iteration, self-learning factor and social learning factor etc.

Step2: Randomly generate binary population Q .

Step3: The fitness (energy cost) of each particle in the population is calculated, and the local optimal value p_{best} of each particle and the global optimal value g_{best} of all particles are obtained.

Step4: Update the position and velocity of each particle according to (9) and (10);

Step5: If the termination condition is met, Output the best individual and its energy cost. Otherwise turn to step 3.

4. Simulation and Results

In order to evaluate the performance of the ENPSO, a simulation program is designed with MATLAB R2014a. Without considering the random factors such as wireless channel interference and signal conflict, the AWSN with sensor nodes randomly distributed in 100m × 100m two-dimensional plane area is considered.

Assuming that some parameters are fixed. The number of data received or sent by each node is 1Mbps bits. The parameter of signal amplification is 100 $pJ / bit / m^2$. The electronics energy parameter is 50 nJ / bit . The number of cluster head nodes is 10% of the total number of sensors.

In ENPSO, the number of particles is 80 and the maximum generation is 100. All nodes cannot move once placed. In order to better show the advantages of the ENPSO algorithm, GWO and SA are simulated and compared by MATLAB. In figure 1 and figure 2, 100 nodes and 500 nodes are randomly located in the monitored area respectively, where the black filled circle is the cluster head and the hollow circle is the common sensor. All of them are located in the in a square of 100 m^2 .

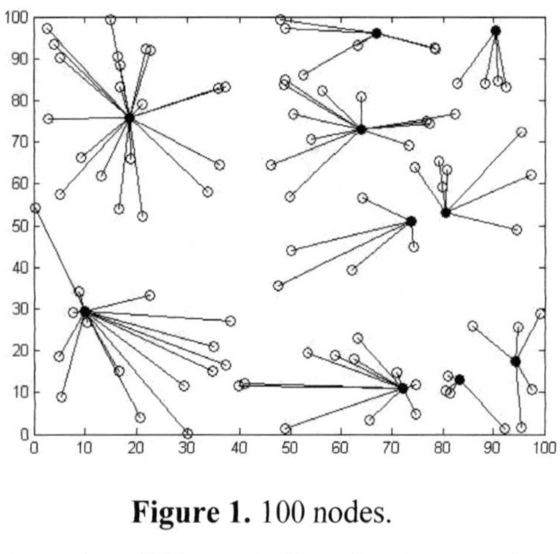

Figure 1. 100 nodes.

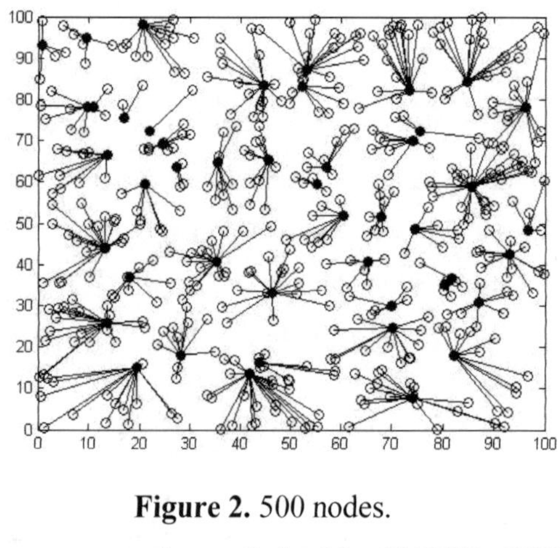

Figure 2. 500 nodes.

Figure 3 and figure 4 show the comparation of energy consumption optimized by ENPSO, GWO and SA with different number of sensor nodes when the amount of cluster heads is 10% of the total number of sensors. In figure 3, there are 100 nodes in the sensing area. At the beginning of the iteration. The optimization speed of the three algorithms is very fast. In the last 50 generations, the convergence speed of the algorithm becomes slow. However, the ENPSO have a higher convergence rate than GWO and SA. After 100 generations, the energy consumption optimized by ENPSO is 70.69J and that optimized by GWO and SA are 78.22J and 87.73J respectively. Results demonstrate that the energy cost optimized by ENPSO is 9.63% and 19.54% smaller than GWO and SA when the number of nodes is 100 with 10% cluster heads.

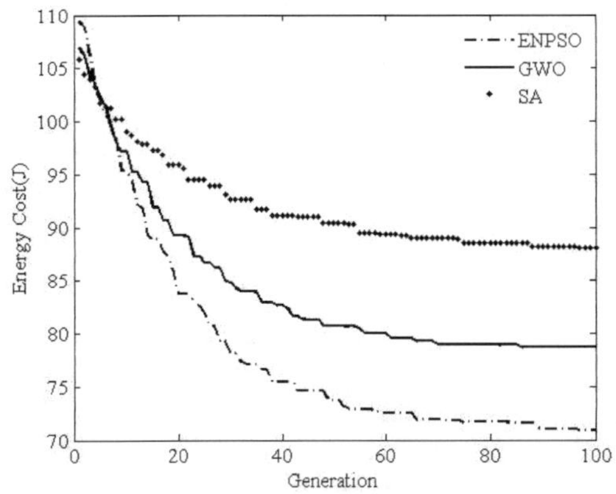

Figure 3. 100 nodes.

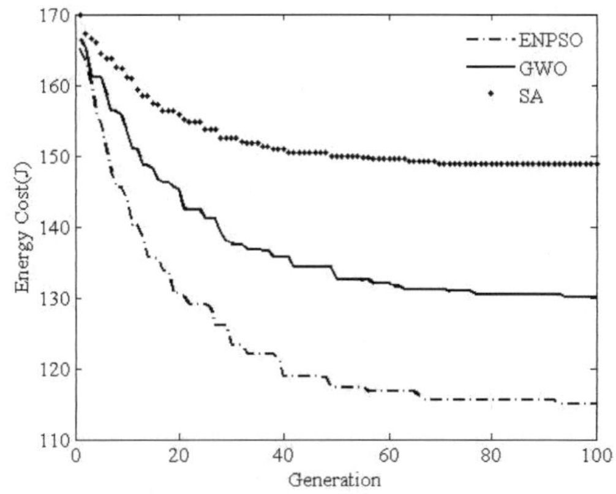

Figure 4. 200 nodes.

In figure 4, when the number of sensor nodes is 200, the energy cost optimized by ENPSO, GWO and SA are 114.64J, 129.72J and 148.60J respectively. It can be seen from the data that the communication quality has been significantly improved.

It is proved that the ENPSO algorithm based on low energy clustering can effectively rationalize the location distribution and the number of cluster heads, and finally reduces energy consumption of sensor and prolongs network lifetime.

5. Conclusions

In this paper, in order to improve the performance and convergence speed, we introduce the main idea of ENPSO and give the implementation process of the algorithm. It combines elite selection and niche sharing mechanism, which can increase the diversity of the population and improve the ability of the ENPSO to avoid falling to the local optimization, and further enhance the global search ability. In the simulation, we compare the energy cost optimized by ENPSO with GWO and SA with different amount of aeronautical sensor nodes. Experiment results show that ENPSO have better effectiveness to reduce the energy cost of transmitting data and receiving data. It has a high convergent rate in finding global optima.

Acknowledgments

This paper is supported by the project of Youth and Middle-aged Scientific and Technological Innovation Leading Talents Program of the Corps (No. 2018CB006), the Major scientific and technological projects of the Corps (2017AA005-04), the project of Promoting Scientific Research Cooperation and High-level Personnel Training with Americas and Oceania. The corresponding author is Jie Zhou.

References

[1] Wang Y, Qiu L and Luo Y J 2020 A piezoelectric sensor network with shared signal transmission wires for structural health monitoring of aircraft smart skin, *Mechanical Systems and Signal Processing,* **141**, 106730.

[2] Amrita G, Subir H and Sajal K D 2020 Distributed on-demand clustering algorithm for lifetime optimization in wireless sensor networks, *Journal of Parallel and Distributed Computing,* **141**, pp 129-42.

[3] Amir Abbas B and Keivan N 2020 HQCA-WSN: High-quality clustering algorithm and optimal

cluster head selection using fuzzy logic in wireless sensor networks, *Fuzzy Sets and Systems*, **389**, pp 114-44.

[4] Ian J, Fakhre A and Manuel Esperon M 2020 Simulation of an aircraft environmental control system, *Applied Thermal Engineering,* **172**.

[5] Salim El K, Rehan Ullah K and Nejah N 2020 MW-LEACH: Low energy adaptive clustering hierarchy approach for WSN, *IET Wireless Sensor Systems*, **10,** pp126-9.

[6] Amir S, Gulustan D and Farzad K 2020 HEEL: A new clustering method to improve wireless sensor network lifetime, IET *Wireless Sensor Systems*, **10**, pp 130-36.

[7] Liu S 2020 Energy-Saving Optimization and Matlab Simulation of Wireless Networks Based on Clustered Multi-hop Routing Algorithm. *International Journal of Wireless Information Networks*, **27**, pp 280-8.

[8] Akshay V, Sunil K and Prateek Raj G 2020 Fuzzy logic based effective clustering of homogeneous wireless sensor networks for mobile sink, *IEEE Sensors Journal*, **20**, pp 5615-23.

[9] Wang W J and Tong G X 2020 Multi-path unequal clustering protocol based on ant colony algorithm in wireless sensor networks, *IET Networks*, **9**, pp 56-63.

[10] Fakhrosadat F and Marjan Kuchaki R 2018 Memetic fuzzy clustering protocol for wireless sensor networks: Shuffled frog leaping algorithm, *Applied Soft Computing*, **71**, pp 568-90.

[11] Shalli R, Syed Hassan A and Ravi R 2020 Dynamic clustering approach based on wireless sensor networks genetic algorithm for IoT applications, *Wireless Networks*, **26**, pp 2307-16.

[12] Gupta G P, Sonu J 2018 Integrated clustering and routing protocol for wireless sensor networks using Cuckoo and Harmony Search based metaheuristic techniques, *Engineering Applications of Artificial Intelligence*, **68**, pp 101-9.

[13] Deepika A, Wasim Muhammad Huzaif Q and Pooja P 2020 GWO-C: Grey wolf optimizer-based clustering scheme for WSNs, *International Journal of Communication Systems*, **33**, e4344.

[14] Sweta P, Shankar T and Rajesh A 2018 Lifetime improvement in wireless sensor networks using hybrid differential evolution and simulated annealing (DESA), *Ain Shams Engineering Journal*, **9**, pp 655-63.

[15] Maria Antony Sahaya S and Rajendran P 2020 Improvement of battery lifetime in software-defined network using particle swarm optimization based cluster-head gateway switch routing protocol with fuzzy rules, *Computational Intelligence*, **36**, pp 813-23.

[16] Li K S, Feng Y and Chen D M 2020 A global-to-local searching-based binary particle swarm optimisation algorithm and its applications in WSN coverage optimisation. *International Journal of Sensor Networks* **32** pp 197-208.

IOP Conf. Series: Materials Science and Engineering **926** (2020) 012025 doi:10.1088/1757-899X/926/1/012025

Optimization of Aviation Wireless Sensor Network based on Discrete Cuckoo Search Algorithm

Rui Yang[1], Mengying Xu[1] and Jie Zhou[1]

[1]College of Information Science and Technology, Shihezi University, Shihezi, 832000, China

jiezhou@shzu.edu.cn

Abstract. Aiming at the problems of multiple coverage blind areas, poor monitoring quality, and low utilization of sensor network nodes in aviation wireless sensor networks, a discrete cuckoo search (DCS) is proposed, and an objective coverage optimization function is established. Aiming at the fact that the previous cuckoo search is not suitable for discrete values, the binary code conversion formula is used to perform binary code conversion on the jump path of each position update of Levy flight, which makes the algorithm converge faster and avoids the limitations of local search. In the simulation, the discrete cuckoo search was compared with ant colony optimization (ACO) and particle swarm optimization (PSO). Simulation experiment results show that the proposed discrete cuckoo search (DCS) has a great advantage in convergence speed, more targets are detected, and the target coverage in the aerial wireless sensor network is effectively improved.
Key words: Discrete cuckoo search, aviation, wireless sensor networks, coverage optimization

1. Introduction

The wireless sensor network (WSN) is make up of many sensors that can sense and check the external world through wireless communication. It is widely used in environmental monitoring, production safety monitoring, intelligent transportation, medical health, military and other fields [1-2]. WSN has the characteristics of miniaturization, low power consumption, less wiring, easy installation, etc. There are practical problems such as heavy cables, difficult wiring, and poor flexibility in the network, which provide a new method of information acquisition and processing for avionics systems [3-4].

In the aviation field, In the aviation field, the coverage is a key issue in the configuration of WSN. In most cases, a good number of sensor nodes are directly and randomly covered to meet the sensor node's coverage requirements of the target area [5-6]. In view of the special area of the sensor and the limited battery power of the sensor, it is often necessary to randomly deploy sensor nodes in the monitoring area on a large scale, which may lead to partial areas covering blind spots or high-density nodes, and the quality of target detection will be directly affected by detection blind spots[7]. When receiving nodes receive, process and transmit data, high node density wastes energy, and too much redundant data will cause data disturbance and block the channel, affect the reliability of the network, and cause a lot of waste of resources in energy consumption and cost. Therefore, it is necessary to adjust the deployment of sensor nodes to make it more evenly distributed and higher coverage in the detection area, so as to more rationally allocate cyberspace resources and better complete perception and information acquisition tasks, which is of great significance for improving network survivability, improving network reliability, and saving network construction costs.

With the extensive application of intelligent algorithms in optimization, most of the researches have used intelligent algorithms to optimize the coverage of WSN. Literature [8] proposed a WSN based on firefly algorithm (FA) to maximize coverage. Due to the consideration of reducing the moving distance, the coverage of the algorithm still needs to be improved; Literature [9] proposed an optimization method for node deployment based on extrapolated artificial bee colony (EABC), the coverage is improved compared with the original algorithm, but its coverage is still not ideal; Literature [10] proposed a coverage optimization strategy based on the improved grey wolf optimizer (GWO), but did not consider the presence of obstacles in actual deployment; Literature [11] proposed an adaptive chaotic quantum particle swarm optimization algorithm to optimize mobile sensor network coverage. Compared with chaotic quantum particle swarm optimization algorithms and particle swarm optimization, the coverage performance of wireless sensor networks has been more effectively improved. Literature [12] aimed at coverage rate and used improved fish swarm optimization (FSO) to optimize the deployment of sensor nodes, which significantly improved the network coverage area. But only for isomorphic sensors, it does not consider heterogeneous sensors and the monitoring environment in complex situations. Although the above method is effective in deploying uplinks on WSN nodes, in order to meet actual application requirements, the coverage and uniformity of WSN still need to be improved.

Cuckoo Search Algorithm (CSA) is an intelligent optimization algorithm which has emerged in recent years with the advantages of easy implementation, few parameters, and strong search capabilities [13]. CSA is widely used and provides a powerful tool for wireless sensor network coverage optimization technology. To this end, in view of the harsh environment and unmanned monitoring scenarios where traditional aviation sensors are located, and the number and monitoring capabilities are limited, a discrete cuckoo search (DCS) is introduced [14-15]. In this paper, a method for optimizing the placement of sensor nodes in aeronautical wireless sensor networks to improve the target coverage rate is proposed, and the DCS is used to solve the target function, which makes up for the deficiencies of some optimization algorithms and improve the efficiency and accuracy of the optimization solution. The results of simulation and theoretical analysis demonstrate that the proposed discrete cuckoo search can effectively increase the coverage of the target area of the mid-sensor nodes of the aviation WSN. At the same time, compared with ant colony optimization (ACO) and particle swarm optimization (PSO), the discrete cuckoo search has stronger convergence, and the number of monitored targets is more.

2. System model

Due to the limited sensor perception capability, only a limited number of targets within the monitoring range of the sensor node can be covered. The coverage relationship matrix M in formula (1) represents the coverage relationship between X monitored targets and Y sensor nodes in the aeronautical sensor network.

$$M = \begin{bmatrix} m_{1,1} & m_{1,2} & \cdots & m_{1,N-1} & m_{1,N} \\ m_{2,1} & m_{2,2} & \cdots & m_{2,N-1} & m_{2,N} \\ \vdots & & m_{x,y} & & \vdots \\ m_{X-1,1} & m_{X-1,2} & \cdots & m_{X-1,Y-1} & m_{X-1,Y} \\ m_{X,1} & m_{X,2} & \cdots & m_{X,Y-1} & m_{X,Y} \end{bmatrix} (m_{x,y} \in \{0,1\}) \quad (1)$$

As shown in equation (1), in the coverage relation matrix M , $M_{x,y}$ means that the coverage relationship between the x-*th* monitored target and the y-*th* sensor node. When $M_{x,y} = 1$, it means that the y-*th* sensor node can monitor the x-*th* monitored target, or that the x-*th* monitored target is within the coverage of the y-*th* sensor node However, when the x-*th* monitored target cannot be detected by the y -*th* sensor node, or the x -*th* monitored target is not within the coverage of the y -*th* sensor node, $M_{x,y} = 0$.

Also because of the limited monitoring capabilities, not all targets within the coverage area can be monitored. The monitoring matrix N represents the monitoring relationship of whether the target to be monitored is monitored by the sensor node, such as equation (2).

$$N = \begin{bmatrix} n_{1,1} & n_{1,2} & \cdots & n_{1,Y-1} & n_{1,Y} \\ n_{2,1} & n_{2,2} & \cdots & n_{2,Y-1} & n_{2,Y} \\ \vdots & & n_{x,y} & & \vdots \\ n_{X-1,1} & n_{X-1,2} & \cdots & n_{X-1,Y-1} & n_{X-1,Y} \\ n_{X,1} & n_{X,2} & \cdots & n_{X,Y-1} & n_{X,Y} \end{bmatrix} (n_{x,y} \in \{0,1\}) \tag{2}$$

In the monitoring relationship matrix N, $n_{x,y} = 1$ means that the x -*th* monitored target is within the monitoring range of the y -th sensor node and the y -th sensor node can monitor the x -*th* monitored target. When $n_{x,y} = 0$ means that the x -*th* monitored target cannot be detected by the y -th sensor node. If the maximum number of monitoring targets in each sensor node within the coverage is S, the constraint condition is as shown in equation (3).

$$\sum_{x=1}^{X} n_{x,y} \le S, y = 1 \ldots Y \tag{3}$$

Suppose a target to be monitored requires at least C sensor nodes to detect at the same time, and a sensor node monitors at most S targets simultaneously, the maximum quantity of monitored targets is maximized as the problem optimization target. The mathematical model of the problem is represented by formula (4)-(6).

Objective: $\max f(n_{11}, n_{12}, \cdots, n_{XY}) = \sum_{x=1}^{X} w_x$

$$\text{among them } w_x = \begin{cases} 1 & \sum_{y=1}^{Y} n_{x,y} \ge C \\ 0 & \sum_{y=1}^{Y} n_{x,y} < C \end{cases} \tag{4}$$

Subject to:

$$\sum_{x=1}^{X} n_{x,y} \le S, y = 1 \ldots Y \tag{5}$$

$$n_{x,y} \le m_{x,y} \tag{6}$$

In equation (5), the objective function is expressed as the total amount of monitored targets. If the x -*th* target can be successfully detected, it is represented by $w_x = 1$, but when $w_x = 0$, it means that the target cannot be successfully detected. Constraint 1 in equation (5) indicates that the monitoring capability of the sensor node is limited and constraint 2 indicates that the monitored target that can be monitored by the sensor node must be within the sensor coverage.

3. Target coverage of aerial wireless sensor networks based on discrete cuckoo search

3.1. Description of cuckoo search algorithm

Cuckoo search algorithm (CSA) is by imitating the cuckoo breeding and special way flight evolved. In nature, cuckoos generally do not have the ability to hatch, and usually lays eggs in other foster birds' nests for hatching. If host bird discovers the cuckoo's behavior, the host bird will directly throw the cuckoo's eggs out of the nest, or gives up the current nest, and choose a new location to build the nest. In order to avoid this, cuckoos generally choose to lay eggs in the nest where the nest bird has just laid eggs, which is not easy to be found by the host bird, thereby reducing the possibility of the host bird moving out of the nest. At the same time, the cuckoo will remove one or more eggs of the host, so as not to be found by the host to increase the number of eggs, but also reduce the competition of the host's juvenile dimension. Once the cuckoo's eggs are preserved, they usually hatch before the host bird's eggs. These new larvae will instinctively push the eggs of the host birds that have not hatched out of the nest to ensure that the host bird can raise the cuckoo larvae.

The CSA is to simulate the cuckoo's behavior of continually looking for a suitable horse's nest to lay his eggs in order to breed offspring, so as not to be discovered by the host, but to be hatched and raised into an adult. Random search, comparison, and judgment can finally solve the optimization problem. The cuckoo algorithm presupposes the following three assumptions: (1) Each cuckoo randomly selects a nest for spawning each time, and can only lay one egg at a time. (2) Keep the best nest for future use. (3) Make sure the amount of offered bird's nest is unchanged, and host may find cuckoo eggs and cause hatching failure. There are two basic search methods for the cuckoo algorithm: one is to use Levy Flight to randomly swim to discover a suitable bird's nest, and second is to discover a brand-new bird's nest location by random drift after deviations are discovered. The specific actions are usually below:

Step 1: Let the maximum amount of iterations be T, the population size be G, and the probability of discovery be p. The bird's nest is initialized randomly, which is the initial solution of the optimization problem.

Step 2: According to the fitness function, evaluate the fitness of the initial bird nest and retain the best quality nest.

Step 3: Follow the Levy Flight random search method to look for the bird's nest position. Use the fitness function to calculate the updated bird's nest quality and compare it with the look for bird's nest quality. Replace poorer bird nests with better ones. The update formula of bird's nest position is (7).

$$X_i^{k+1} = x_i^k + \alpha \oplus L(\lambda) \tag{7}$$

In equation (7): X_i^k and X_i^{k+1} are the location of the *i-th* bird's nest in the *k-th* and $k+1$ iterations respectively ; $\oplus$ is point-to-point multiplication; α is the step size control parameter; $L(\lambda)$ is the search path, $L(\lambda) \sim u = t^{-\lambda}$.

Step 4: Generate a random number r and compare it with the discovery probability P_a. If $r > P_a$, the selected bird's nest position is discarded, then use the random walk search method to find the new bird's nest location. The bird's nest position is updated by formula (8).

$$X_i^{k+1} = X_i^k + r \times (X_j^k - x_i^k) \tag{8}$$

In equation (8): r is the scaling factor, $r \in [0,1]$; X_j^k is the bird's nest near X_i^k.

Step 5: Compare the improved bird's nest with the previous best bird's nest, and keep bird's nest with better quality

Step 6: Perform steps 3 to 5 until the previously set number of iterations is completed, and the best quality bird nest is output.

3.2. Discrete cuckoo search algorithm

CSA is a new type of natural element-based heuristic algorithm proposed by British scholars Yang and

Deb on the basis of swarm intelligence technology in 2009. CSA has strong global search ability, few selection parameters and strong robustness. But the original cuckoo search algorithm is only appropriate for resolving continuous values, not for discrete values. Ouyang et al. proposed a discrete cuckoo search (DCS) in order to resolve discrete problems.

The DCS requires a binary code transformation of the jump path for each position update of Levy flight. The transformation formula of the binary coding is as follows:

If $rand() \le p_r$

$$Sig(Levy) = \frac{1}{1+\exp(-Levy)}$$

$$w_x^{g+1} = \begin{cases} 1 & rand \le Sig(Levy) \\ 0 & otherwise \end{cases}$$

Else

If $Levy \le 0$

$$Sig(Levy) = 1 - \frac{2}{1+\exp(-Levy)} \tag{9}$$

$$w_x^{g+1} = \begin{cases} 0 & rand \le Sig(Levy) \\ w_x^g & otherwise \end{cases}$$

Else

$$Sig(Step) = \frac{2}{1+\exp(-Step)} - 1$$

$$w_x^{g+1} = \begin{cases} 1 & rand \le Sig(Step) \\ w_x^g & otherwise \end{cases}$$

End If

End If

Where rand is the generated random number; p_r is the binary coded control parameter and satisfies $p_r \in [0,1]$. The larger p_r, the stronger the global diversity of the discrete cuckoo algorithm; the smaller p_r, the stronger the convergence of the discrete cuckoo algorithm.

4. Simulation

In the experiment, set the monitoring range to 500×500, the target to be monitored is 200, and the positions of the sensor nodes and the target to be monitored are randomly distributed. Within the coverage area, each sensor node can monitor up to 4 targets, and the number of sensor nodes that need to be monitored at least for each target to be monitored is 3 at the same time.

Set cuckoo algorithm parameters: population number $N = 40$, maximum evolution number $T = 100$, and initial probability of discovery $p_r = 0.25$. To be able to prove the effect of the algorithm, DCS, PSO and ACO were used to optimize the target coverage method of the WSN with the same parameters, then the target coverage rate of the three algorithms was obtained.

Figure 1 shows the target coverage rate when the number of sensor nodes is 200 and the sensor sensing radius is set to 60 meters and 80 meters respectively. In Figure 1, (a) is a perception radius of 60 meters, and (b) is a perception radius of 80 meters. It can be seen from Figure 1 that the DCS has faster convergence speed and successfully avoids the phenomenon of evolutionary stagnation. Under different radii, DCS improved the target coverage rate better than PSO and ACO. The target coverage rates reached 82% and 95% when the radii were 60 meters and 80 meters, respectively.

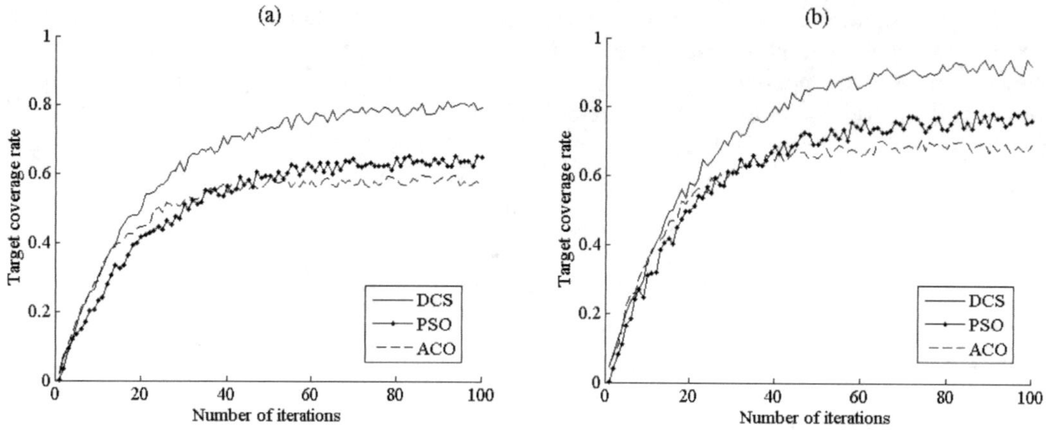

Figure 1. The change of target coverage rate with the number of algorithm iterations

Figure 2 is a line chart of the number of targets monitored by the sensor node. In Figure 2, the number of sensor nodes is 200 and the sensor sensing radius is set to 60 meters and 80 meters respectively. It can be seen from the simulation result graph that as the amount of sensor nodes raises, the number of detected targets also increases. Compared with the two algorithms of PSO and ACO, the number of targets monitored by DCS has obvious advantages.

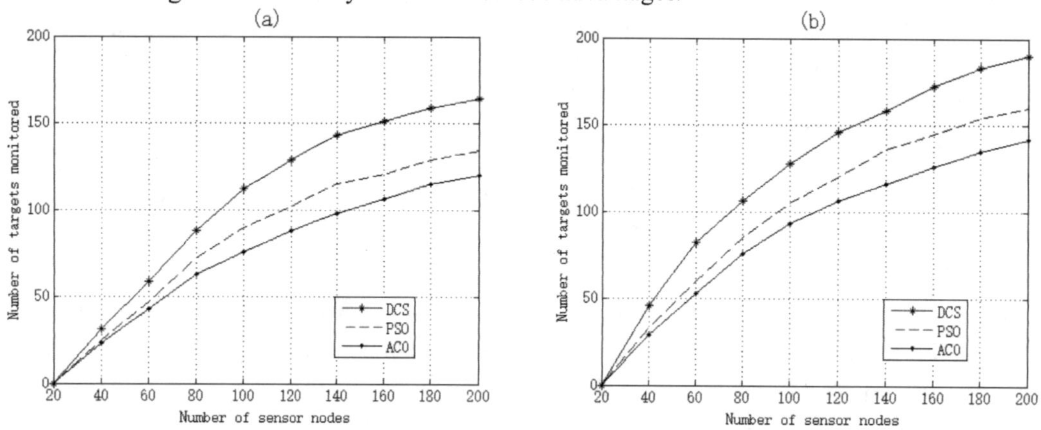

Figure 2. The number of successfully detected targets changes with the number of algorithm iterations

5. Conclusion

Aiming at the problems of limited monitoring capability, large redundancy, short life span, and high cost of the detection process in the traditional aviation wireless sensor network, a discrete cuckoo search is proposed. In the simulation and experiment, the DCS is compared with PSO and ACO. The outcomes demonstrate that the proposed discrete cuckoo search algorithm offers faster convergence speed and stronger global search capabilities, which can effectively increase the target coverage of aviation wireless sensor networks.

Acknowledgments

This paper is supported by the project of Youth and Middle-aged Scientific and Technological Innovation Leading Talents Program of the Corps (No. 2018CB006), the Major scientific and technological projects of the Corps (2017AA005-04), the project of Promoting Scientific Research Cooperation and High-level Personnel Training with Americas and Oceania. The corresponding author is Jie Zhou.

References

[1] Y Nishikawa *et al* 2018 Design of stable wireless sensor network for slope monitoring, *2018 IEEE Topical Conference on Wireless Sensors and Sensor Networks (WiSNet)*, Anaheim, pp. 8-11.

[2] L Zhang, J Qu and J Fan 2016 Topology Evolution Based on the Complex Networks of Heterogeneous Wireless Sensor Network, *2016 9th International Symposium on Computational Intelligence and Design (ISCID)*, pp. 317-320.

[3] M U H Al Rasyid, F A Saputra and M R Ismar 2015 Performance of multi-hop networks using beacon and non-beacon scheduling in Wireless Sensor Network (WSN), *2015 International Electronics Symposium (IES)*, Surabaya, pp. 195-199.

[4] R K Verma, K K Pattanaik and S Bharti 2015 An adaptive mechanism for improving resiliency in Wireless Sensor Networks, *2015 IEEE 10th International Conference on Industrial and Information Systems (ICIIS)*, Peradeniya, pp. 525-530.

[5] P Tsai, L Kong, S Vaclav, J Pan, V Istanda and Z Hu 2016 Utilizing Cat Swarm Optimization in Allocating the Sink Node in the Wireless Sensor Network Environment, *2016 Third International Conference on Computing Measurement Control and Sensor Network (CMCSN)*, Matsue, pp. 166-169.

[6] F Shahzad, T R Sheltami and E M Shakshuki 2016 Multi-objective optimization for a reliable localization scheme in wireless sensor networks, in *Journal of Communications and Networks*, **vol. 18,** pp. 796-805.

[7] L Peng, C Guifen and G Ruijuan 2019 Research on wireless sensor network location based on Improve Pigeon-inspired optimization, *2019 IEEE/CIC International Conference on Communications Workshops in China (ICCC Workshops)*, pp. 228-232.

[8] Bingnan Pei, Hao Zhang, Tengda Pei and H. Wang 2015 Firefly algorithm optimization based WSN localization algorithm, *2015 International Conference on Information and Communications Technologies (ICT 2015)*, pp. 1-5.

[9] V Neagoe and C Neghina 2018 An Artificial Bee Colony Approach for Classification of Remote Sensing Imagery, *2018 10th International Conference on Electronics, Computers and Artificial Intelligence (ECAI)*, pp. 1-4.

[10] H Xu, X Liu and J Su 2017 An improved grey wolf optimizer algorithm integrated with Cuckoo Search, *2017 9th IEEE International Conference on Intelligent Data Acquisition and Advanced Computing Systems: Technology and Applications (IDAACS)*, pp. 490-493.

[11] H Yu, X Tiantian and H Pu 2014 An improved quantum particle swarm optimization and its application in system identification, *The 26th Chinese Control and Decision Conference (2014 CCDC)*, pp. 1132-1134.

[12] M S R Nalluri, T SaiSujana, K H Reddy and V Swaminathan 2017 An efficient feature selection using artificial fish swarm optimization and svm classifier, *2017 International Conference on Networks & Advances in Computational Technologies (NetACT)*, pp. 407-411.

[13] F Chibane, A Benammar and R Drai 2018 Parameters Estimation of Ultrasonics Echoes using the Cuckoo Search and Adaptive Cuckoo Search Algorithms, *2018 26th European Signal Processing Conference (EUSIPCO)*, pp. 2415-2418.

[14] D Jung, C Eom and C Lee 2019 Discrete Cuckoo Search Algorithm for MIMO Detection, *2019 34th International Technical Conference on Circuits/Systems, Computers and Communications (ITC-CSCC)*, pp. 1-4.

[15] L Zhang, N Yin and S Wang 2017 The Adaptive Discrete Cuckoo Search for Parallel Flow Shop Problem, *2017 10th International Symposium on Computational Intelligence and Design (ISCID)*, pp. 68-71.

IOP Conf. Series: Materials Science and Engineering **926** (2020) 012026 doi:10.1088/1757-899X/926/1/012026

Design of a New Type of Deployable Bridge

Li qiyu[1,*] , Yuan jiehong[1] , Sun haitao[1] ,Zhou shiming[1] and Peng yuxing[2]

[1] College of Aerospace Science and Engineering,National University of Defense Technology, Changsha 410072

[2] College of Electronic Science and Technology,National University of Defense Technology, Changsha 410072

[*]1041737365@qq.com

Abstract. In order to ensure that the rescue workers can quickly cross the natural obstacles such as ditches and valleys to enter the disaster area when the traffic facilities are destroyed Seriously by natural disasters, a new type of developable bridge is designed. The design of the bridge is based on the theory of shear hinge and the one-dimensional shear hinge element is used as the basic element. Under the function of dead-weight, the bridge can be unfolded by rope traction. It can be locked into service state after being unfolded to the design position. In this paper, through kinematic simulation and finite element static analysis, the developability and bearing capacity of the developable bridge are verified. The results show that the deployable bridge can meet the design requirements.

1. Introduction

The key characteristic of a deployable structure is that is has two different stable state: the fully folded state and the fully unfolded state. When the deployable structure is in the fully folded state, it is small in size and easy to transport and store. When it is fully deployed, the structure is stable and can bear loads.

Because of its advantages of being easy to transport and store in the fully folded state and being able to bear loads in the fully unfolded state, the deployable structure has been applied to the common articles of daily use, such as telescopic doors and umbrellas. In civil engineering, aerospace engineering and other fields the deployable structure also has a good application prospect.

In the field of bridge engineering, the developable bridge has a good application prospect because of its advantages of convenient transportation and fast erection speed.In 1926,a inventor in the Netherlands, L. Deth had invented a kind of emergency deployable bridge which is the prototype of the large deployable bridge[1].In 2007, the University of Wisconsin designed a deployable bridge system of fiber reinforced concrete[2]. There is a "deployable roller bridge" in London, England[3].The whole bridge can be folded into octagonal structure by hydraulic system, which is convenient for ships to pass through the river smoothly. In 2013, Hiroshima University[4] in Japan designed a movable bridge with a scissor-hinged structure for deployment.

In the 1960s, the Planning and Design Institute of the Ministry of communications of China designed a detachable "321" portable steel bridge on the basis of the original British Bailey bridge, and further developed the "200" portable highway steel bridge[5] In 2010, Sun Guangxu[6] of National Defense Science and Technology University designed a new type of deployable bridge section based on the task requirements of detachable beam bridge. In 2011, Guan Fuling of Zhejiang

University and her team[7] designed a kind of light military deployable bridge using scissor hinge and sleeve deployable mechanism.In 2016, Xiong Haibei et al.[8] of Tongji University designed a kind of deployable arch bridge with the triangle shear type element as the basic deployable component.

When natural disasters such as earthquakes and floods occur, a large number of conventional bridges will be destroyed, there is an urgent need for a convenient transportation, rapid emergency bridge to cross the obstacles, and the developable bridge can achieve such a goal.

In this paper, a new type of deployable bridge structure is proposed with the shear hinge element as the basic element.In this paper, the author used SolidWorks mechanical design software to design the components of the developable bridge. After that, the author uses SolidWorks to carry on the simulation assembly and the interference inspection to the developable bridge.The mechanical system simulation software ADAMS is used to simulate the deploying process of the deployable bridge. Using SolidWorks simulation finite element analysis module to analyze the stress and deformation of deployable bridge structure.

2. Bridge structure design

2.1. The design goals of the deployable bridge

Depending on the scenario,the deployable bridge designed in this paper should meet the following requirements:first of all,when the bridge is fully extended, the span should more than 22m and the bridge height should less than 3m.Secondly,in terms of bearing capacity, it can be used for large vehicles with a weight of 30 tons to pass through, and the material will not yield when passing through the bridge.Thirdly, as an emergency bridge with span less than 50, the mid span deflection under the design load shall be less than $L/120$, where L is the bridge span.At last, the deployable bridge should be flexible to be closed and unfolded, and the time of closing and unfolding should be less than 10min.

2.2. Conception of the project of the developable bridge

The main structure of the developable bridge designed in this paper adopts the shear hinge structure because that the deployable principle of the shear hinge structure is simple, and it is convenient to store and transport. Another reason for the design to chose the shear hinge structure is that the shear hinge structure can achieve different configurations easily by adding different constraints which is easy for designers to choose the type of bridge.

When the construction member only bears the axial force, the material bears the most uniform force and can give full play to the performance of the material, while the reasonable arch structure has the characteristics of only compression and not bending, so the developable bridge type of this paper is arch bridge.In consideration of the convenience of production and assembly, the arc-shape is selected for the arch axis.At the same time, the arc-shaped arch axis is selected, and the span of the developable arch bridge can be realized by increasing or decreasing the number of shear hinge structures.

In order to improve the stability of the structure on both sides of the developable arch bridge, diagonal supporting bars and tie bars along the driving direction of the developable arch bridge are proposed.The main structure, supporting bar and tie bar of the developable arch bridge form a triangle so as to improve the stability of the structure. The horizontal thrust of the arch end is borne by the tie bar, which reduces the requirements of the geological conditions of the arch bridge.The concept of the developable arch bridge is shown in the Figure 1.

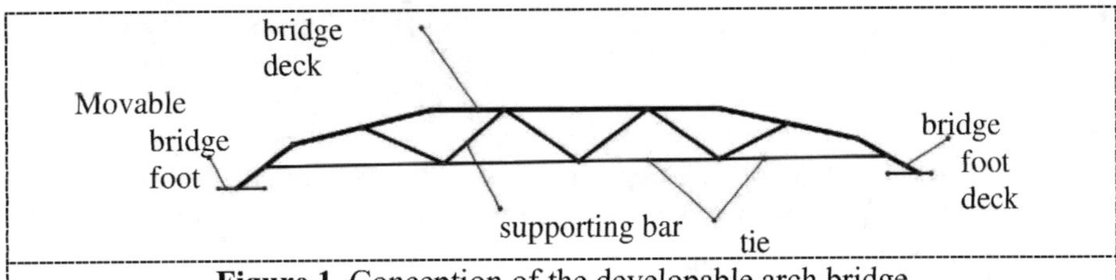

Figure 1. Conception of the developable arch bridge.

IOP Conf. Series: Materials Science and Engineering **926** (2020) 012026 doi:10.1088/1757-899X/926/1/012026

2.3. Scheme design of main structure of deployable arch bridge

2.3.1. The developable condition of shear hinge structure. The developable arch bridge designed in this paper is based on shear hinge structure and the smallest element is shear hinge element. Figure 2 is the schematic diagram of shear hinge unit. When the two points of the shear hinge element (point A and C) move in opposite direction, the shear hinge structure can be folded. It can be imagined that when the shear hinge structure is folded to the final state, point A, B, D and point C, B, D will be almost in a straight line. At this time, the distance between two points B and D is the maximum. Therefore, it is easy to obtain the developable condition of shear hinge structure:

$$AB+AD=BC+CD \tag{1}$$

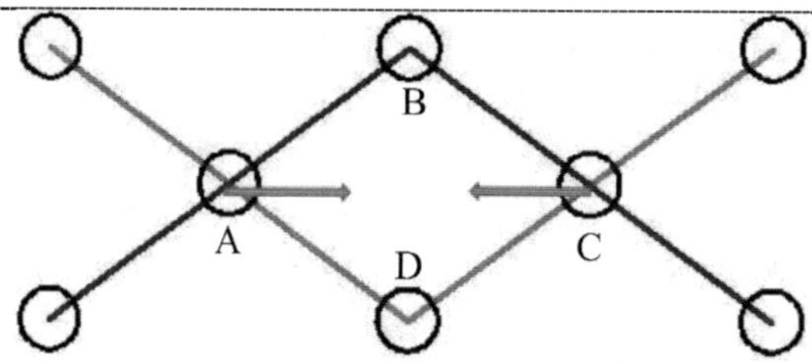

Figure 2. Shear hinge element.

As shown in Table 1, by adjusting the length of the member, the shear hinge structure with different shapes can be formed.

Table 1. The formation of shear hinge structure with different shapes.

Member length relationship	The shape of shear hinge structure
AB=AD=BC=CD	Rectangle
AB=BC,AD=CD, AB≠AD	Arch

*2.3.2 The main structure design of arc arch.*Figure 3 is for the structural diagram of circular arc arch shear hinge.Now it is known that the radii of the three arcs are R_1, R_0, R_2 and the degree of the central angles δ corresponding to each scissor hinge element.According to the cosine theorem, we can know the length of each of bar:

$$l=(R_1^2+R_2^2-2R_1R_2\cos\delta)^{1/2} \tag{2}$$

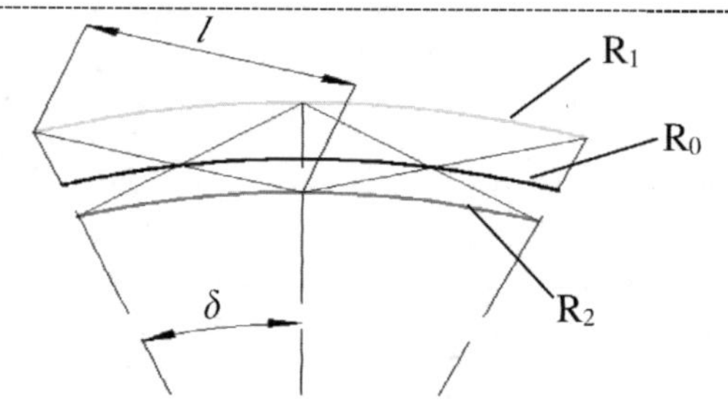

Figure 3. The shear hinge structure of circular arc arch

In order to meet the design span requirements, the arch axis of the extendable arch bridge designed in this paper is an arc with a radius of 45.5m and a center angle of 30°.According to the developable principle of shear hinge and the design of arch axis, the dimensions of the final designed main component is shown in Figure 4.

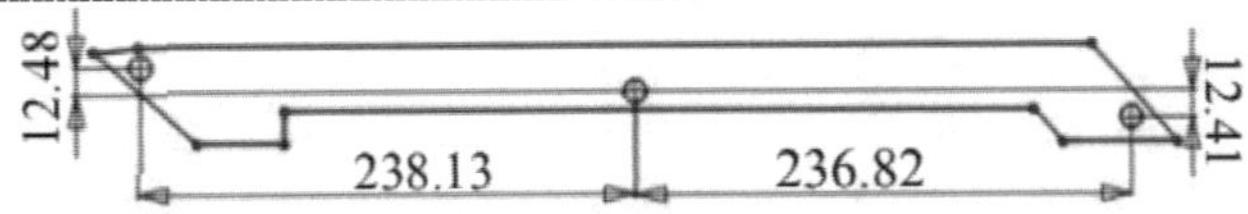

Figure 4. Dimension drawing of main component.

2.4. The overall scheme design of the developable arch bridge

The whole structure of the developable arch bridge designed in this paper is rut structure, which consists of two single span bridges.Single span bridge is composed of bridge deck, bridge foot deck, bridge foot, pin shaft, support bar, tie bar and locking hook.The bridge deck, bridge foot panel and bridge foot are made of aluminum alloy materials, and the rest of the components are made of stainless steel.The main structure of the arch bridge is composed of four groups of deck, two groups of deck and bridge foot.The supporting bar and tie bar are used to assist the bearing of the bridge body. There are six groups of supporting bars, which are respectively installed at the bridge deck interval on both sides of the bridge body.There are two kinds of tie bars, sixteen in all,which are arranged on both sides of the bridge body, and the two ends are hinged.The pin shafts are used to connect the various components.The function of the locking hooks is to keep the deployable arch bridge structure stable and working normally.In the SolidWorks software, the simulation assembly is carried out, and interference inspection is carried out for the inspection of the developable arch bridge. The result shows that there is no interference.

As the deployable arch bridge to fully deployed state,the span of the developable arch bridge is 24m, the width of the bridge deck is 1.4m, and the rise height is 1.6m.The Fully expanded graph of the developable arch bridge is shown in Figure 5.

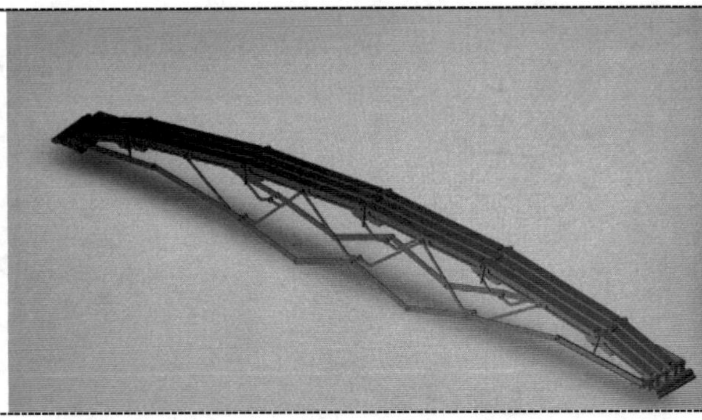

Figure 5. Fully expanded graph.

3. Simulation of the unfolding process of the developable arch bridge

In order to check the unfolding feasibility of the developable arch bridge, this section carries out kinematic simulation of the developable arch bridge through ADAMS,the mechanical system simulation software.

3.1. Principle of developable arch bridge

As shown in Figure 6, it is necessary to hoist the solid line arrows B and C in the drawing with steel cable, then use the moment M(Figure 7)in formed by the staggered distribution of tension and gravity to expand the developable arch bridge.Similarly, the deployable arch bridge can be folded when the dotted arrows A and D are hoisted.

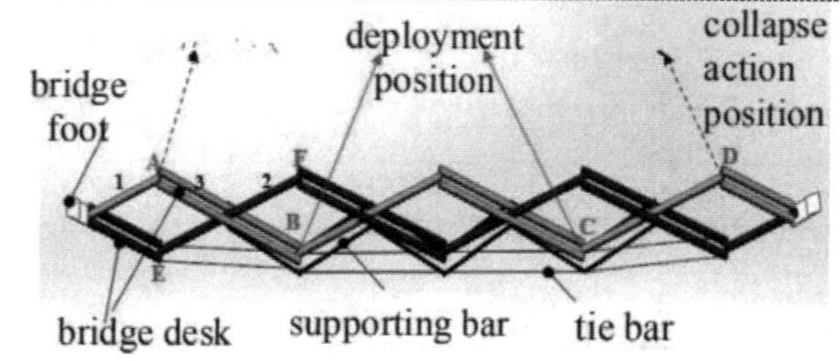

Figure 6. The principle of unfolding and closing of deployable arch bridge.

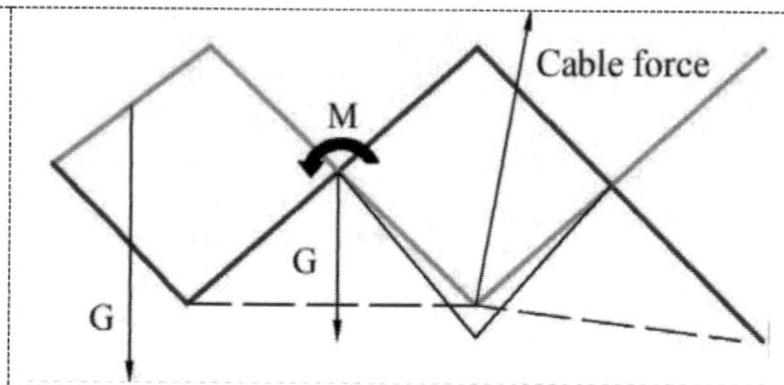

Figure 7. Detail drawing of unfolding and closing of deployable arch bridge.

3.2 ADAMS simulation

The kinematics simulation of deployable arch bridge is carried out by using Adams software, which is divided into three steps:

Firstly, the geometric model of developable arch bridge should be established, but because the modeling ability of ADAMS is not well,this paper adopts building model in SolidWorks and then import into ADAMS.In the process of importing the model from SolidWorks, the geometric model of the developable arch bridge is reasonably simplified that the hinge shafts with small mass and independent of the developability are removed.

Secondly, the physical model is established by adding kinematic pairs, boundary conditions and material properties.After the physical model is established ,the author check the model and find that the model includes 50parts, 74revolutes, and 2sliding pair.The system freedom of the deployable arch bridge is zero, which shows that under the driving action, the model can make certain movement.

Finally, the author verified the expansibility of the designed deployable arch bridge by simulation. The process of model folding can be observed as Figure 8 and 9.

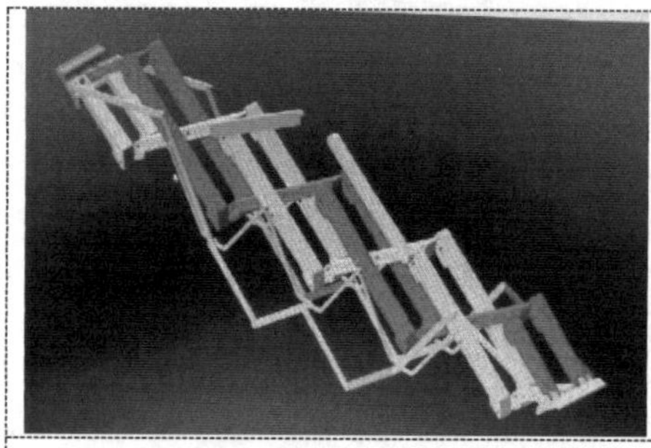

Figure 8. Retraction process of deployable arch bridge.

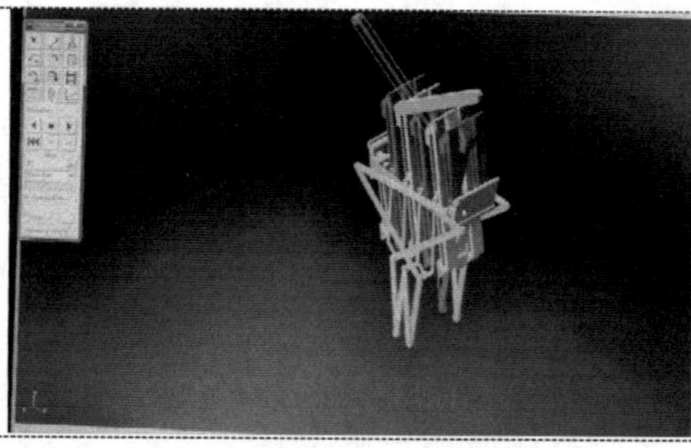

Figure 9. Retraction state of deployable arch bridge.

4. Analysis of bearing capacity of deployable arch bridg

In this section, the author mainly uses the simulation module of SolidWorks mechanical design software to analyze the deformation and internal force of the developable arch bridge under the action of gravity and design load when it is fully expanded.

4.1 The establishment of the finite element model of the developable arch bridge

This section establishes the finite element model by using the model designed in SolidWorks software in the section1.

4.1.1 Material properties of finite element model. In the global model, only elastic analysis is needed. If the stress of the structure exceeds the yield strength, the structure needs to be strengthened. Table 2 is for the material properties of each component of the developable arch bridge.

IOP Conf. Series: Materials Science and Engineering **926** (2020) 012026 doi:10.1088/1757-899X/926/1/012026

Table 2. The material properties of each component.

	Material	Density(kg/m^3)	Poisson's ratio	Modulus of elasticity(Pa)
Bridge deck **Bridge foot deck** **Bridge foot**	Aluminium alloy	2700	0.33	6.9×10^{10}
Pin shaft **Support bar** **Tie bar** **Locking hook**	Stainless steel	7800	0.28	2.1×10^{11}

4.1.2Analysis type and boundary conditions. The static analysis method can be adopted for the model, so the static stress analysis is selected when setting the type of calculation example, which mainly focuses on the stress and deformation of each component of the structure.

People can connect the bridge foot with the ground well on both sides of the bridge, so the boundary condition is set to restrict the 6 degrees of freedom of the bridge foot module on both sides, so that the bridge foot is firmly connected with the ground. Boundary conditions are set in the fixture and fixed geometry is selected.

When using the finite element analysis module of SolidWorks, the connection between the parts defined in the previous assembly is only the position relationship, not the connection between the parts. Before the finite element analysis, it is necessary to establish the connection between the components. The basic developable element of the developable arch bridge designed in this paper is the shear hinge element, so the connection between the components is hinge.There are many kinds of joints and contact ways in SolidWorks. Because the connection between the parts is hinge joint, the pin joint is selected.Joint property is fixed ring (no translation) so that the two parts of the pin connection can rotate relative to each other, but not translate relative to each other.The author has checked the freedom of the developable arch bridge. After checking, the developable arch bridge has been completely constrained.

4.2 Deformation and internal force of deployable arch bridge under dead weight

The deformation cloud chart under the action of self weight is shown in Figure 10. It can be seen from Figure 10 that the deformation of the main body of the span module of the developable arch bridge under the action of self weight is not large, the maximum deflection in the middle of the span is 3mm, and the maximum deformation occurs at the tie bar. Figure 11 is the stress cloud chart under the self weight of the deployable arch bridge. The maximum stress in the middle part of the whole small span of the folding bridge is 20MPa, and the maximum stress appears in the bending part of the support bar is 70MPa .

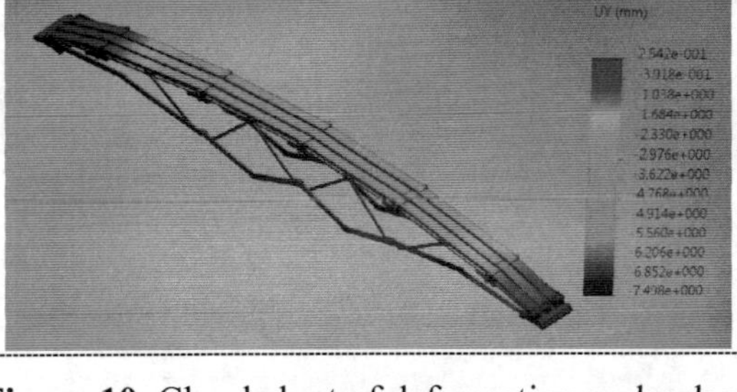

Figure 10. Cloud chart of deformation under dead weigh of deployable arch bridge.

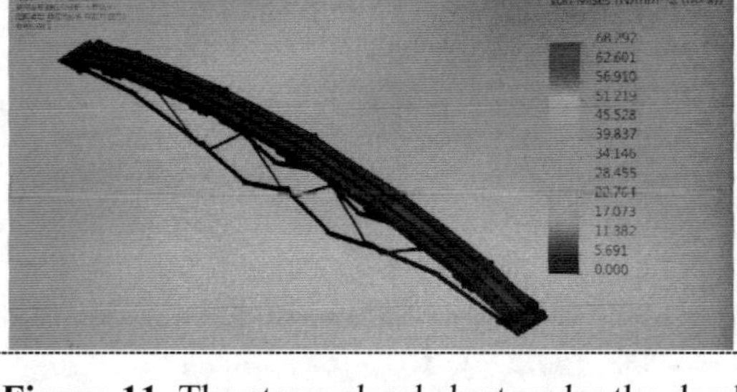

Figure 11. The stress cloud chart under the dead weight.

4.3 Deformation and internal force of deployable arch bridge under design load

It can be known from the overall scheme design part of section I that the design of the deployable arch bridge is jointly borne by two bridges when bearing the design load. When taking one of the bridges for analysis, the design load is halved.

Mid span loading is the most unfavorable condition for the stress of the deployable arch bridge.Therefore, two 5meters long and 0.7meters wide cuboids are placed in the mid span module, and 15 tons of pressure is applied on them to simulate the design load.

The deformation nephogram under the design load is shown in Figure 12. It can be seen from Figure 12 that the maximum deformation of the deployable arch bridge under the design load is 125mm in the middle of the span, which is less than $L/120$ of the bridge span.This means that the expandable arch bridge can meet the deformation requirements under the design load

It can be seen from Figure 13 that the maximum stress of the bridge deck under the design load is 190MPa, which appears in the middle of the span. The maximum stress of the deployable arch bridge appears at the foot of the bridge, which is 240MPa.

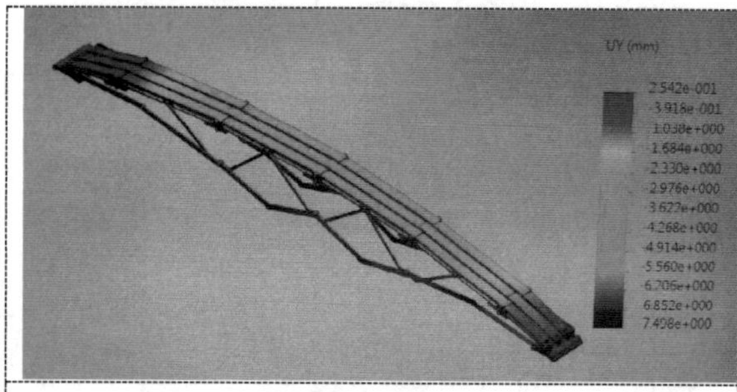

Figure 12. Cloud chart of deformation under design load.

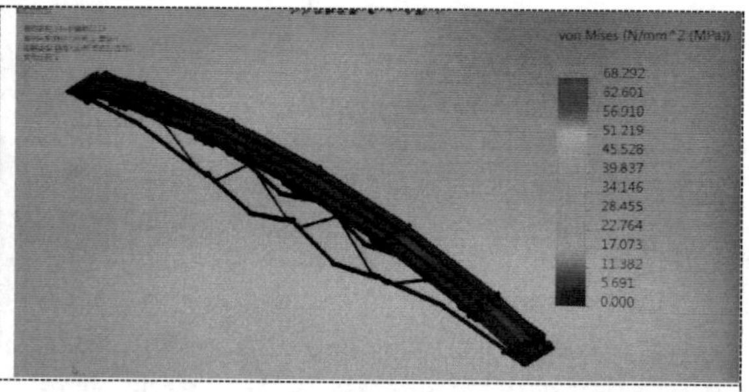

Figure 13. The stress cloud chart under design load.

From the above finite element analysis, it can be seen that the developable arch bridge designed in this paper can meet the use requirements under the design load.

5. Conclusion

Based on the requirement of high-mobility engineering support for rapid deployable emergency bridge, a new design scheme of the deployable arch bridge is presented in this paper.The folding process of the deployable arch bridge is simulated to ensure that the deployable arch bridge can be unfolded and folded according to the design. At the same time, the stress and deformation of the deployable arch bridge under the design load are analyzed, and it is found that the bearing capacity of the deployable arch bridge can meet the design requirements.

The deployable arch bridge designed in this paper has novel structure, simple configuration and can be manufactured in factory.The deployable arch bridge has a good adaptability, according to demand can be reasonable increase or decrease the number of bridge span modules to meet the requirements of crossing different obstacles.It is very convenient to transport and unfold the arch bridge designed in this paper.The problem of large deformation of deployable arch bridge under design load can be improved in the future structural design.

In conclusion, the expandable arch bridge designed in this paper can preferably meet the requirements of rapid emergency bridge for buliding the emergency rescue channel in non military emergency activities such as disaster relief.

Reference

[1] XIA Zhiyuan, DONG Jun, CHI Pei. Application and development of deployable Bridge *J*, *World Bridge*,2013.**(06)**

[2] HANUS J P. Investigation of a deployable military bridge system with a fiber reinforced concrete deck *D*.Madison:College of Military Studies,University of Wisconsin-

Madison,2007:93-9

[3] Fantastic bridge design *J Encyclopedic knowledge*, 2011**(05)**: F0002-F0003.

[4] ARIO I, NAKAZAWA M, TANAKA Y, et al.Development of aprototype deployable bridge based on origami skill *J Automation in Construction*, 2013,**32**:104-111.

[5] XU Guanyao, YU Zhong quan Present situation and development trend of fabricated steel bridge *J. Steel structure*, 1999**(03)**:47-49.

[6] SUN Guangxu Scheme design and dynamic analysis of deployable bridge section *D*. National University of Defense Technology, 2010

[7] GUAN Fuling, ZHOU Yijun, KUANG Qi, Qian Lifeng. Design and dynamic analysis of a light deployable military bridge *J*, *Journal of Southwest Jiaotong University*,2012,47**(05)**:735-740+747..

[8] XIONG Haibei, SONG Yijie. Exploration of a new type of developable arch bridge *J*, *Journal of Hunan University (Natural Science Edition)*,2018,45**(01)**:19-25

IOP Conf. Series: Materials Science and Engineering **926** (2020) 012027 doi:10.1088/1757-899X/926/1/012027

Optimal Design of a Typical Fuselage Cross Section for a Certain Type of Civil Aircraft

Zhenwei Zhang

Aircraft Integration Department, Shanghai Aircraft Design and Research Institute

zhangzhenwei@comac.cc

Abstract. Optimal design of a typical fuselage cross section for civil aircraft is one of the most important aspects of overall aircraft design. In this paper, a typical cross section of a certain type of civil aircraft (referred to as model A) is taken as an example and compared with a similar model B of strength calculation and analysis of deformation, bending moment, stress and other structural bearing efficiency. The principle of the difference is explored and the principle of engineering design of a typical fuselage cross section is summarized. Then the design of the original cross section is optimized by integrating structural load-bearing efficiency, space arrangement and other factors. Strength evaluation is also carried out. Finally, the optimization plan of a typical fuselage cross section of model A is given.
Keywords: fuselage cross section; structural load-bearing; optimal design

1. Introduction

For the civil aircraft, the selection of a typical cabin cross section is extremely important for the performance, economy and comfort of the aircraft. Realizing the optimal design of the typical cross section of the fuselage is an important measure to improve the overall index of the aircraft. Actually, the design of a typical cabin cross section reflects a trade-off between the comfort level of the cabin and the related parameters such as the weight cost and size increase required to meet that comfort level [1]. In general, for aircraft with pressurized cabins, most of the fuselage cross sections are designed to be circular. However, from the perspective of effective space utilization, the circular cross section will waste space with a cross section diameter that is not particularly large[2]. Therefore, in order to meet the requirements of passenger seats and containers layout for aircraft with smaller diameter, multi-circular cross section is often adopted, which is composed of multiple arcs and smooth transition curve coordinated with it. The arc radius and center position can be flexibly changed according to the needs of cabin layout to achieve full utilization of cabin space [3]. The design of model A fuselage cross section was carried out in the form of a multi-circular section. The fuselage section of similar model B is also shown in figure 1. Compared with model B, the lower half of the typical fuselage section of model A has a smaller radius and a smaller transition radius, but a larger transition area.

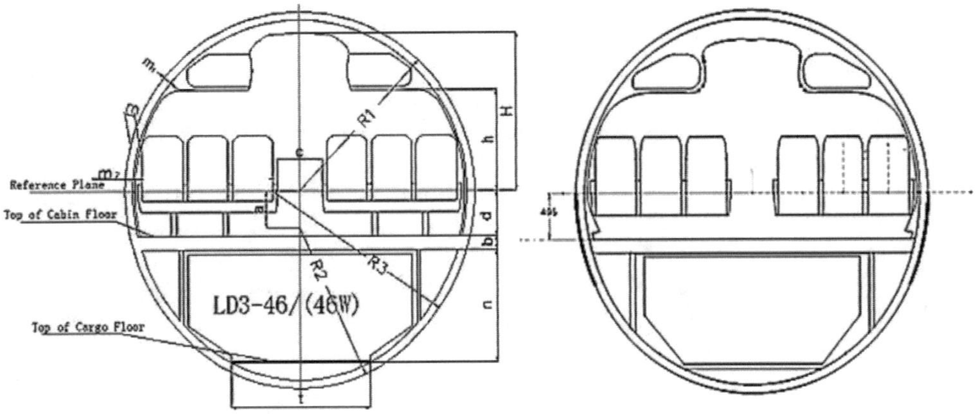

Figure 1. Typical fuselage cross section of model A and model B

2. Problem analysis

According to the difference between the above two models in the typical cross-sectional shape of the fuselage, the strength analysis of the structural bearing efficiency such as deformation, bending moment and stress is carried out. The same finite element model is established, and the same pressurized load is applied. The load balances itself inside the fuselage[4]. Three-point static constraint is adopted during the calculation, and the element model is modeled by 13 frames of straight sections. The element attributes are assigned according to the real structure or corresponding parameters [5], as shown in figure 2.

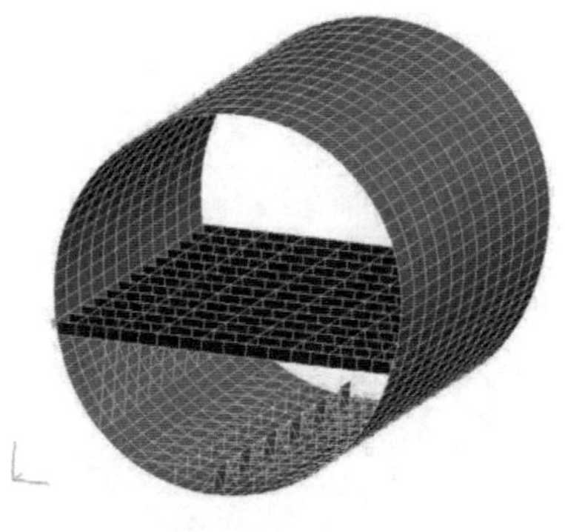

Figure 2. Finite element model of straight frame fuselage

2.1. Deformation

Take the middle frame of the straight section for deformation analysis. The deformation results in this paper are all displayed at a magnification of 10 times. The frame of the straight section of model A is deformed outward most in the left and right sides of the cargo hold. The middle of the cabin floor beam is downward, and large displacement of the cargo floor occurs upwards together with the bottom wall panel; The straight fuselage frame of model B has the largest outward deformation in the left and right side areas above the cabin floor, and the displacement occurs toward the inside in the left and right side areas of the cargo, while the cabin floor beams remain basically horizontal. Therefore, the deformation state of the cabin floor beam of the straight section of model B is better than that of model A, and it is more in line with the design expectations. The deformation is shown in figure 3.

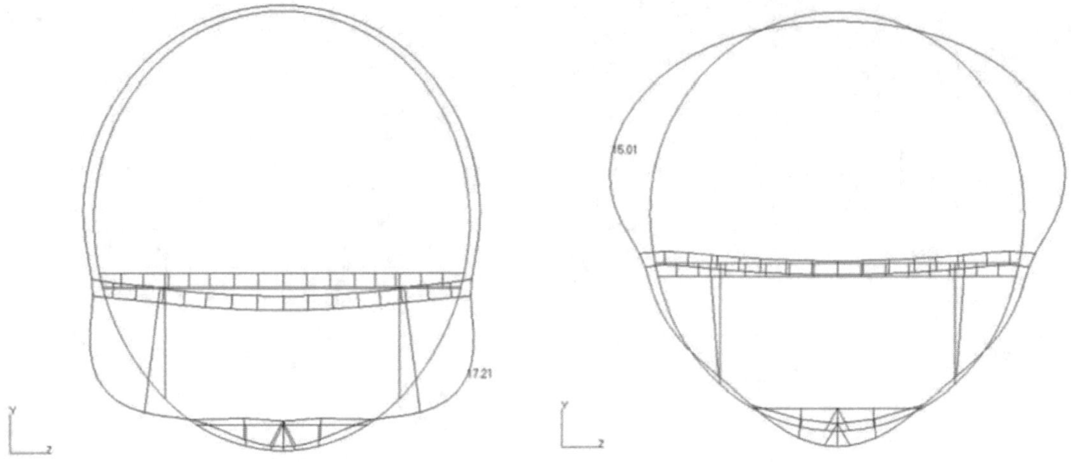

(a) (b)

Figure 3. Deformation of typical fuselage cross section of model A and model B

2.2. Bending moment distribution

The bending moment distribution of the inner end and the distribution of the floor beam of the left frame of the middle straight fuselage frame is compared. The bending moment distribution of the straight fuselage frame of model B is more uniform than that of model A, and the bending moment of the floor beam is much smaller. Therefore, the bending moment distribution of the straight fuselage frame and floor beams of model B is more reasonable than that of model A. The bending moment distribution is shown in figure 4.

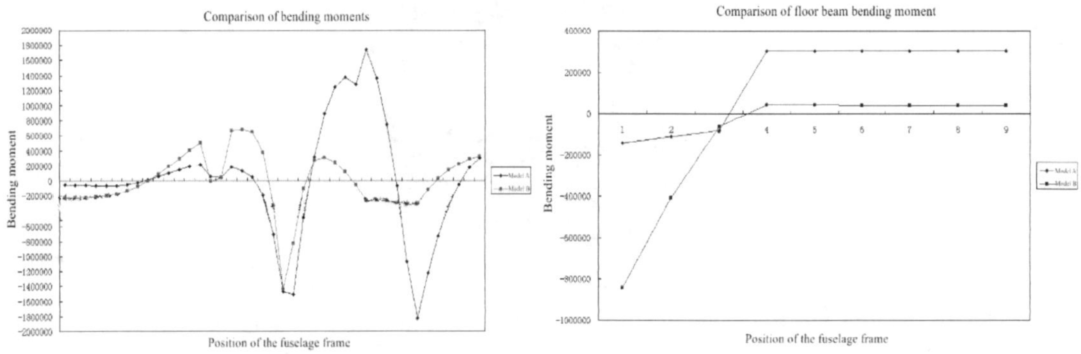

(a) (b)

Figure 4. Comparison of bending moment distribution of fuselage frame and floor beam of typical cross section between model A and model B

2.3. Stress distribution

The stress analysis on the inner and outer edges of the left side frame of the middle straight fuselage frame is carried out. The stress distribution on the edges of typical straight fuselage cross section of model B is more uniform and reasonable than that of model A, and the load is smaller. The stress distribution is shown in figure 5.

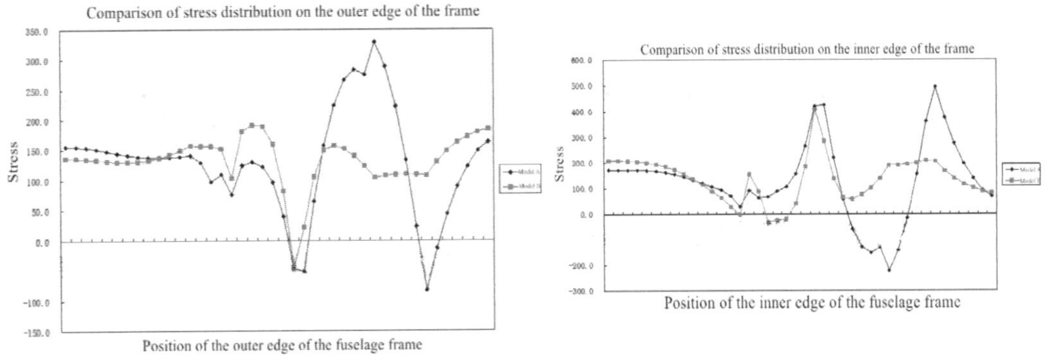

(a) (b)

Figure 5. Comparison of stress distribution on the inner and outer edges of the left side frame of typical cross section between model A and model B

2.4. Influence of structural weight gain

If the cross section of the fuselage remains unchanged, the local structural parameters between the cabin floor and the cargo floor of typical cross section of model A are strengthened in accordance with the design of equal stress level, so that the stress level of the straight section of the fuselage of model B can be reached, and the resulting weight is as high as 108.7 kg. Comparison of stress on the inner and outer edge of typical fuselage cross section frame of model A and model B after partial reinforcement with twice the supercharged load is shown in Figure 6.

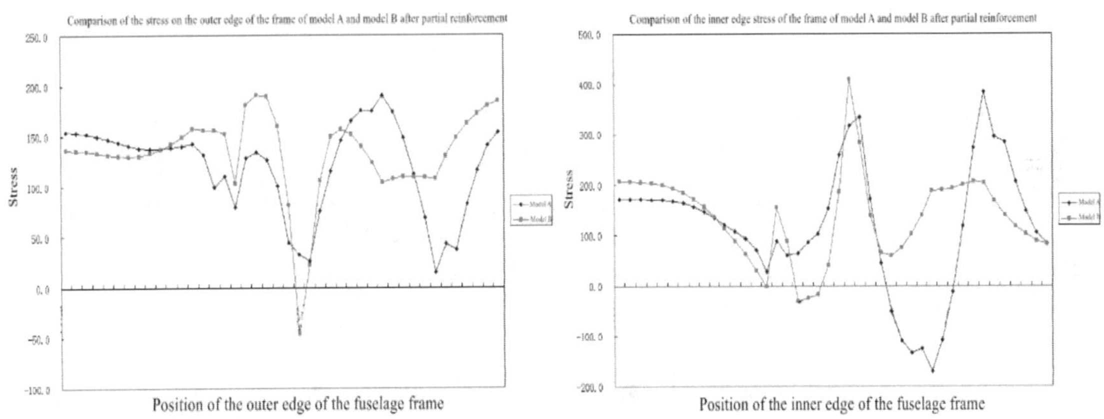

(a) (b)

Figure 6. Comparison of the stress distribution on the outer and inner edge of typical fuselage cross section frame of model A and model B after partial reinforcement

2.5. Conclusion

It can be seen from the above that when the same structural parameters are used for design, the typical cross section parameters of the straight fuselage of model B are more reasonable, and the structural bearing efficiency is higher. If the straight fuselage of model A is partially reinforced to achieve a considerable stress level, an additional weight of about 108.7 kg is required. At the same time, the area of the triangular section of the fuselage of model A is smaller than that of model B, which increases the difficulty of system arrangement in the triangle. Therefore, the existing fuselage cross section of model A needs to be optimized.

IOP Conf. Series: Materials Science and Engineering **926** (2020) 012027 doi:10.1088/1757-899X/926/1/012027

3. Optimal design

3.1. Principle research

To better illustrate the effect of differences in fuselage cross-sectional shape on frame load-bearing efficiency, three 1/4 arcs of different diameters were tangentially docked to one 1/4 arc, as shown in figure 7. Uniform normal load is applied, and it is found that the distribution of the axial force at the tangent point is not much different, but the bending moment is abrupt, and the greater the radius change, the greater the abrupt gradient.

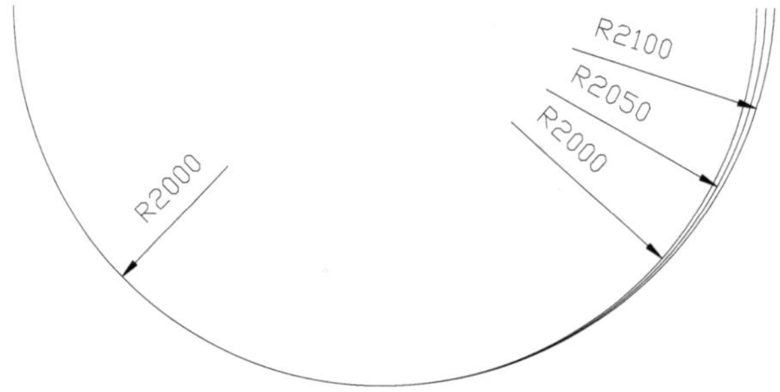

Figure 7. Schematic diagram of theoretical analysis and comparison of the cross section

It can be inferred that the difference between the radius of the upper and lower semicircle of the straight fuselage cross section of model A is large and the transition area is long, so the frame load between the cabin floor beam and the cargo floor is too large.

3.2. Cross section data analysis of model B

This article makes an in-depth study on the geometric data and design rules of the typical fuselage cross section of model B. It is found that the transition arc of the upper and lower circles of model B is short, and the upper and lower end points of the transitional arcs are equal in distance to the surface of the floor. At the same time, on the premise that the radius and total height of the upper half of the cross section remain unchanged, change the distance between the center of the upper and lower circles, the absolute difference between these two values are showing an increasing trend. It can be inferred from this: a) Model B has undergone detailed optimization design in the design of the transition arc length; b) the midpoint of the transition arc in height is near the upper surface of the floor, and the structural load-bearing efficiency is the highest .

3.3. Cross section data analysis of model A

By comparing the geometric data, it is found that the transition circle of the fuselage cross section of model A is too long, and the absolute difference between the upper and lower end points of the transition arc and the upper surface of the floor is too large. Compared with model B, the lengths distribution of transition arc and lower arc of model A has changed drastically, the cross-sectional area is reduced by 0.016%, and the triangle area is reduced by 4.695%. The reduction of the triangle area leads to the difficulty of system arrangement. The relevant data comparison is shown in table 1.

Table 1. Comparison between typical cross section of model A and typical cross section of model B

	Relative value to model B				
	Triangle area	cross section area	Transition arc length	Lower arc length	Cross section circumference
Model B	0%	0%	0%	0%	0%
Model A	-4.695%	-0.016%	53.4%	-23.45%	0.002%

3.4. Plan optimization

Based on the above analysis, at the same time, in order to minimize the amount of change and the impact of the cross section change, the following optimization principles are determined:

a. The arc radius of the cross section shape should be as close as possible, and the length of the transition arc should be shortened as much as possible, so as to benefit the bearing efficiency of the structure;

b. Keep the radius of the upper circle on the original cross section unchanged;

c. Keep the total height of the original cross section unchanged;

d. Optimize the upper and lower center distances of the design variables (i.e. changing the radius of the lower circle) to make the transition arcs evenly distributed near the floor plane;

e. strive to make the cross section circumference as small as possible to facilitate drag reduction.

Based on this, this paper developed two new typical fuselage cross sections.

Plan one: reduce the center distances of the upper and lower circle, so that the midpoint of the transition arc in height is near the upper surface of the floor; Plan two: further reduce the center distances of the upper and lower circle, so that the center distances are equivalent to that of model B .The optimization plans are shown in figure 8.

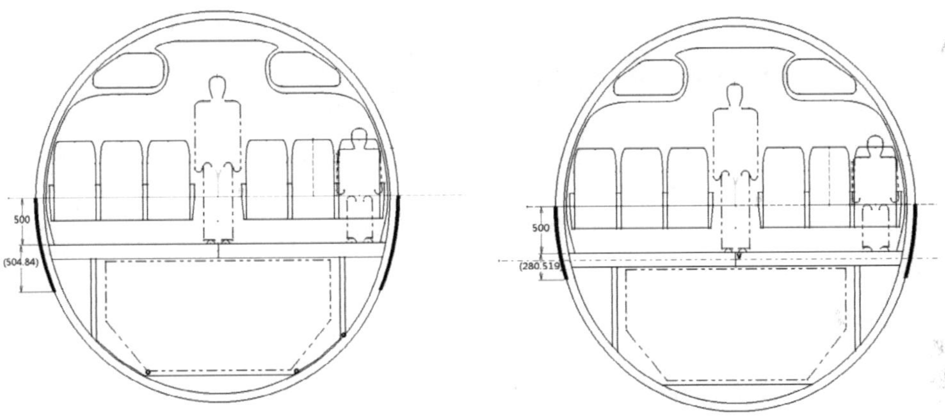

(a) (b)

Figure 8. Improvement plan one and plan two of typical fuselage cross section of model A

The data comparison of improvement plan one, two and model A, model B is shown in table 2.

Table 2. Data comparison of model B, improvement plan one, plan two and original plan of model A

Model	Relative value to model B				
	Triangle area	cross section area	Transition arc length	Lower arc length	Cross section circumference
Model B	0%	0%	0%	0%	0%
Model A	-4.695%	-0.016%	53.4%	-23.45%	0.002%
Improvement plan one	3.176%	0.453%	0.461%	0.267%	0.230%
Improvement plan two	2.890%	0.463%	-6.705%	3.334%	0.235%

Using the same finite element analysis method as in Chapter 2 of this article, the structural load analysis of the two improvement plans is carried out. The deformation and bending moment distribution are shown in figure 9-10.

IOP Conf. Series: Materials Science and Engineering **926** (2020) 012027 doi:10.1088/1757-899X/926/1/012027

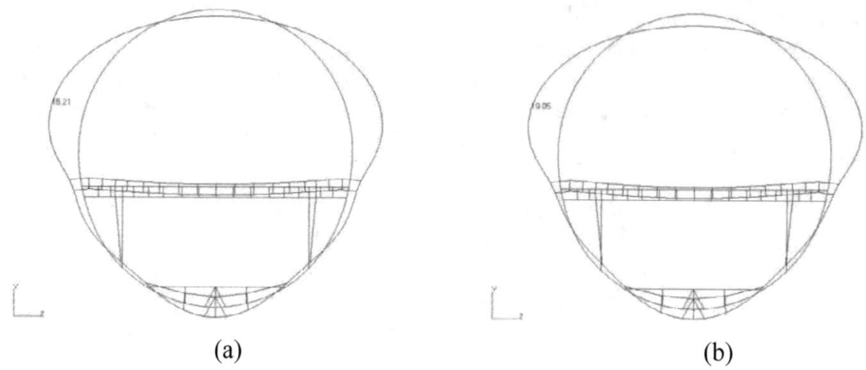

Figure 9. The deformation of fuselage section of model A for improvement plan one and plan two

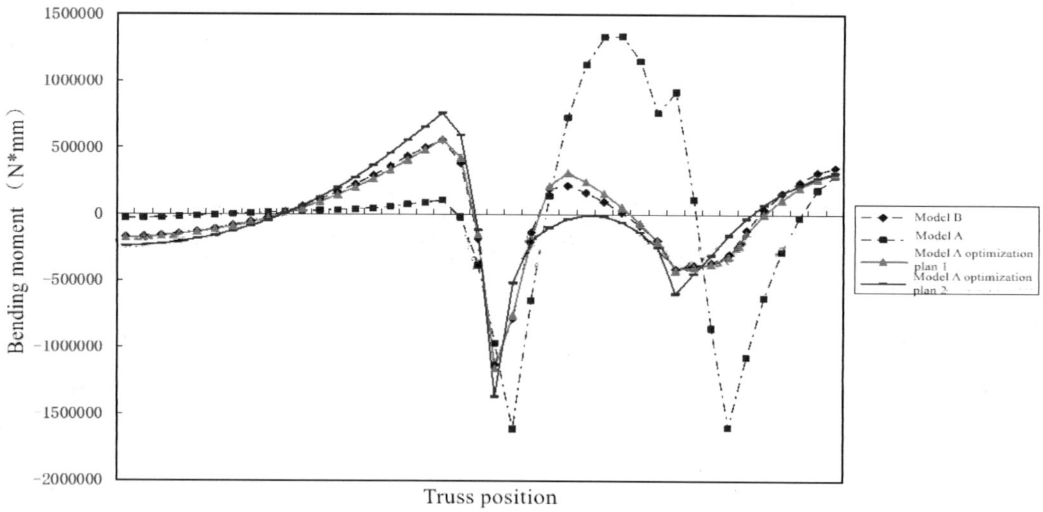

Figure 10. Comparison of the distribution of bending moments of the straight frame fuselage frame

4. Conclusion

The distribution of bending moment and deformation of the fuselage frame of the plan one and plan two have been greatly improved compared with the original plan. The peak and valley values of the bending moment decreased significantly. The stress distribution of the fuselage frame is more uniform, which improves the structural efficiency and realizes the weight reduction of the structure. The distribution of bending moment and deformation of the fuselage frame of plan one is equivalent to those of model B. Although the cross-sectional area is increased by 0.453% compared to model B, the triangle area is increased by 3.176%, Which is beneficial to solve difficulties in system layout.

It is determined that improvement plan one is adopted as a typical fuselage cross section for the engineering implementation of model A. The comparison between plan one and the original cross-section is shown in figure 11. It can be seen that compared with the original cross-section, the roundness of the cross section of plan one is better, which is conducive to structural bearing.

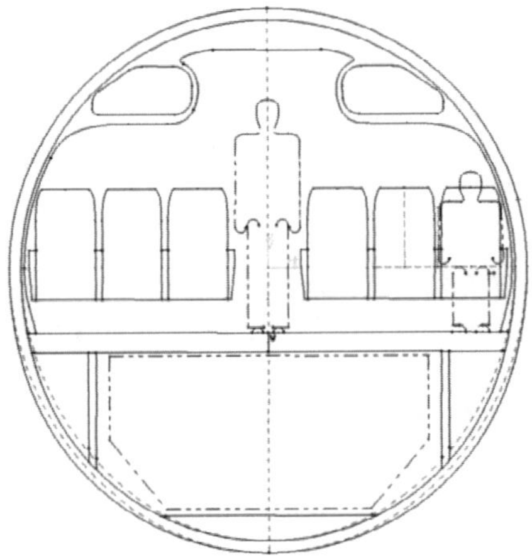

Figure 11. Comparison of optimization plan one with the original cross section of model A（Black solid line is the improvement plan one, red dotted line is the original cross section）

By optimizing the design of the typical fuselage cross section of model A, the principle of cross section design is explored, the influencing factors and design principles of cross section design is mastered, which accumulated good experience for the subsequent design of aircraft.

References

[1] 2011 Civil aviation regulation of China (CCAR) *Part 25-Airworthiness Standards: Transport Category Airplanes.* R4.
[2] WuYang, Zhang Zhenwei. Research on Dual-channel Civil Aircraft Cross Section Design. 2014.09. *Science & Technology Vision*, 58-59.
[3] 2005. *General design of civil aircraft Aircraft Design Handbook, Volume 5*(Beijing: Aviation Industry Press).
[4] 2001. *Load, strength and stiffness Aircraft Design Handbook, Volume 9*(Beijing : Aviation Industry Press).
[5] Zhiping Q, Xiaojun W. 2011. *Aircraft structural strength analysis and design basis* (Beijing: Beijing University of Aeronautics and Astronautics Press).

Creative Design of Streamline Groove Mechanical Seal Based on TRIZ Theory

Wang Heshun[1,*], Zhou Lei[1], Liu Xiaoming[2], Zhang Chening[2], Wang Zeping[2] and Zhu Weibing[1]

[1]School of Mechanical Engineering, Xihua University, Chengdu Sichuan 610039, China;
[2]Sichuan Nikki Seal Co. Ltd., Chengdu, Sichuan 610046, China

[*]E-mail: wangheshun@mail.xhu.edn.cn

Abstract. In view of the problems of poor stability and large leakage of dynamic pressure non-contact mechanical seals in a specific working environment, through the analysis of the mechanical seal principles and end structure, based on the TRIZ theory, choose the "structure stability", "adaptability and versatility" as the improve elements, the "harmful emission" and "foreign harmful factors" as the deteriorate elements. The principle numbers of the invention are obtained from the contradiction matrix. According to the principles of the invention, a new type of streamline groove mechanical seal is designed. Through the technology of making groove on the end face, a fluid film with a certain stiffness is obtained, and the end face is sealed. Through improve the stability and adaptability of the mechanical seal operation, it can operate safely and stably in a specific working environment and meet the requirements of the seal.

Keywords: seals; mechanical seals; TRIZ theory; streamline groove

1. Introduction

Mechanical seal is a kind of sealing device that at least has a pair of end surfaces perpendicular to the rotational axis, under the action of the compensation element and the dielectric pressure to prevent leakage of fluid, also known as end face seal. It is indispensable part in fluid machinery and power machinery [1]. From the patent of the first face seal in the UK in 1885 to the present one hundred years, the mechanical seal technology has developed greatly. The mechanical end face seal has the characteristics of reliable operation, small leakage, long service life and low power consumption, so it is suitable for petrochemical industry and water pump repair industry[2].According to the size of the seal face clearance, mechanical seals are divided into contact seals and non-contact seals, and non-contact seals are divided into hydrostatic and hydrodynamic pressure types. From the perspective of the development of sealing technology at home and abroad, fluid dynamic pressure mechanical seals with macro grooves on the surface can better meet the conditions of high PV values, and have received greater attention[3].This article focuses on the research and design of non-contact mechanical seals. The grooved mechanical seal on the sealing surface has become an advanced sealing technology in the field of non-contact mechanical seals. With its unique dynamic pressure effect and good sealing performance, it has been widely used [4]. Among them, the most widely used grooves are: spiral grooves, "T" grooves, arc grooves, straight grooves, etc. With the rapid development of petrochemical, energy, aerospace and other fields, higher requirements are placed on the performance, service life and reliability of mechanical seals under extreme conditions (high temperature, high pressure, high speed or ultra-low temperature, vacuum, etc.)[5].Based on the analysis of mechanical seals, this paper uses TRIZ theory to improve the sealing performance and applicability of mechanical seals, and carries out an innovative design.

2. TRIZ theory

TRIZ theory is an innovative method theory proposed by G. S. Altshuller and his research team on the basis of analyzing and studying more than 2.5 million patents in various countries and integrating multiple disciplines and field principles, which means the theory of solving invention problems. Generally speaking, the process of solving problems in TRIZ theory is as follows: first, the actual problem to be solved is transformed into a problem model; then, the solution model is obtained by using TRIZ tools; finally, the solution model is applied to the specific problem to obtain the solution[6].There are four basic models of TRIZ problem: technical contradiction model, physical contradiction model, functional model and physical field model[7].Technical contradiction model is widely used in engineering and technical problems, and contradiction matrix is an important tool for analyzing technical contradictions. The contradiction matrix is composed of 39 general engineering parameters and 40 invention principles refined by G. S. Altshuller and his research team. The 39 general engineering parameters are arranged horizontally and vertically. Horizontal represents the deteriorating parameters, vertical represents the improving parameters. There is a contradiction between the improved parameters and the deteriorated parameters. The number in the grid where the engineering parameters cross is the number of the recommended principle of invention[8].With the development of technology and economy, the number of general engineering parameters has increased from 39 to 48, the number of invention principles has also increased to 77, the contradiction matrix table has changed from 39×39 to 48×48, and the diagonal positions have been filled[9].In the mechanical design industry, there is also a basic process of convenient invention principle solution method[10].

3. Innovative design

3.1. Problem description

In the traditional seal face groove design and research process, a groove shape is generally designed first, and then some sealing performance indexes are selected to optimize the groove structure and parameters in order to obtain the best end face fluid film characteristics. The idea is to give the groove type first and then correlate the groove type parameter with the seal. This makes the groove configuration unable to actively and well match the characteristics of the end face fluid film.so it is difficult to obtain the best dynamic pressure effect of the end face fluid, and thus cannot obtain greater fluid film stiffness and opening force.

3.2. Extracting contradictions and parameters determining

In the design process of mechanical seal, the leakage rate is an important parameter to determine the seal performance. For a given slotted mechanical seal, there is a corresponding minimum leakage. In certain workplaces (such as nuclear power plants), the leakage requirements are so strict that some grooves cannot meet the requirements, so new grooves need to be designed. However, the new groove design takes a long time and may not be suitable for emergency situations. At the same time, consideration should be given to increasing the stability of the mechanical seal in the face of overload in order to better adapt to the workplace. According to the above analysis, the engineering parameters determined are "adaptability", "structural stability", "harmful emission" and "external harmful factors".

3.3. Check the contradiction matrix

According to the analysis, the improvement parameters are "adaptability" and "structural stability", and the deterioration parameters are "harmful emission" and "external harmful factors". Find the contradiction matrix table to get the serial number of the invention principle, as shown in Table 1.

Table 1. Contradiction matrix table (partial)

improvement parameters		…	30	…	40
deterioration parameters			harmful emission		external harmful factors
…					
21	structural stability		1,15,24,21,12,35		40,35,31,17,11,24,18,30
…					
32	adaptability		31,9,30,14,18,29,3		35,24,31,3,40,11,19,14,17

3.4. Invention principle

From "structural stability" and "harmful emission", the invention principle number can be obtained: 1, 15, 24, 21, 12, 35. No. 1 is segmentation, and the description is: a. Divide the object into separate parts; b. Make the object detachable; c. Increase the division of the object. This principle does not contribute to the solution of the problem.No.15 is dynamicity, described as: a. The change of the characteristics of the object (or external medium) should be optimal at each working stage; b. Divide the object into several parts that move relative to each other; c. Make the immobile object moving. According to this principle, the mechanical seal can be designed as a rotating form. The dynamic pressure type non-contact mechanical seal relies on the dynamic pressure effect of the fluid to achieve the sealing effect.No.24 is intermediary which is described as: A. The use of intermediate objects that can be migrated or transported; B. To temporarily attach another (easily separable) object to an object. According to the principle, the sealing medium can be used to solve the problem. The sealing device relies on the sealing medium to form a fluid film between the sealing rings to achieve the sealing.No.21 is reducing the harmful effect time, No.12 is equipotential, No.35 is physical or chemical parameter changes, none of these principles contribute to the solution of the problem.

From "structural stability" and "external harmful factors", the invention principle number can be obtained: 40, 35, 31, 17, 11, 24, 18, 30. No.40 is composite material. According to this principle, composite material can be selected for the seal ring to improve the adaptability and stability of the seal.No.31 is porous materials, No.17 is one-dimensional variable multidimensional, No.11 is pre-protection, No.18 is mechanical vibration, none of these principles contribute to the solution of the problem.No.17 is flexible shell or film, according to this principle, a film can be placed between the seals to prevent leakage.

From "adaptive" and "harmful emission", the invention principle number can be obtained: 31, 9, 30, 14, 18, 29, 3. No.9 is in advance reaction, according to this principle, a compensating device can be installed on the sealing device to cope with small axial movement.No.14 is the curvature increasing, This principle does not contribute to the solution of the problem.No.29 is pneumatic or hydraulic structure, according to this principle, sealing structure can be achieved according to hydraulic or pneumatic principle.No.3 is local changes, according to this principle, the sealing ring torus can be grooved to produce hydrodynamic pressure effect.

From "adaptation" and "external harmful factors" the invention principle number can be obtained: 35,24,31,3,40,11,19,14,17. No.19 is periodic action, according to this principle, the sealing grooves can be evenly distributed on the sealing ring.

Based on the above principles, the groove shape on the seal end face is designed according to the flow linear motion track of the seal fluid on the seal end face. He sealing principle is to make the sealing fluid squeeze and shrink gradually from the opening of the groove to the root of the groove, and finally form a strong shearing and squeezing effect at the root of the groove, thus forming a significant dynamic pressure effect of the fluid. The basic shape of the boundary line of the streamline groove is determined by the fluid track of the end seal fluid film. The particles of the sealing fluid move in an approximately uniform speed along the circumferential direction, the sealing fluid particles move with approximately uniform velocity in circumferential direction and in radial direction with the regularity that the velocity is inversely proportional to the radius. The synthesis of circumferential and radial motion is the basic streamline shape of sealed fluid particles. As shown in figure 1, the relationship between the polar Angle and the polar diameter of the streamline groove marginal line (12,13) is as follows:

$$\theta = \theta_0 + \frac{\pi h \omega (r_0^2 - r^2)}{Q} \tag{1}$$

where:

θ_0 is the initial polar Angle of type line edge starting at the opening of the groove;
r_0 is the radius of the starting point of the groove (11) at the opening of the groove (11);
h is the average clearance between two opposite axial sealing faces;
Q is the leakage of medium volume in sealing clearance;
ω is the angular velocity of relative rotation of the two sealing end faces.

4. Streamline groove seal

As shown in Figure 1~2, shows the structure of mechanical seal device with a streamline groove end face seal ring, and the streamline groove is located on the outer diameter side of the rotary ring seal end face. The mechanical seal device is composed of a static ring 1, a rotating ring 2, a static ring seat 3, a shaft sleeve 4, a transmission pin 5, a tolerance ring 6, a compression sleeve 7, a push ring 8, a spring 9, an anti-rotation pin 10, etc. The coaxial lines of the static ring 1 and the rotating ring 2 are arranged face to face, and the opposite end faces form a sealing surface. The static ring 1 is axially supported by the push ring 8 and spring 9 and positioned by the anti-rotation pin 10. Therefore, the static ring 1 can only float freely along the axial direction. The moving ring 2 is expanded on the cylindrical surface of the shaft sleeve 4 radially by the tolerance ring 6, contacts with the step surface of the shaft sleeve 4 axially and is compressed axially by the compression sleeve 7, and is fixed together with the shaft sleeve 4 by the driving pin 5 in the circumferential direction, so the moving ring 2 can rotate synchronously with the shaft sleeve 4 and the rotating shaft.

On the seal face of the moving ring shown in Figure 1 and Figure 2 (it can also be a static ring, or the seal face of the rotating ring and the static ring at the same time), a streamline groove 11 is provided, which is divided into 8-18 numbers and generally no more than 25 evenly arranged streamline grooves by the same form isolation part 14 (weir area) with the end plane height.

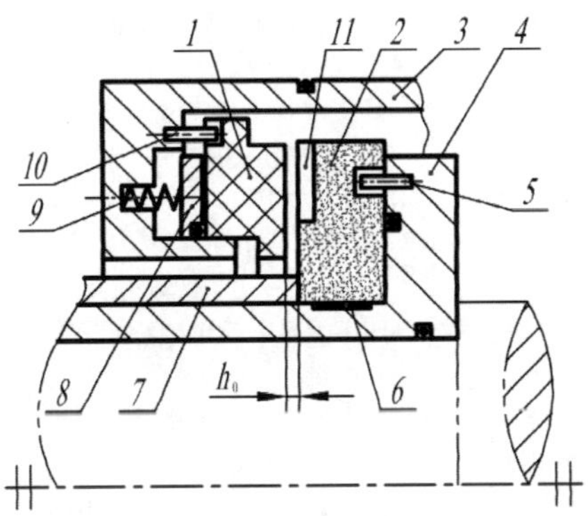

Figure 1. streamline groove seal

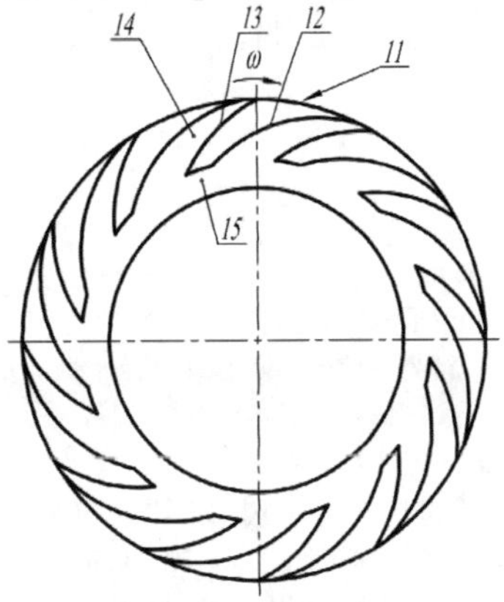

Figure 2. streamline groove face

As shown in Figure 2, the ratio of the radial length of the ring area with groove part on the sealing end face to the radial length of the ring area without groove part is 0.3~5, preferably 0.6~1.5.The ratio range of the arc length of the opening along the circumference of the sealing ring and that of the adja-cent isolation part along the circumference of the sealing ring is 0.2~5, and the preferred ratio range is 0.8~1.2.The depth of the type groove(h_g) can generally be 0.002~10 mm, preferably 0.002~0.030 mm. The number of grooves on the seal end face may be determined according to different working condi-tions and requirements. Under normal circumstances, along the circumference of the sealing end face, there are less than or equal to 25 grooves. In terms of the driven pressure effect and the processing technology, the number of grooves on the seal end face can be optimized as 8~18.The shape of the seal ring is shown in Figure 2.

The clearance of the groove-free area of the two sealing end faces is 3μm, the groove depth is 6μm, the inner diameter is 120 mm, the radial groove length is 12mm, the blocking structure (15) radial length is 10 mm.The above data were taken to calculate and analyze the bearing capacity of the fluid film in logarithmic spiral groove and "T" shaped groove respectively. Firstly, the optimized structure parameters of logarithmic spiral groove and "T" groove are obtained. Then, the designed groove should be consistent with the first two types in terms of the number, depth and ratio of the dam (block structure 15) except for the borderline, to enhance the comparability of the results. The results show that the bearing capacity of the fluid film in logarithmic spiral groove is 620N, that in "T" groove is 560N, and that in streamline groove is 667N. By contrast, the new type of streamline groove mechan-ical seal fluid film has a greater bearing capacity

5. Conclusion

In this paper, TRIZ theory is applied to carry out innovative design of dynamic pressure non-contact mechanical seal. Problems encountered in practical application are analyzed, technical contradictions are found, engineering parameters are determined, and the serial number of invention principle is ob-tained through contradiction matrix table. By analyzing the principle of invention, a new type of streamlined groove mechanical seal is designed Innovatively. Compared with the current sealing ring in the form of logarithmic spiral groove or "T" shape groove, the sealing device with the above design structure seal ring can have more superior fluid dynamic pressure effect. Under the same conditions, the bearing capacity of the fluid film on the end face of the designed streamline groove is increased by 7.6% compared with that of the logarithmic spiral groove, and increased by 19.1% compared with that of the "T" groove, which obviously has better opening force, stiffness and bearing capacity, and also has better dynamic pressure effect of end face fluid.

Acknowledgments

The project was supported by the key R&D projects of Science and Technology Department of Si-chuan Province(2019YFG0351), and the Key Laboratory of Fluid and Power Machinery, Ministry of Education.

References:

[1] Gu Yongquan. Mechanical seal practical technology. Beijing: China Machine Press,2001.

[2] Liu Shuangyu. Analysis of mechanical seal technology and its application. China Plant Engi-neering, 2020(04):175-176.

[3] Zhang Jinfeng, Yuan Shouqi, Cao Wuling. Present research status and prospect on mechanical face seal, Fluid Machinery, 2004(10):26-31.

[4] Wang Yuming, Liu Wei, LIU Ying. Current research and developing trends on non-contacting mechanical seals. Hydraulics Pneumatics & Seals,2011,31(02):29-33.

[5] Wang Yan, YU Xuemei, Lu Long, He Yiqiang. Current research on slotting technology of non-contact mechanical seal. Hydraulics Pneumatics & Seals,2016,36(11):1-6.

[6] Wang Xiaojun. Structural optimization for internal leakage of water-cooling jacket tube for CDQ based on TRIZ theory. Fuel&Chemical Processes,2020,51(03):10-13.

IOP Conf. Series: Materials Science and Engineering **926** (2020) 012028 doi:10.1088/1757-899X/926/1/012028

[7] Yan Huigeng. Introduction to technological innovation methods[M]. Chemical Industry Press,2003.
[8] LIU Guitao, Yuan Chunjing, Dai Yi gui. The application of technology contradiction solve matrix based on TRIZ. Machinery Design & Manufacture,2007(08):63-65.
[9] Peng Huijuan, Cheng Siyuan, LI Suyang, Xiang Mengqun. The Overview of TRIZ's Theoretical Systems. Machinery Design &Manufacture,2013(10):270-272.
[10] Zheng Min. Discussion of Solving Machinery Problem Using TRIZ Theory. Development &Innovation of Machinery&Electrical Products,2015,28(05):22-24.

IOP Conf. Series: Materials Science and Engineering **926** (2020) 012029 doi:10.1088/1757-899X/926/1/012029

Creative Experiment on Decorative Products for Condominium using Responsive Technology

Natcha Sirikhvunchai, Aran Wanichakorn and Ravitep Musikapan

College of Creative Industry, Design Innovation, Srinakharinwirot University, 114, Sukhumvit 23 Khlong Toei Nuea Subdistrict, Wattana District Bangkok, Thailand 10110.

ntchnew@gmail.com

Abstract. Creative experiment on decorative products for condominiums using responsive technology has the purpose of creating decorative products using responsive technology which can improve the interior atmosphere which can be controlled by the user and adjusted according to their needs. The constraint of the design is the interaction between the two emotional states of normal (heart rate of 60-80bpm) and high stress level (heart rate of 80-100bpm). Another constraint of the design is the technological constraint, which can control the lighting, noise level, and temperature according to the users' needs. This research is conducted based on the previous one called *"A Study of Interactive Media to Design and Develop Decorative Product in Condominium"*. The knowledge gained from the previous research about incorporating technology into product design is utilized in this research to be experimented on decorative products with responsive technology. The results of this research can be divided in two parts. The first part is summarizing user behaviors in which the results are obtained from both primary and secondary methods. The results are then used to create product design constraints. The second part is the resulting product design which is aesthetically satisfying, suitably for the physical environment and the three factors which can be controlled by sensors include lighting, noise level and temperature lighting, which is controlled by, noise sensor and humidity sensor respectively. There are three resulting designs which have three levels of intelligence according to different usages of living room, dining room, and bedroom.

1. Introduction

Nowadays, the way of living of people around the world has changed. Whether it is caused by the development of technology, values, or culture that has changed according to the era. The increase and decrease in population of depends on the physical and biological environments that are related to human life. In addition, the world is shifting towards globalization, things are linked together logically, can work together and communicate with one another using the characteristics on internationalization whether humans, environments, buildings, or even various products.

We may notice a general overview of human characteristics and their lifestyles. For example, accommodations are more urbanized with greater population density. As a result, it leads to more shared spaces. It means the area that used to be a detached house was transformed to a residential building in the form of shared building. There is a cohabitation among many households in one single building such as a condominium building or a small residential unit to meet the residents' needs to live in, not a typical house. Today, most families are much smaller than they used to be. Many people would prefer to stay single, live on their own, and tend to raise pet animals like dogs, cats, hamsters and so on. The world has become smaller as a result of Internet of Things and the power of Big Data that can be processed and used anywhere, anytime. Whether it is use of teaching, marketing, various business including, construction, maintenance, architecture, interior design of the building as well. Humans are not the only creatures that can communicate and interact with technology, but also it is architecture, places, including products and appliances as if they were living. It is good for increasing the convenience of users anytime and anywhere. A new living culture will occur in the future in which humans can interact with objects. On the other hand, things can communicate and recognize emotions and human feelings. The researchers are interested in creative experiment on decorative products for condominiums using responsive technology which can be explained further from Diagram shown in Figure 1, research conceptual framework.

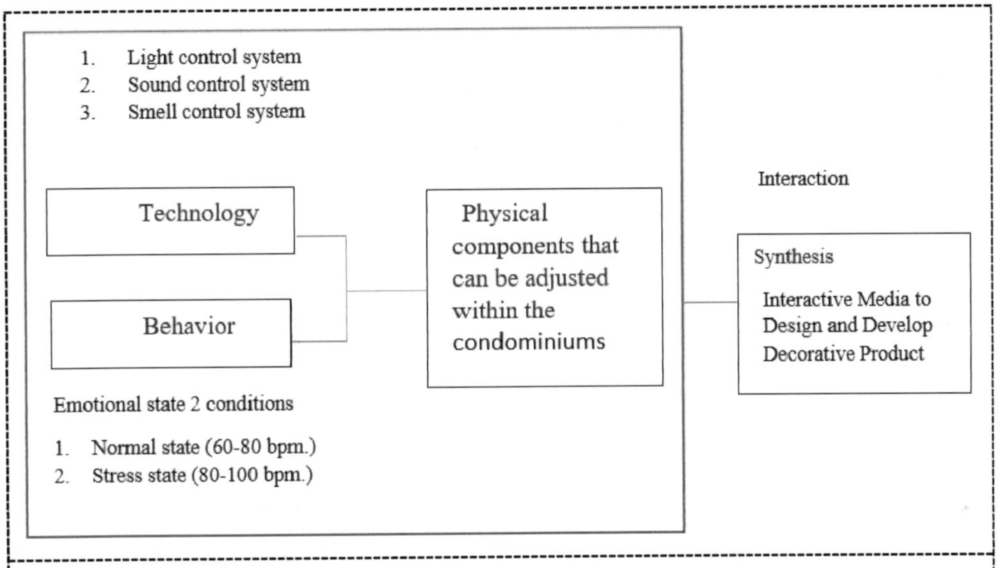

Figure 1. Research conceptual framework.

From the research conceptual framework can be explained the creative experiment on decorative products for condominiums using responsive technology caused by the design under the second part design specifications as follows; 1) physical components that can be adjusted within the condominiums 2) emotional behavior of users. Technology can be divided into 3 systems consisted of; light control system, sound control system, and smell control system which respond to the emotional state of the users consisting of normal and stress states.

2. Methodology

The methods used in this experiment is a Creative Research consisting of 7 steps, which can be explained as follows:

2.1. Analyze Behavior

Analyze behavioral conditions. The researcher studied the emotional expression and response to emotions which separated the emotions from the heart rate [1]. The researchers divided the emotional state into 2 conditions which are normal state at the heart rate of 60-80 bpm and the stress state at the heart rate of 80-100 bpm.

2.2. Analyze Technology

Analyze conditions in the application of technology by studying the physical characteristics that can be adjusted within condominiums which consists of light, sound, and temperature [2] by using 3 types of technology, namely the light control sensor system, sound control sensor system, and humidity control sensor system.

2.3. Education
Education component working plan and the proportion of usage areas with 3 systems technology which are selected for emotional interaction with users, including the light control sensor system, sound control sensor system, and smell control sensor system. The researcher creates a working plan which the response system designs the external appearance of the product.

2.4. Interview
Interview real estate experts. Data collection used both primary data and secondary data. Data were analyzed by create image design concepts.

2.5. Conceptual Design
Present the concept of Mood Board design in 3 concepts consisted of: design ideas with geometric shapes, design ideas with shapes inspired by pets, and design concept with free shape.

2.6. Inquiry and Trial Design
Make an inquiry an expert to choose the best way to create a prototype. Then create trial design and checking the operation of technology.

2.7. Production
Create interactive decoration products, prototypes to be used in other to design applications in the future.

All process can be explained further by a diagram shown in Figure 2.

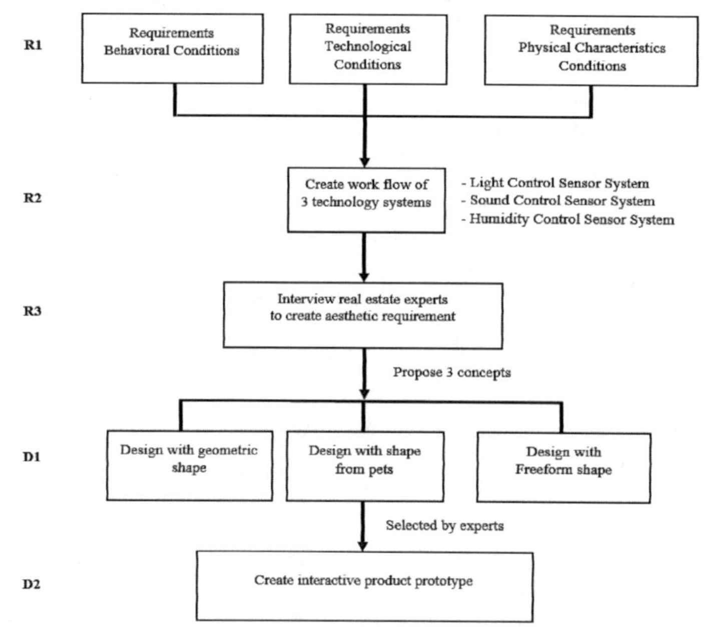

Figure 2. Work Process Diagram.

IOP Conf. Series: Materials Science and Engineering **926** (2020) 012029 doi:10.1088/1757-899X/926/1/012029

3. Results

The research results are divided into 2 parts that is, the first part is gathering knowledge about behavior from primary and secondary data, which leads to the creation of product design specifications. The second part is the design of interaction decoration products that meet the needs of beauty and appropriate for the physical space of condominium.

Section 1: Knowledge used to design specifications

3.1. Study of usage behavior and elements that create the condominiums atmosphere and the physical characteristics of condominium can be explained with Table 1.

Table 1. Shows usage behavior, atmosphere composition, and physical characteristics of condominiums.

Behavioral conditions	Technological conditions	Physical of condominium conditions
Emotional state with heart rate 2 conditions 1. Normal state (60-80 bpm.) 2. Stress state (80-100 bpm.)	3 Technology systems 1. light control sensor system 2.sound control sensor system 3. humidity control sensor system	Divided the minimum usage space of condominium into 3 parts 1.living room 2.dining room 3.bedroom

From the table can be explained that preliminary design specification study results divided into 3 parts as follows; behavioral condition: a study of factors that indicate the mood of the user [1] that is heart rate which can be divided into 2 conditions such as normal state at the heart rate of 60-80 bpm., and stress state at the heart rate of 80-100 bpm. In addition, technological conditions; the indoor environment can be adjusted by users [3] consisted of light, sound and temperature leading to the selection of 3 systems technology, namely light control system, sound control sensor system, and humidity sensor system that used in decoration of the condominium interior consists of 3 parts, which are the living room, dining room and bedroom.

3.2. The study of consumption characteristics, tastes and preferences of real estate consumers. Primary data from the interview with 3 real estate experts and secondary data about the Central Business District real estate project can be explained with table 2.

Table 2. Data from the interview 3 real estate experts / 3 companies [4-6].

Questions	Answers
Suite type	There are studio room, 1 bedroom suite, and 2 bedroom condominium
Exterior design and the interior	No relationship due to the exterior design is the design of the architect but the interior design depends on the residents which have different identity.
Area	25-56 Sqm.

Questions	Answers
Consumer group and favor	Group 1 People aged 40-50 years are high-end consumers, interested in spacious and comfortable condominium with large areas. It meets the needs of long stays and suitable for the elderly. The selection of furniture emphasize individuality preference to be used for a long time. Group 2 People aged 22-39 years whose high incomes of parents, interested in luxury and comfortable residential units. Large areas are not necessary because most of the accommodation is only to sleep at night. For this group should focus on the common area, activity area because they tend to be bored easily therefore they often change furniture and decorations.
Price / square meter	150,000-300,000 Baht / sqm.
Opinions about the application of technology in design and decoration.	Interesting and very much agree. At present, the real estate business in Thailand focus on technology using in terms of system management and building inspection, security and convenient. If technology usage is expanded to decorate the building will be regarded as adding aesthetic benefits to the project and able to motivate consumers to become more interested in buying.

3.3. Design and create working plans of technology in 3 systems to 3 different decorative products by describing the 5 types of features which are the usage area interaction characteristics. Core patterns of technology using and the level of each technology intelligence divided into 3 levels which are lowest, moderate and maximum.

Table 3. Working plan 3 technology styles to 3 decorating products

Various features	Model 1	Model 2	Model 3
Space	Living room	Dining room	Bedroom
Interaction characteristics	Response to the emotions	Strengthen the atmosphere	Respond to the environment
Core technology	1. Heart rate sensor 2. Light control sensor	1. Sound control sensor 2. Bluetooth, MP3 3. Wireless charging system	1. Humidity control sensor 2. Light control sensor
Pattern of use	It is a decoration that responds to user mood through heart rate measurement which divided into 3 color light in 3 phases such as; red at the initial stage, green for normal mood (60-80 bpm.), and blue for stressful mood (80-100 bpm.)	It is a decoration that enhances the living atmosphere. There are2 main functions consisted of wireless mobile charger and speaker. The inside consists of bluetooth and Mp3 allowing users to control music playback.	It is a decoration that responds to the environment which accommodate sleepy atmosphere. When power on, it will humidify and will be a bright lamp automatically. When the light is off at night, it will be an automatic bright lamp with blue light to create a relax atmosphere.
Levels of intelligence	Maximum	Lowest	Moderate

All of Technology Systems inside shown in Figure 3.

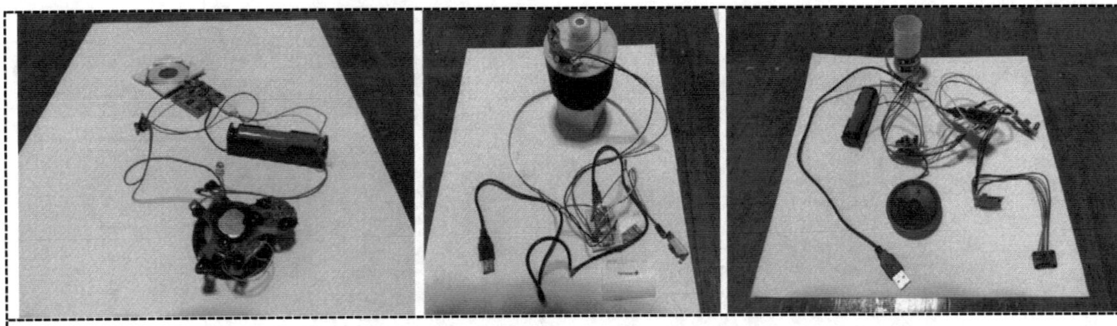

Figure 3. Technology Systems Inside.

Section 2: The design of interactions decoration products with responsive technology for condominium.

This section is an explanation of the exterior design work caused by the extension of use from Table 3. After presenting 3 design concepts, namely geometry shape that has been inspired by pets and free shape. The data from discussing with the experts, it was found that design using shapes that are inspired by pets is the most suitable method for further development as prototypes because consistency and visual which linked to the deep inner feelings of each person. The researcher designed by using the concept of pet into 3 products that was inspired by dogs, cats and fish. The results of the design and interaction experiments shown in Figure 4-8.

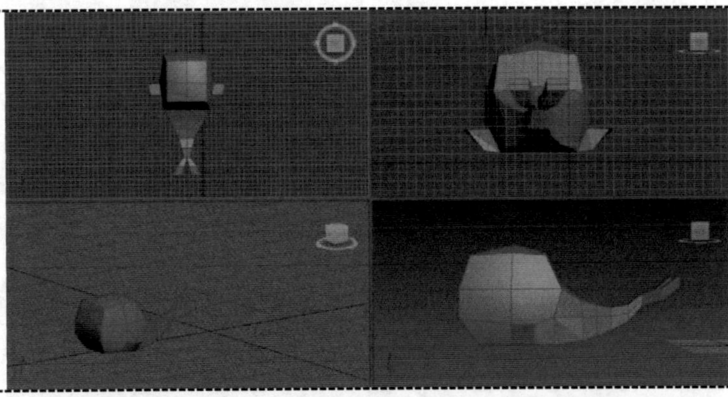

Figure 4. Form Design Model 1 using 3DSMax.

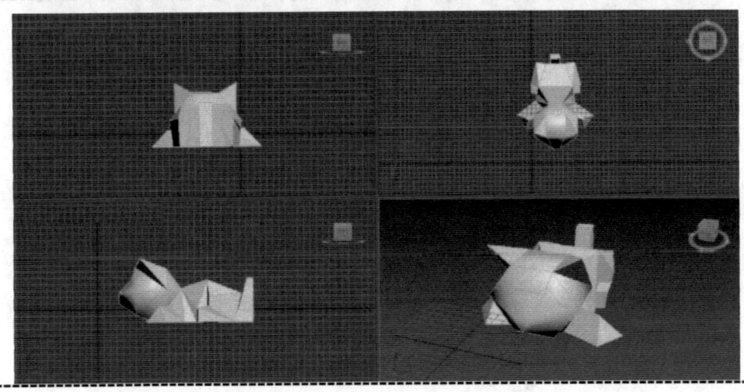

Figure 5. Form Design Model 2 using 3DSMax.

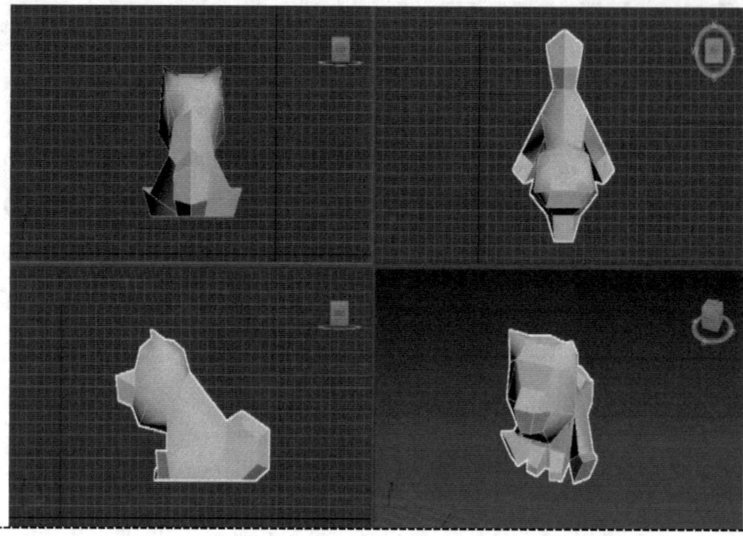

Figure 6. Form Design Model 3 using 3DSMax.

Figure 7. Decoration Product Prototypes.

Figure 8. Perspective Example When Decorate in Condominium.

Acknowledgments
The researchers feel deeply grateful Assist. Prof. Dr. Aran Wanichakorn, Assist. Prof. Dr. Ravithep Musikapan and undergraduate students Product Design Program Faculty of Arts Srinakharinwirot University for cooperation and continuous support. Including real estate experts consisting of Mr. Pichakorn Siriwong, Mr. Yuthachai Phatanasirimongkol and Ms. Sasitha Lorsakul. Thank you for this opportunity.

References

[1] Sopharak, Akara. (2003). Emotional State Classification Using Heart Rate Signal. (Master's Thesis), Graduated School, Mahidol University, Nakorn Prathom.

[2] Sirikhvunchai, Natcha ; Wanichakorn, Aran and Musikapan, Ravitep. (2019). A Study of Interactive Media to Design and Develop Decoration Product in Condominium. Paper presented at the 2019 International Art & Design Symposium, Ewha University, Korea.

[3] Srihirun, Kraweekrai. (10 October 2019) Responsive Technology and Architectural Design/Interviewer: Natcha Sirikhvunchai. Graduated School, Srinakharinwirot University, Bangkok.

[4] Siriwong, Pichakorn. (28 October 2019) Real Estate Development/Interviewer: Natcha Sirikhvunchai. Graduated School, Srinakharinwirot University, Bangkok.

[5] Phattanasirimongkol, Yutthachai. (1 November 2019) Real Estate Development/Interviewer: Natcha Sirikhvunchai. Graduated School, Srinakharinwirot University, Bangkok.

[6] Lorsakul, Sasitha. (5 November 2019) *Real Estate Development/Interviewer: Natcha Sirikhvunchai*. Graduated School, Srinakharinwirot University, Bangkok.

IOP Conf. Series: Materials Science and Engineering **926** (2020) 012030 doi:10.1088/1757-899X/926/1/012030

Exploring the Attractive Attributes and Consumer Preferences for Dream Cake

CHIA-PAO HUNG [1], CHUN-CHIH CHEN[2,*], SHU-MING WU[1] AND PING-LAN HUNG[3]

[1]Tungfang Design University, Kaohsiung, 82941, Taiwan
[2]National Kaohsiung Normal University, Kaohsiung, 80201, Taiwan
[3]Fu Jen Catholic University, New Taipei City, 24205, Taiwan

[*]Corresponding author's e-mail:ccchen@nknu.edu.tw

Abstract. The study is mainly to understand consumer preference toward the charm of cake decoration via in-depth interview. Through the interview of 37 participants, we could understand the awareness of consumers to dream cake decoration and provide a physical and theoretical charm factor analysis via evaluation grid method. From the experimental results, it is known that the major factors constituting the charm of dream cake are style, technique, theme, colour, while the abstract image and physical features and properties constituting these four factors are known as well. From abstract aspect, the dream cakes are considered lovely sweet and fun. As to physical aspect, the dream cakes are featured at brilliant skill, matured technique and exquisite. After subsequent verification, the results may be used as references and guidelines for dream cake decoration in the future.

1. Introduction

Following the progress of world and cultural enhancement, foods are no longer to ally human's hunger but to feel and enjoy the gourmet foods. Lu [1] indicated that the foods and beverages are gradually used to stress the visual, smell, touch, taste and sound such five senses for marketing. That is, the "cultural content", "experiencing process" such entire sense joyfulness of eating are valued more and more. In recent years, people love both traditional bread and big cake in early age and exquisite cake and stylish donut now. The diversified styles, colours and flavours of baking products may stimulate people's sense and further arouse the desire for eating. Among them, the cake is the top choice. Through stylish creativity, technical presentation, the cake may lead the trend and attract consumer to check in and share.

All kinds of stylish cakes were introduced to the market successively. From common fruit or chocolate to cartoon character and fantasy cakes aiming at the wedding or even the "customized" stylish cake according to different needs of customers, they all gradually change the traditional consumption ways of consumers in Taiwan [2]. Tseng [3] pointed out that charm is an important enzyme to promote civilization, social development, energetic life and emotional soul, which is closely related to human's felling and important content and motivation in life. Charm itself has unique feature generating different charm awareness to different groups. To attract specific users, you must touch them to arouse the sympathy [4]. Therefore, the study takes extremely charming edible craft -dream cake as example to discuss the perception and the preference factors to dream cakes. The main purposes of study are as follows:

(1) The perception and evaluation of consumers toward dream cakes.

(2) The study of charm property preferences of consumers to dream cakes.

2. Literature Review

2.1 Dream cake

The cake decoration not only provides bakers a stage showing their professional technique to advance the appearance of cake, but also includes the purposes of adding flavor of cake and extending the preservation period of cake [5]. With fresh cream decoration, simulation via chocolate and sugar such materials, current cakes overturn the traditional fresh cream style and gradually form the trend in Taiwan. The dream name of cake is mainly derived from the details of cake that may highlight the taste. The cakes with fine and delicate embellishment symbolize the hope, happiness and eternity. The hands of current bakers have skills to create the extremely similar cakes like magicians. Meanwhile, the characters played by the cakes are not just simple one. The wedding cakes only have country, jewellery, flower, colour painting styles. The themes of cakes not just mean the happiness, but also the beautiful scenery in wedding. When the wedding cake encounters with beautiful flowers running into a romantic and elegant time, the entire cake may look simple and elegant even there is a flower only. The cakes may please the eyes through different decoration skills or adding other elements like strawberry, matcha, chocolate blending in the pastry, allowing them be more colourful. In addition to increase the business value, different bakers or cake operators have their own exclusive recipe appearing the myriads of flavours [6]. Nevertheless, the main purpose is to get eye-attracting cake style.

2.2 Miryoku engineering and evaluation grid method

"Miryoku engineering" was the study initiated by Japanese scholar Masato Ujigawa altogether with several scholars for the purpose of "the technology and knowledge of creating charming products and space [7]. Lu [8] mentioned that Miryoku engineering is no longer limited to physical product development flow, but the study from the connection between human sense and product to the conversion of people's feeling against industrial experiences. In recent years, there are many researches start using Miryoku engineering to extract industrial experiences or local charm factors. Along with increasing of consumption awareness, the product development trend was turned to consumer-oriented. Consumers not only require for basic functions, the look, comfort and individual demand are all important considerations of consumers as selecting the products. The "evaluation grid method" in Miryoku engineering is an important study approach. It is an approach originated from psychology to catch personal knowledge and summarize the ideas. It summarizes individual characters of target subject after discussing the similarity or differences among the subjects mainly by using personal interview. By applying evaluation grid method the reasons why consumers or users would be attracted are discussed via in-depth interview to further lead the participants to convert their ideas to more specifically to physical reasons and abstract senses and then summarize the true thinking of participants. The study expects to catch the charm of cake from participants through preceding features of evaluation grid method under Miryoku engineering abovementioned to facilitate the analysis.

2.3 Food design

In general, it is hard to imagine that mode or tools design and production may be a part of food design. In fact, not just eating is related with food design, all the patter or appearance of food mode produced for purpose or request, e.g. the egg cakes loved by both adults and children with various modes in shape of rooster, gun, Patrick Star, flower or tools allowing smoother eating or cooking process are all included in "Design for Food", i.e. the more familiar "Product Design" [9]. The food based design

IOP Conf. Series: Materials Science and Engineering **926** (2020) 012030 doi:10.1088/1757-899X/926/1/012030

elements and materials to change the texture, appearance, colour or flavour for the purpose of passing a thought, message or adding five-sense experiences are
categorized as Food Design.

3. Research method

3.1 Subjects of in-depth interview

The study has invited 37 participants aged 20-35 to interview their perception and reasons for favourable dream cake decoration. It is to understand their perception to dream cakes and conclude the charm factor evaluation grid drawing through in-depth interview.

3.2 Work selection

The study collects 50 pictures of dream cakes produced by famous baking fondant classroom teachers. All teachers have cake related certificates and led students for competition abroad and served as judges. The works are exquisite and ingenious examples with high standards. 5 other baking professional teachers and 5 consumers select final 15 samples through discussion of focus group as shown in table1.

Table 1. Dream cake pictures

No.1 Angel's secret angel place	No.2. European image	No.3. Purple magic house	No.4. Run wild	No.5. Pinky party

No.6. Alice in wonderland	No.7. Fantasy carousel	No.8. Baby only	No.9. Splendid years	No.10. Journal time

No.11. Magic time	No.12. Sweet date	No.13. Affection	No.14. Greenhouse lady	No.15. Sweet talk
				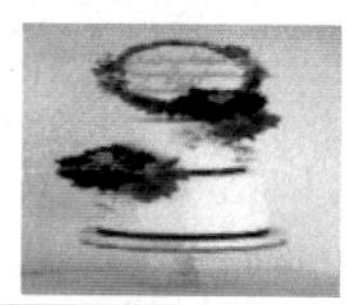

3.3 Interview steps

The study asks interviewees to select 5 pictures representing dream cake most. Meanwhile, the interviewees are requested to explain the reason for selecting such pictures such as technique, creativity, charm etc. It is to understand the perception and evaluation of consumers toward dream cakes via the interview, and conduct the study of charm preference at the same time. The interview questions are as shown in table 2.

4. Research results

4.1 Dream cake preference research

The research aiming at the consumer preferences to dream cake, The ranking of dream cakes is shown as table 3.

Table 2. Interview Questions

No.	Question
1	Pick the top five which you feel most dreamlike among the 15 pictures (write down the No.)
2	Why would you like these five cakes? (is it look adorable? attractive? vivid color? or…)
3	How do you feel about the technique (skill) of these five cakes? (exquisite? absolutely lifelike? or…)
4	What do you think about the creativity of these five cakes? (Similarity? have story? or….)
5	What is the event that you think is most appropriate to use these five cakes? Why? (E.g.: wedding, because it is symbol of romance and dream or formal event because the fine technique makes the cake like the art). Please tell your own opinions

Table 3. Ranking of Dream Cakes

Rank	No.	Picture	Votes	%
1	No.7. Fantasy carousel		28	15%
2	No.11. Magic time		24	13%
3	No.6. Alice in wonderland		19	10%
4	No.9. Splendid years		15	8%
5	No.8. Baby only	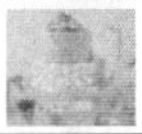	13	7%
6	No.10. Journal time	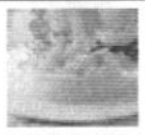	13	7%
7	No.2. European image		11	6%
8	No.15. Sweet talk	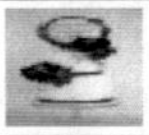	11	6%
9	No.3. Purple magic house		10	6%
10	No.1 Angel's secret angel place		9	5%
11	No.4. Run wild		8	5%
12	No.5. Pinky party	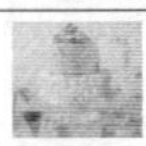	8	4%
13	No.4. Greenhouse lady		8	4%
14	12. Sweet date		6	3%
15	13. Affection		2	1%

The score of No. 7 fantasy carousel is the highest, accounting for 15%, followed by No. 11 magic time 13%, and third place No. 6 Alice in wonderland 10%, fourth place No. 9 splendid years 8%, fifth place No. 8 baby only 7%.According to analysis of ranking, the reasons of some participants for picking top (1) Theme: the colour combination of cake is adorable and sweet, and very skilful while

IOP Conf. Series: Materials Science and Engineering **926** (2020) 012030 doi:10.1088/1757-899X/926/1/012030

the butter decoration technique is super, very creative and full of story. The creation aiming at the theme will be very beautiful, graceful in style and show due respect in the table 3. (2)Colour: colours are vivid and riot, bringing happiness feeling to people! It is fine, exquisite and romantic. (3)Technique: I think the style and colours of this cake is extremely skilful. The technique is covered in the craftsmanship is stunning and breaks general thinking of cake. Its unique creativity may highlight the distinctive taste and style of host in private event. (4)Style: the pink cake with flower for decoration is very beautiful in layers. It is very delicate and romantic, making people falling in love. It is very suitable for the wedding because who can resist the fine and romantic style! It is known from the explanation of interviewees that they value the aesthetics prevented by the theme, colour, technique and style. Similarly, if the colour of cake is not vivid enough, the thematic expression is not clear and less layering or the technique is common in term of exquisite and fineness, the preferences of interviewees would be lower.

4.2 Evaluation grid method

The summary of interview content is concluded in the charm factors of dream cake for the evaluation grid method map, shown as figure 1. According to the interview results, the charm factors of dream cakes are mainly divided into style (appearance), technique, and theme and colour such four factors.

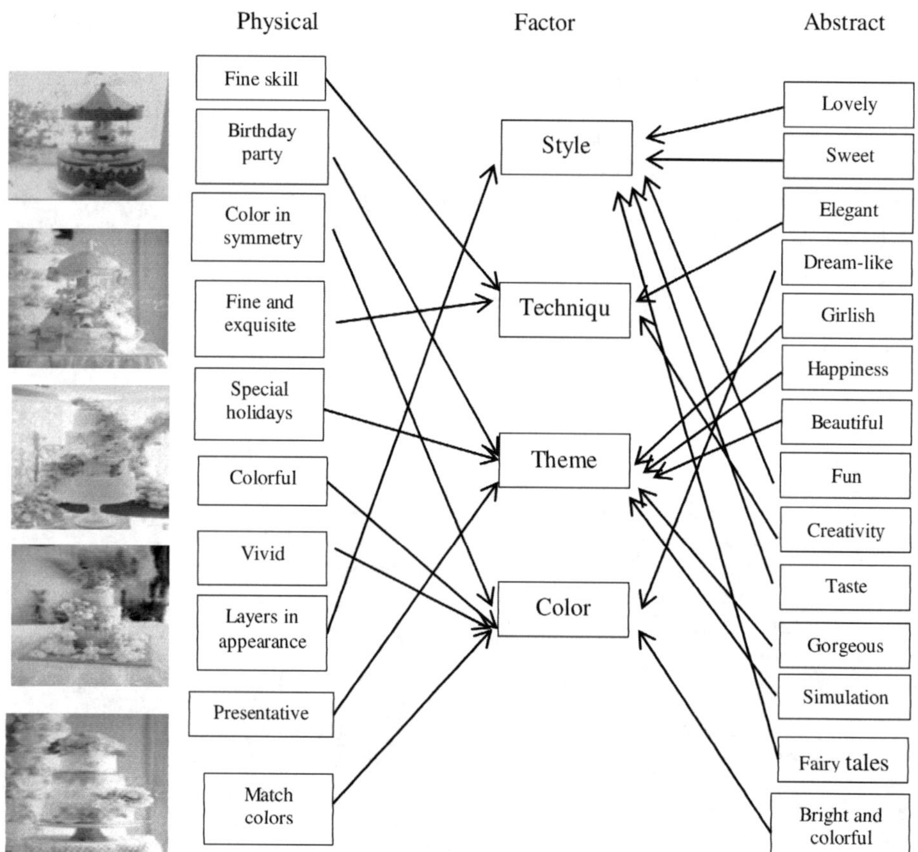

Figure 1. Evaluation grid method map of charm factors for dream cake

Through the description of interviewees for the charm of dream cake, the charm of cake could be from physical factors such as – colour allocation, layering style, fine skill, special holidays like birthday party, wedding, anniversary activities etc. or from abstract factors such as – existence like in fairy tale, gorgeous, elegant, beautiful, similarity, dreamlike and happiness feeling etc. It could be seen that charm has inside and outside factors. Just as Tseng [3] mentioned that charm is an important is enzyme to promote civilization, social development, energetic life and emotional soul, which is closely related to human's felling and important content and motivation in life. That is, charm is a kind of attraction and effect. Meanwhile, with proof of evaluation grid method, the top five dream cakes have features of charm factors above-mentioned, e.g. dreamlike, vivid colour, beautiful, romantic, clear layers such factors. Relatively, the cakes in last five places are relatively lack of such characters and therefore are fewer selected. The study may determine what is the request of interviewees for the presentation and appeal of cakes.

5. Conclusion

According to preceding results, the map of evaluation grid method is established accordingly and four original charm factors style, theme, technique, colour, 10 physical charm factors and 15 abstract charm factors are concluded. In addition, it is found from the arrangement of interviews, among the original charm repeatedly mentioned by different interviewees, the style, theme, technique and colour are mostly identified by the interviewees. In respect of physical charm factors, the most frequently mentioned factors are vivid colour, followed by colourful, skilful, exquisite and fine, representative; the most frequently mentioned abstract factor by the interviewees is romantic, followed by dreamlike, story, fairy tale, happiness and beauty. To satisfy the market and customer demands, identify the emotional requirement of consumers and provide services more satisfactory to economic effects and sensible by customers, it is hoped to increase the number of interviewees, use integrated charm evaluation grid of cake decoration as basis for subsequent questionnaire design, use extensive samples to verify the charm factors of cake decoration to confirm the reliability and validity of charm factor. Meanwhile, find the definite charm and identify the differences of customer demands through more complete research results as basis of references for dream cake decoration design of businesses in the future.

References

[1] Lu Ying-Ti 2005 *The Influences of Colours on Appetite Arousal—The Example of Cake Tray* master's thesis Graduate School of Visual Design National Yunlin University of Science and Technology (Yunlin)

[2] Yu Pei-Shan 2016 *The study of stylish cake business model in Taiwanese market, taking stylish cake workshop as example* master's thesis Graduate School of Food and Beverage Management University of Kang Ning (Tainan)

[3] Tseng Li-Dan 2007 The evaluation of Taiwanese local cultural and creative industry –taking Yilan International Children's Folklore and Folkgame Festival", 12th seminar essay collection of Chinese Institute of Design – pioneer design; the integration of technology and culture No. 268

[4] Lin Feng-Han 2001 *Charming quality and product design*, master's thesis Graduate School of Industrial Engineering National Tsing Hua University (Hsinchu)

[5] Hsu Hua-Chiang, Huang Dern-Shun and Gu De-Tsai 1979 *Cake and pastry* New Taipei City: China Grain Product Resaerch and Development Institute

[6] Mao Show-Zi 2012 *The study of motivation and demand of matured generation to cakes*, master's thesis Department of Applied Science of Living Chinese Cultural Univeresity (Taipei)

[7] Ma Min-Yuan 2010 Briefly talk about KANSEI technology of new product development in Japan Industrial Materials Vol.280 pp.3

[8] Lu Zong-Zer 2010 *The study of charm elements and pleasant feeling sustainability of pleasant products* master's thesis Graduate School of Industrial Design National Cheng Kung University (Tainan)

[9] Zhan Bo-An 2017 Six major fields of food design! Let eating being fun https://vocus.cc/food-design/5a121864eceaed97b402617a/

ATDMAE 2020 IOP Publishing
IOP Conf. Series: Materials Science and Engineering **926** (2020) 012031 doi:10.1088/1757-899X/926/1/012031

A Study on the Deep Learning based Prediction of Production Demand by using LSTM under the State of Data Sparsity

Ho-Seok Jung[1] and Sungbum Park[1,*]

[1] Graduate School of MoT, Hoseo University, 20, Hoeo-ro 79 beon-gil, Baebang-eup, Asan-si, Chungcheongnam-do, Republic of Korea 31499

[*]Corresponding author's e-mail: parksb@hoseo.edu

Abstract. Rapid market condition changes and growing uncertainties require accurate product demand forecasting to reduce the risks of business operations. So far, recurrent neural networks (RNNs) have been introduced as the representative models for forecasting time series data. However, it showed limited performance when learning time pattern occurring over a long period of time. This study performs deep learning-based production forecasting analysis with Long-term short-term memory (LSTM) and dropout algorithm as the alternative for the conventional time series analysis methods and the exponential smoothing method. Our result provides the reference for sales target setting, facility investment and production planning, inventory control, supply chain management, and marketing strategy establishment.

1. Introduction

Accurate product demand forecasting enables companies to strategically implement the key elements of overall corporate management, such as supply chain management (SCM), facility investment planning, and maintenance of appropriate inventory levels [1].

Recently, recurrent neural networks (RNNs) have been introduced as models for forecasting time series data such as written language and speech sound, and product demand. However, conventional RNN model showed limited performance when learning the specific patterns in time series data occurring over a long period of time. To overcome this problem, long short-term memory (LSTM) has been proposed. The concept of memory introduced by LSTM enables patterns recognition in time series data over a long period, which has led to improvement in the RNN performance.

However, prediction using LSTM based deep learning model has hardly been successful so far, either. This is due to the fact that time-series data for event forecasting of interest generally has a sparsity problem. For example, for data occurring once a year, there exist only five samples for five years. In addition, the demand patterns five years ago are highly likely to differ from the present demand patterns. So, the representativeness of the sample also decreases.

Meanwhile, various learning methods have been proposed along with LSTM in order to increase the forecasting ability. Among the algorithms, the most commonly used one is dropout. The dropout rules out some nodes in a deep network with the starting probability (for example, 0.6) during the learning process. Dropped-out nodes are randomly selected each time. The output value for the current connection strength (weight) is calculated with dropped-out nodes. Once completing learning, the output value for new data is calculated by multiplying the output values of all nodes by 1 – previous probability without omitting the nodes. This approach improves the stability and accuracy of learning [2].

In this idea, this study attempts to apply the LSTM with dropout method with the aim of verifying its applicability on the demand forecasting. The result of this study shows how dropout actually works and how corporations utilize the experimental results through demonstration.

2. Theoretical review

2.1. Changes in the methods of time series analysis

Deep neural networks (DNNs) use more intermediate layers than conventional neural networks to improve the capacity to represent data. Currently, representative models of deep learning include convolutional neural network (CNN) models and recurrent neural network models, and recurrent neural networks are known to be suitable for learning and reasoning for time series data.

A recurrent neural network (RNN) is a neural network structure used for modelling sequence data. The basic idea underlying the RNN model is that since each data after the present data in the sequence provides new information, the information is used to update the present state of the model, When a sentence of a text is read, the data of the present state is updated with each new word, and this state is dependent on the previous words as well as the newly introduced word.

RNN models are based on the concept of a chain structure, and there are various types of RNN models depending on how data are maintained and updated. As the word 'recurrent' in the term suggests, RNN is composed of a kind of 'loops.'

Although RNN processes the current input data in connection with the previous input data, it has a limit in remembering the input data that was processed very long time ago. According to Bengio et al. [1994], when learning for long sequences of input data is performed using BPTT (Backpropagation Through Time) in a RNN model, the vanishing gradient problem or the exploding gradient problem may occur. These problems occur because the gradient value vanishes or increases exponentially because of the recurrent weights that are repeatedly multiplied during the process of learning of even the data of the distant past.

The methods to solve the vanishing gradient problem, extended RNN models such as Long Short-Term Memory (LSTM) or gated recurrent units (GRU) have been proposed as fundamental solutions. LSTM was first introduced in 1997 and is now one of the most widely used models in natural language processing. GRU, which was proposed in 2014, is a simplified version of LSTM. Both models were introduced to solve the vanishing gradient problem and are known to handle long sequence data effectively.

2.2. Long Short-Term Memory (LSTM)

Gradient descent, which is a learning method of RNN, uses the change of gradient. One strength of RNN is that it connects the present task to previous information. However, RNN has a limitation in considering long-term dependency of data due to the vanishing gradient problem, which refers to the loss of major information as it goes through several time steps when processing long sequences in the learning process. Therefore, studies on LSTM have been conducted to solve the long-term dependency problem of RNN.

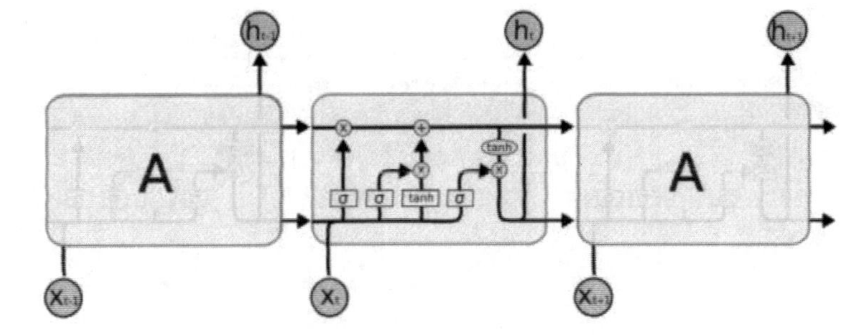

Figure 1. LSTM Architecture [3]

LSTM is composed of four layers with a recurrent structure as shown in Figure. 1. The core of LSTM is a continuous cell state coming through the gate. The continuous cell state is also called the conveyor belt. The information coming through the conveyor belt is transmitted without any changes. LSTM can add or delete information through the input gate, forget gate, and output gate. In short, the gates perform the role of selectively transmitting data and thereby remove previous data to continue learning. LSTM performs calculation using gate vectors.

ft is the forget gate vector and serves as the weight to remember the previous cell state, and it is the input gate vector and serves as the weight to acquire new information. On the other hand, ot is the output gate vector, and performs the role of selecting output candidates. x t is the input vector and, ht is the output vector. ct is the cell state vector, W and U are parameter matrices, and b is a vector, respectively. ft, it, ott are gate vectors. Two types of activation functions are used in LSTM: σ signifies a sigmoid function, and tanh represents the hyperbolic tangent function [4].

2.3. Recurrent neural network and dropout

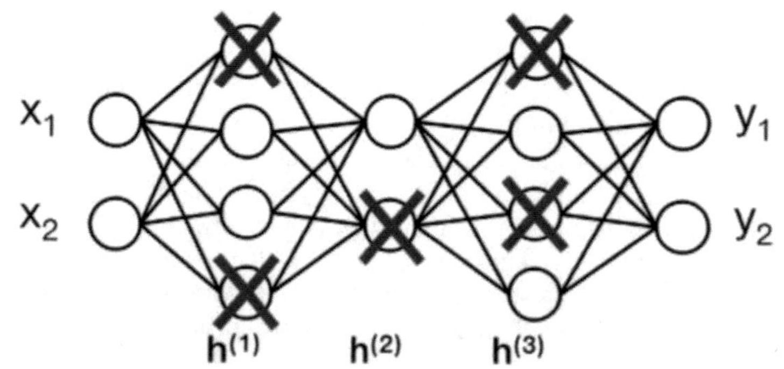

Figure 2. Dropout application in DNN model [2]

Among various algorithms that enhance the learning capabilities of deep networks, dropout is the algorithm that has received the most attention and is currently commonly used.

In the initial version of dropout proposed for RNN and LSTM, recurrent connections are maintained as they are and dropout is applied only to non-recurrent connections (Figure 2 and Figure 3). This method is to omit some part of connection lines instead of dropping out nodes as in the conventional dropout. As a result, this technique led to improved performance in tasks such as handwriting recognition, but showed poorer performance in phoneme recognition, compared to other normalization techniques such as weight noise injection [2].

Dropout for RNN and LSTM, that is, rnnDrop distinguishes it from existing dropout algorithms for RNNs is to apply dropout not to connections but to nodes themselves. RnnDrop is also different from existing dropout methods for RNNs in that rnnDrop does not generate a new dropout mask for each time frame, but a dropout mask is generated once for the entire sequence and the same mask is applied to all time frames [2].

This method is more natural than conventional dropout methods for RNNs in that one dropout mask is used for one sequence, just as one random dropout mask is used for one image. Moreover, in order to train an RNN model, the recurrent layer is generally unfolded to apply backpropagation (backpropagation through time or BPTT), as shown in Figure 4, and at this time, unlike other dropout methods, rnnDrop allows the same neural network connection structure to be maintained in the unfolded neural network at each time step even after its application.

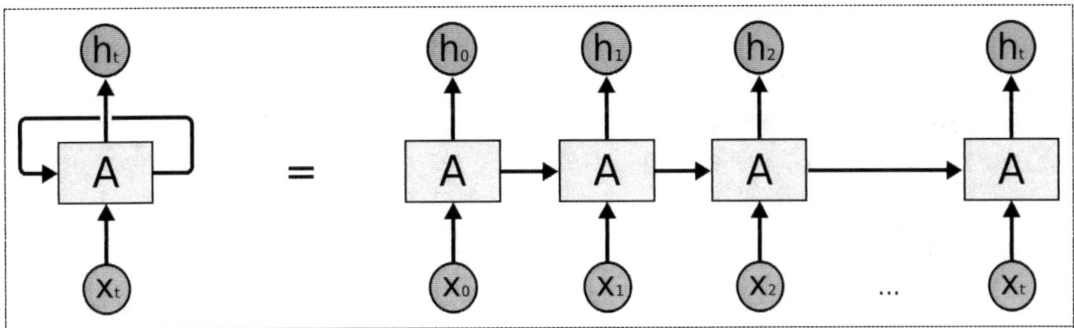

Figure 3. Unfolding in Recurrent layer [3]

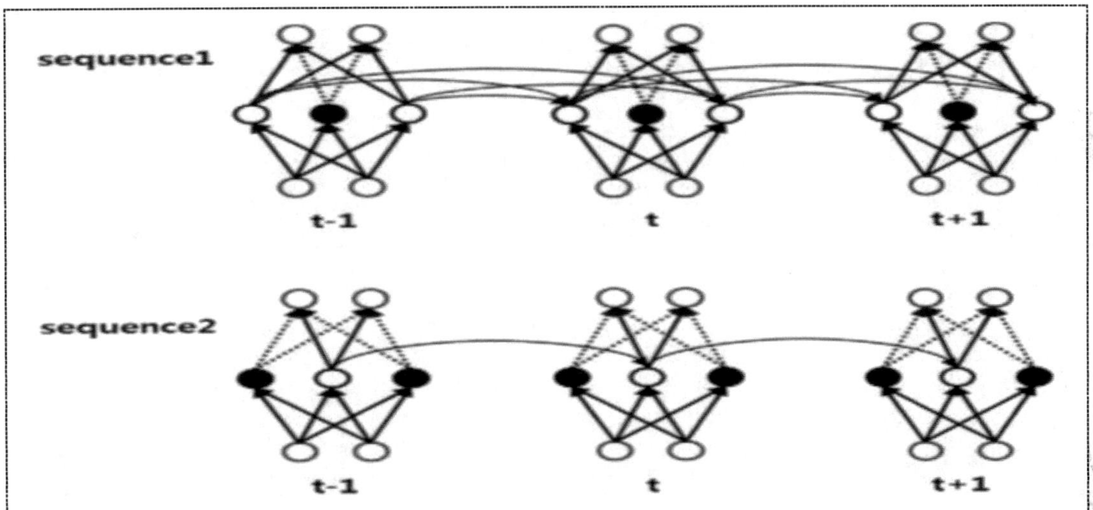

Figure 4. Dropout for Recurrent layer [1]

3. Methodology and results

3.1. Date collection and preprocessing

Corporation H makes available the data of production plans, the previous day's production performance, and the inventory quantity to its partners through its website. The production plan refers to the planned production quantity that Corporation H plans to produce on the retrieval date and the next day (the company shares the hour-by-hour updates of the production plan, but this study reflected the total daily quantity). The production performance of the previous day represents the amount of materials used for production of Corporation H on the previous day. The inventory quantity is the quantity of materials that Corporation H has in its possession. The production plan, inventory quantity, and the previous day's production performance distributed by Corporation H on the retrieval date were collected from January to June 2018. In addition, the shipment quantity sent from D company to the customer company, which is the target variable in this study, was collected from January to June. The data was sorted and arranged by product category to be analyzed as shown in Table 1.

Table 1. Research variables and statistics

Variable	Category	Description	Min	Mean	Max	Kurtosis	Skewness
Month	numeric	Month of occurrence	1	3.01	5	-1.31	-0.02
DAY	numeric	Date of occurrence	1	15.62	31	-1.19	0.01
ENGINE	numeric	Major category	0	4.08	8	-1.48	0.01
Product	numeric	Specific item	0	25.50	51	-1.20	0.00
Demand	numeric	Quantity demanded by Corporation H	0	261.05	4,792	11.07	3.35
Act	numeric	Quantity supplied by D company to Corporation H, considering Demand, BeAct, stock, etc.	0	276.23	4,536	9.88	3.11
BeAct	numeric	Stock of Corporation H Before Act	-6	268.37	4,356	9.82	3.08
Stock	numeric	Stock of Corporation H	-558	151.01	3,808	22.30	4.16
SDemand	numeric	Cumulative Demand	0	20,708.65	362,166	13.03	3.39
SAct	numeric	Cumulative Act	0	19,610.98	358,312	14.71	3.49
SBeAct	numeric	Sdemand – Sact	-132,246	1,097.67	72,836	32.63	-0.65

Assuming that D company is notified by Corporation H today that the next day's Demand is 170, Act represents D company's performance of delivering parts to Corporation H's warehouse in consideration of BeAct and stock as well as the Demand. With respect to Act, BeAct and stock which occurred because D company did not produce products according to Demand even though Corporation H distributed the information of Demand, were also taken into consideration.

3.2. Data analysis

In general, Corporation H has independent factories for each type of engine. A production line of each factory manufactures several products of the same engine product category. Since products in the same product category are expected to have a strong impact on each other in terms of demand, this research set product category as the unit of analysis. Accordingly, during data pre-processing, data was divided according to the product category, and demand forecasting was performed separately for each product category.

The input layer was composed of Demand, BeAct, stock, SDemand, SAct, SBeAct, and Act, and the hidden layer was set as a single layer considering that events were sparse and there was small data available for use. DNN models were trained by randomly sampling 80% of the response-to-demand data, and the experiments were repeated 52 times to derive verification statistics by applying trained models to the tests of the remaining 20% of data [5].

Table 2. Hyper parameters for LSTM model

Option	Setting
Input Dimension	6
Output Dimension	1
Sequence Length	151
Hidden layer	1
Loss function	Mean squared errors (MSE)
Optimizer	AdamOptimizer
Activation function	Rectified linear unit (ReLU)
Dropout ratio	0.0

Figure 5 shows the flowchart of the LSTM model for time series forecasting.

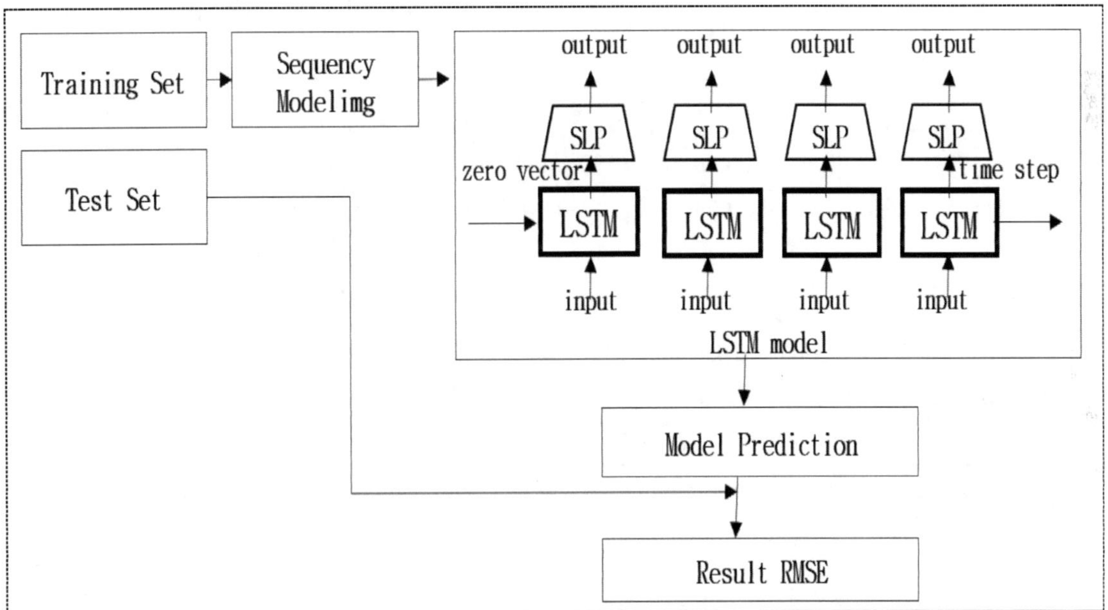

Figure 5. Flow chart of the proposed prediction LSTM model

The Google TensorFlow library was used and the DNN model was set through the optimization process. The LSTM model was trained by randomly sampling 80% of the 153 matchup datasets, and the trained model was applied to the remaining 20% of the matchup data to calculate the amount of evapotranspiration, and the obtained value was compared with the ground observation data. These steps were considered to comprise a single experimental procedure, and this experiment was repeated 200 times to derive verification statistics [5].

As shown in Table 2, Adam (Adaptive Moment Estimation) was used as the optimizer to minimize the loss function and improve the learning speed. Also, ReLU (Rectified Linear Unit) was used as the activation function that solves the vanishing gradient problem of the loss function that can occur in the backpropagation process (Agarap, 2018). In this study, the DNN model was built by optimizing the weights and bias set using the backpropagation algorithm, and then, dropout was performed again to improve optimization [2] (Choi & Min, 2015). Table 3 shows the result. In the situation where data sparsity exists, it can be seen that the prediction accuracy of the model improves as the dropout is not applied to the LSTM model.

Table 3. LSTM analysis result according to various dropout ratio

	Dropout ratio					
Prod No	0	0.1	0.2	0.3	0.4	0.5
1	0.759	0.474	0.308	0.109	0.000	0.000
3	0.553	0.527	0.521	0.479	0.439	0.445
4	0.511	0.502	0.482	0.470	0.483	0.465
5	0.942	0.600	0.259	0.152	0.000	0.000
6	0.655	0.619	0.551	0.513	0.459	0.354
7	0.614	0.590	0.554	0.551	0.494	0.443
8	0.482	0.485	0.477	0.478	0.451	0.459
9	0.282	0.278	0.285	0.288	0.278	0.300
10	0.432	0.430	0.432	0.395	0.383	0.375
...	...	...	...	...	...	...
49	0.532	0.277	0.000	0.000	0.000	0.000
50	0.890	0.409	0.080	0.000	0.000	0.000
51	0.906	0.467	0.244	0.000	0.000	0.000
52	0.627	0.541	0.501	0.547	0.429	0.368
Average	0.668	0.452	0.314	0.240	0.182	0.155

Figure 6 shows the accuracy when the LSTM without dropout model was derived by randomly selecting 39 data from a total of 52 product time series data, and the test was performed using the remaining 13 product data. The LSTM without dropout model shows an accuracy of 66.8%.

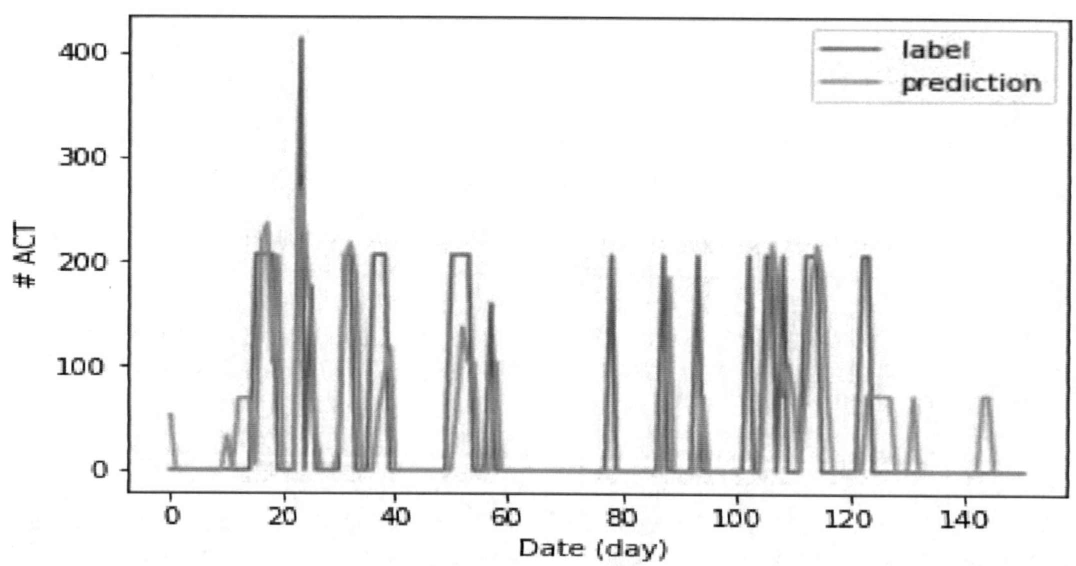

Figure 6. LSTM model without dropout

4. Conclusion

The results of the study indicate that LSTM, which connect the present task to the previous data using neural network methods to predict the results, is a suitable technique for demand forecasting than simple exponential smoothing. Also, dropout was applied to prevent complex co-adaptation of training data and reduce overfitting in neural networks. As shown in Table 3 and Figure 6, no dropout leads to more accurate forecasting compared to the conventional LSTM method.

Dropout has been shown to be useful in the case of relatively large deep network models with a lot of training data. However, it is less effective in the case of models with sparse data. Therefore, there is a need to conduct research on the effectiveness of dropout algorithms under conditions of data sparsity.

Acknowledgments

This research was supported by the MSIT (Ministry of Science and ICT), Korea, under the ITRC (Information Technology Research Center) support program (IITP-2020-2018-0-01417) supervised by the IITP (Institute for Information & communications Technology Promotion).

References

[1] Sunyoung Cho. 2010 Time Series Prediction Using Recurrent Neural Network. Chungju University p 7-11

[2] Heeyeon Choi, Yunhong Min. 2015 Understanding Dropout Algorithms. Communications of the Korean Institute of Information Scientists and Engineers, 33(8), 32-38. p 1,4

[3] http://colah.github.io/posts/2015-08-Understanding-LSTMs/

[4] Samkeun Kim, Takil Oh. 2018 Real-time PM10 Concentration Prediction LSTM Model based on IoT Streaming Sensor data, Journal of the Korea Academia-Industrial cooperation Society, 19(11), p 310-318

[5] Sujin Lee. 2018 Deep Learning-based Estimation and Mapping of Evapotranspiration in Cropland using Local Weather Prediction Model and Satellite Data. The Korean Cartographic Association, 18(3), p 105-116.

IOP Conf. Series: Materials Science and Engineering **926** (2020) 012032 doi:10.1088/1757-899X/926/1/012032

Develop Creative Brick Kiln Cultural Products by Powder Metallurgy Process Technique

Yann-Long Lee[1], Yu-Ching Hung[2], Wen-Chih Chou[1], Chuan-Po Wang[1] and Tzu-Hsua Peng[1]

[1]Dept. of Industrial Design, Chaoyang University of Technology,168, Jifeng E. Rd., Wufeng District, Taichung, 41349 Taiwan
[2]Dept. of Faculty of Mechanical and Manufacturing Engineering Technology, Universiti Teknikal Malaysia Melaka Jalan Hang Tuah Jaya, 76100 Durian Tunggal, Melaka, Malaysia

yannlee@cyut.edu.tw

Abstract. Brick is one of the most common traditional building materials in the world. Located at the foot of the Bagua Mountain in Changhua, Taiwan, Qiaotou community residents make good use of mountain clay to develop the brick kiln industry, and once created a glorious history. With the increase in building materials regulations, the chimneys of brick kilns no longer smoke, and the industry has gradually declined. This study hopes that the traditional brick kiln industry will be improved through technological technique and input a new generation of innovative thinking. The general brick kiln cultural products will be re-granted to new life from the decline and showing the creative thinking in the local culture. The materials test method was used to study the processing technology of brick powder. It is expected to improve the precision of brick products and reduce the production cost through the application of the powder metallurgy process technique. And applied to the brick creative culture product design to create unique merchandise.

1. Introduction

1.1. The history of brick kiln in Taiwan

During the Japanese occupation period, Taiwan's local brick production was earthen and less quality. Therefore, the mechanized brick kiln technology was introduced from Japan. Specially made bricks for the Japanese to instead Taiwanese local bricks. During the Showa era, there were as many as 80 folk brick kiln industries in Kaohsiung at that time, producing bricks used by the public (Chang, 2012)[1] .

The Qiaotou community is located under the Bagua Mountain Range[2] [3] .It is rich in natural earth and stone, especially the clay layer is the best material for making bricks. So in the 1960s, it began to develop into an industrial brick kiln area. At most, there were more than 20 brick kiln factories. The production of red bricks is well known throughout the Taiwan province, creating a regional economic development miracle for more than 30 years. However, in recent years, due to the advancement of building materials, it has been replaced by concrete, which is more fast, strong and durable.

IOP Conf. Series: Materials Science and Engineering **926** (2020) 012032 doi:10.1088/1757-899X/926/1/012032

After the red brick market was sluggish, the brick kiln in the community was transformed into a cultural teaching style in 1999 and create a special small brick and cultural product design. And actively revived the lost brick carving skill, to become a local characteristic craft. Community residents and brick kiln factory integrate brick carvings into their lives and are everywhere in the community. Step by step to create the "Red Brick Art Village" which combines with the history, crafts, and culture.

1.2. Commercial product analysis of brick culture products

At present, the collection and analysis of brick's culture products are shown in Table 1. Most of them are mainly formed by compression molding, and the price is between NTD. 500-2000.

Table 1. Commercial product of brick's culture products (http://www.sanhetk.com.tw/product_list.asp)

Brick culture products	Function and description	Brick culture products	Function and description
	Fireplace: Small brick for decoration. Promote red bricks in DIY activity, so that users can build their own fireplace.		Stove: Burning stove. Hand-made, although its unique, but the stitching is not easy, the price is relatively unfriendly.
	Little doll: Decorations. Although it has a hand-made texture, but it is not refined enough and the price is expensive.	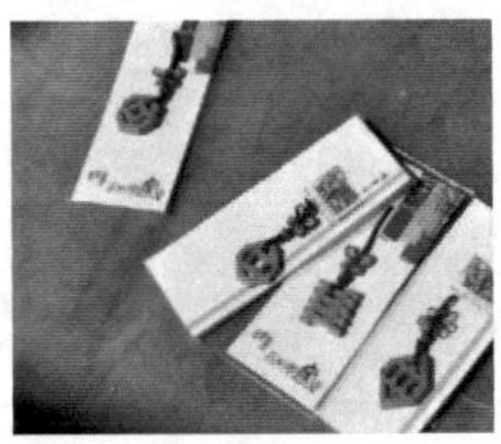	Window grille: Necklace or charm. Designed with a window grille as a charm. Simple and light, but the quality of the production is rough.

In terms of commercial products, the study found that the application of other materials has developed rapidly and has opened the market (Lu, 2001; Lu et al., 2013) [4] [5] . Such as cement cultural products, The 22 Design Studio from Taiwan [6] , including cement ring, cement clock, cement stationary group and so on. The popular watch "The 4th dimension" is the focus of attention.

These cements have successfully developed a variety of creative cultural products, and have been recognized by consumers (Lee, 2018; Hsiao, 2018)[7] [8] . The same building materials of the red brick should also catch up with the market trend as soon as possible.

1.3. The possibility of powder metallurgy molding technology

This study found that brick molding and powder metallurgy molding have many similarities, as shown in the following table 2:

Table 2. Powder metallurgy forming and brick forming process analysis

Step	Powder metallurgy molding	Brick molding
1	Metal powder production	Take soil, Sieve soil.
2	Stirring: mixing of metal components	Pug mill: mix of brick powder and other ingredients.
3	Compression molding (mould without draft)	Extrusion (mould without draft)
4	Sintering (Temperature: 1200-1300°C)	Into the kiln (Sintering temperature: 900-1100 °C)
5	Secondary processing: vibration grinding, surface treatment, heat treatment, machining.	Surface removal of sharp angle.
6	Finished processing--- Quality inspection, cleaning, finishing.	Finished processing - graded and stacking.

For the brick kiln industry, powder metallurgy molding is extremely similar to brick forming (Liao, 2003)[9] . Can brick powder replace metal powder? If it can be directly processed? It is a worthy issue of study. If it can, its production costs will drop 1/3 dramatically. It will help reduce personnel costs and the price and improve product competitiveness.

Facing the difficulties of traditional industrial, this subject boldly assumes that tradition can be innovation in other industries? Can two traditional industries have new opportunities? Can two traditional industries create together? There is no gap in generations, just lacking a bridge. The project tried to find the bridge and let the design become an opportunity for co-creation.

2. Methods

Material testing and characterization is a key function in assessing and ensuring the quality, properties, and behavior of a polymer or composite, and is thus an important step in guaranteeing the quality and success of finished product.

This research hopes to bring the creative cultural products of the red brick through powder metallurgy molding technology. Powder metallurgy molding technology is similar to the brick burning process. This research uses brick powder as raw material, combined with powder metallurgy technology, to shape the red brick products. The case study is conducted on the bricks in central Taiwan.

3. Results

3.1. Material testing process

The experimental test focuses on the feasibility of molding. The first step is to test with an alternative simple mold. The results are: Using water as an adhesive can be replaced by extrusion mold. But it is also easy to disintegrate because of the reduced water. However, the press molding can be more detailed of the brick surface.

The second step is to use the existing mold for the molding test. Fist, discuss the molding problem with the powder metallurgy practitioner, set the molding pressure, mix the appropriate proportion of brick powder and iron powder. Select the appropriate mold for testing, and then perform sintering after the test. The test situation is shown in the following table 3, the final results show that the brick power metallurgy is quite successful.

Table 3. The material testing process of brick power metallurgy

 1. 3D modeling	→	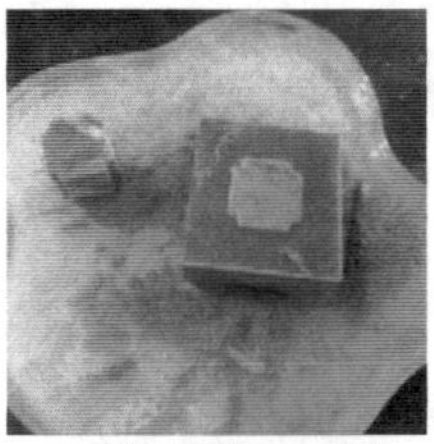 2. Simple mold	 5. Press forming	→	 6. Pressurized product
 3. Discussion	→	 4. Brick powder feed	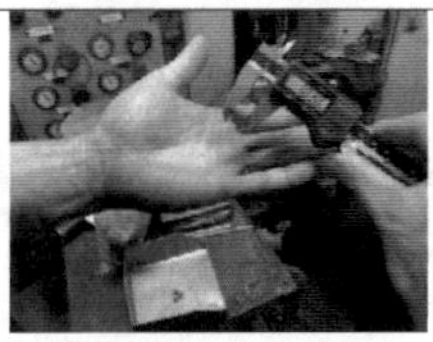 7. Dimensional measurement	→	 8. Sintered product

3.2. Design positioning

This study uses innovative molding methods to allow bricks to escape the image of the past. Not only the shape becomes more delicate and degenerate, but it also shrinkage the capillary pore of the bricks. We also found that the improved small brick is very suitable as a raw material for the diffused stone.

The principle of the diffused stone is mainly to absorb moisture and adjust the humidity by many tiny pores. When the humidity of the diffused stone is lower than the humidity in the air, the moisture is absorbed into the air; conversely, when the humidity of the diffused stone is higher than the humidity in the air, the moisture is discharged to the air. Therefore, when the incense stone drops on the essential oil, it is diffused by the adjustment of humidity.

Through the combination of diffused stones and accessories, many problems with perfumes can be solved. Perfume is usually sprayed on the back of the ear, neck and other parts, through the higher temperature of the body, so that the fragrance can spread more effectively, but the fragrance also disappears quickly only about two to three hours.

Some users are allergic to the ingredients of the perfume, so some people will spray the perfume on the clothes to make the fragrance last longer and not directly touch the skin. But another disadvantage is that some perfumes are prone to dyeing light-colored clothing.

The diffused stone brick jewelry designed through this study allows users to spray perfume on the jewelry to create the "brick" jewelry that belongs to their own fragrance and memories. It can also effectively solve the above problems, and reduce the amount of perfume. When you feel that the aroma is insufficient, you only need to sprinkle a small amount of water on the surface to continue to spread the fragrance.

The flow of the study from analysis, sketching, 3D drawing to questionnaire survey and mold opening is shown in the table 4 below.

Table 4. The flow of the study

 1. Design analysis	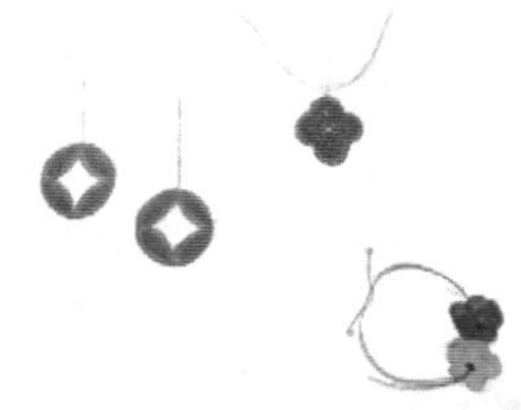 2. Sketch drawing	 3. 3D drawing
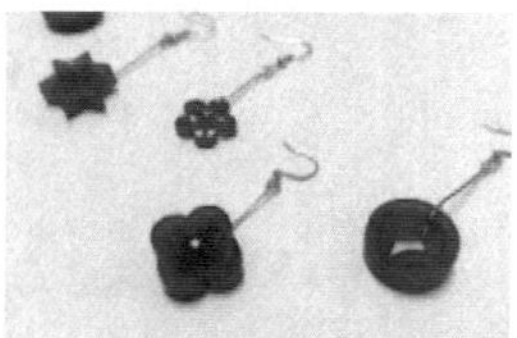 4. 3D printing	 5. Questionnaire	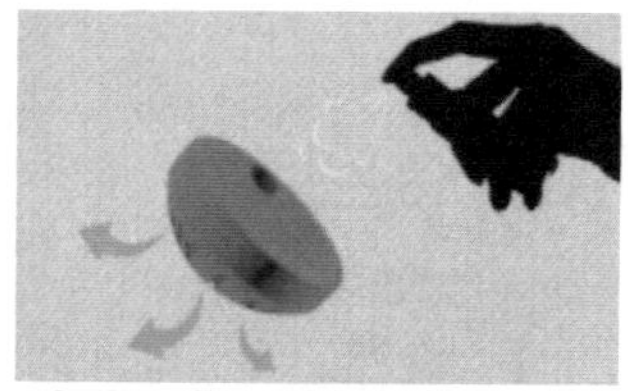 6. Confirm product style
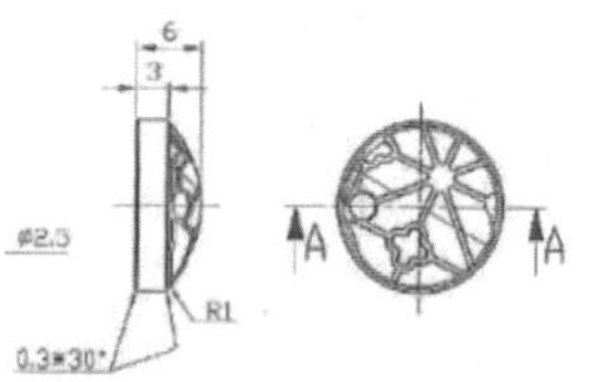 7. Mold design	 8. Mold opening	 9. Mass production

3.3. *Design results*

Using the above-mentioned brick powder metallurgy technology, this study designed several bricks of cultural products, mainly necklaces, and earrings, as shown figure 1.

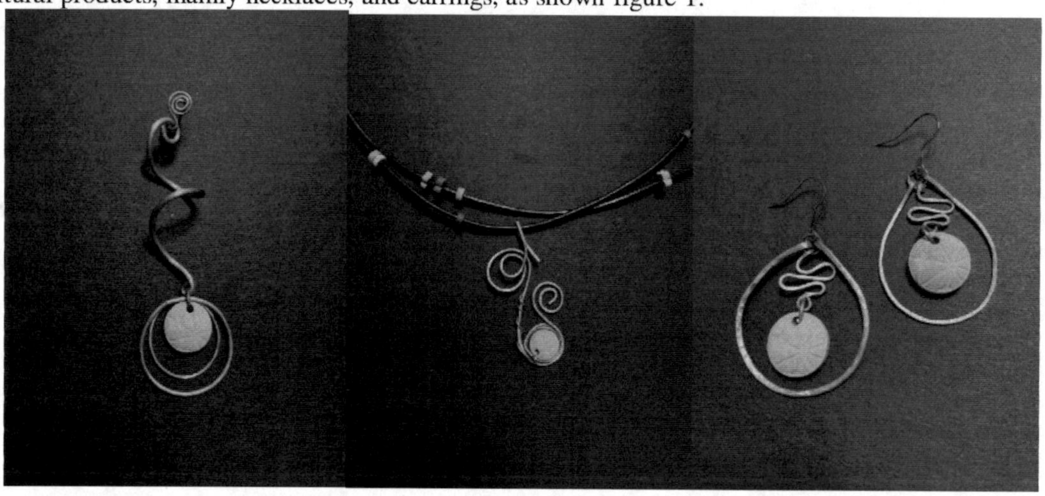

Figure 1. Creative brick kiln cultural products design

4. Conclusion

After the import of the powder metallurgy process technique and design project, it helps to activate the brick kiln industry. And obtained the following results.

- The surface roughness is greatly improved by brick powder metallurgy pressing.

- Reduce manual labor and reduce costs significantly.
- Promote the combination of different industries and create new opportunities for brick cultural product design.
- Bricks made of powder metallurgy, with high public acceptance.
- Innovative formulations and processes have been proven to be viable.

Acknowledgments
Authors are special thanks to the assistance provided by the Soil and Water Conservation Bureau, Council of the Agricultural, Executive Yuan and the Ministry of Education, Taiwan.

References

[1] Chang Hsin-Fang *"A Study on the Development of Brick and Roof Tile Industry in Ciaotou Area, Huatan Town, Jhanghua County."* Taipei National University of the Arts publishing. Master's thesis, 2012.
[2] Shun-Da Kiln Industry, https://www.facebook.com/sdyourbrick
[3] Changhua Huatan Qiaotou Community Development Association. https://www.facebook.com/groups/575609532514242/
[4] Hao Lu *"A Case Study Research on Developing Model of Brick Cultural Products from Jin Lian Shing Factory"*, National Yunlin University of Science and Technology press, 2001.
[5] Chi-Chang Lu, Chun-Chien Wang *"Research on Using the Image of Taiwan's Red Brick in Household Product Design"*, 2013 Craft Creation and Creative Design Conference, National Taiwan University of Arts press, 2013.
[6] The 22 Design Studio, https://www.pinkoi.com/store/22designstudio
[7] Yann-Long Lee, Ming-Yu Hsiao, Chia-Wen Tsai and Shih-Jung Tan *"A New Design Method of Grafting Method in Furniture Design and its Creations"*, Creative Industry International Conference 2018, 2018.
[8] Ming-Yu Hsiao, Yann-Long Lee, Pei-Chun Hsueh, Yung-Ling Tseng *"A New Creative Method "The Cultural Gene Triangular" application in Cultural Products Design"*. 1st IEEE International Conference on Knowledge Innovation and Invention 2018, 2018.
[9] Ji-Qiao Liao *"Powder Metallurgy Experimental Technology"*, Central South University Press, 2013.

IOP Conf. Series: Materials Science and Engineering **926** (2020) 012033 doi:10.1088/1757-899X/926/1/012033

Somatosensory Interaction of the Science Exhibition Based on Flow Theory

Yin Chao[1], Chen Yang[1,*], Wang Jian-wei[1] and Shi Ruo-hao[1]

[1]School of Mechanical Engineering, Guizhou University, Huaxi District, Guiyang City, Guizhou Province,550025, China

[*] E-mail: 892168711@qq.com

Abstract. In order to improve the immersion and pleasure of users from science exhibition, a design method based on flow theory is proposed. Based on the flow theory, it proposes to set up clear science goals and sense of substitution for somatosensory interactive science exhibits, to maintain the unity of user skills and task difficulty, to maintain timely and effective feedback, to promote the guidance layer to maintain the continuation of the flow and other design methods, and Use jack simulation software for environment optimization research and analysis. Verification by the flow experience Likert scale proves that this method can greatly increase the probability and duration of the user's flow state, and effectively compensate for the shortcomings of neglecting the user experience in the design of traditional somatosensory interactive science exhibits. The example proves that the design of the somatosensory interactive science exhibits based on the flow theory can effectively improve the user's science experience and enhance the user's immersion and pleasure. This method has good universality.

1. Introduction

With the upgrading of technology, the popular science experience has been extended to a higher dimension, and its hardware and interaction methods have begun to appear in a variety of choices. In the design of somatosensory interactive science exhibits, how to embody the interactive advantage, whether the user's pleasure can increase with the change of technology, whether the experience process is linear, whether the environmental planning is reasonable, whether the evaluation system is established, lack corresponding methods to guide them. The flow theory is based on the psychological theory of cognitive psychology about mental state and interactive experience. By balancing challenges and user skills, it guides users to achieve the characteristics of high immersion and pleasure. Based on flow theory, the design of somatosensory interactive scientific exhibits, combined with Jack software, can quickly and accurately set up interactive process, build an interactive environment, keep user flowing and improve the popular science experience. Taken altogether We providing a new design method for somatosensory interactive science exhibits by this paper.

2. Exhibit structure

The digital science exhibits consist of three parts: information content, interaction mode, and user environment, as shown in Figure 1.

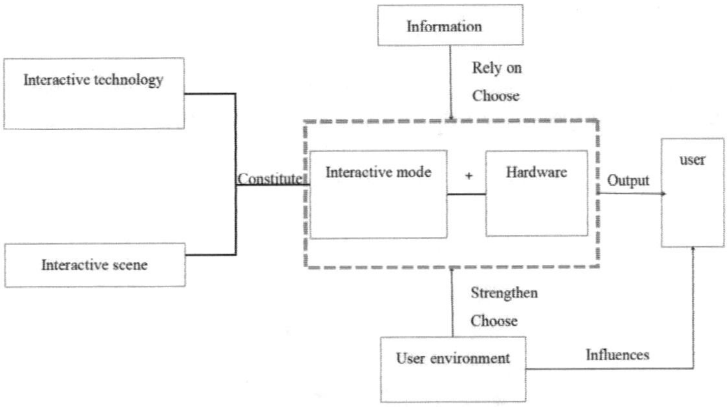

Figure 1. Structure of digital science and technology exhibition.

A large number of experiments have been conducted at home and abroad to combine emerging technologies and products, and it has been proved that the interaction mode composed of interaction scenarios and interaction technologies determines the user's direct experience. In the design of interactive scenes, Cabero [1] and others proposed to judge the educational situation, enrich the educational content by creating AR scenes to improve the immersion of students and strengthen the learning effect; Cai S [2] uses AR and Kinect to build the teaching content, through gestures Real-time interaction with the magnetic field can effectively improve learning efficiency. Experiments prove that the update of interactive scenes can stimulate the short-term interest of young user groups and enhance the efficiency of instant learning. However, Cabero and Cai S did not distinguish and verify multiple interactive technologies in the same scene. The advantages of visual stimulation after the scene were established also showed deficiencies in maintaining the user's high immersion, pleasure and control.

At the same time, domestic researchers tried to experiment with interactive technology as an entry point. Ma Huan [3] proposed a design method based on the integration of dynamic somatosensory interaction technology with multi-channel projection display, to achieve interactive freedom and dynamicity, and to improve the participant' s immersion; Li Hui [4] and others proposed a child attention evaluation system based on Kinect somatosensory interaction, which believed that somatosensory can effectively restore high immersion and natural Interactive mode, the test results under this condition tend to be true, positively determine the role of somatosensory technology in improving sensory sensation, behavior, etc. [5]; natural interaction mode is the main design direction, assisting interactive scene selection and Design is a way to effectively improve the immersion and pleasure of users in the learning and cognitive process.

The interaction method affects the user's perception and acceptance of the content. The design method based on the flow theory connects the content and the interaction technology without affecting the volume of the content, reconstructs the interaction scene, optimizes the user environment, and enhances the science experience.

3. Flow theory and experience

The flow theory is a psychological theory proposed by the famous psychologist Csikszentmihalyi about human psychological state and interactive experience [6]. Flow experience is an important concept of flow theory, which refers to a feeling of putting personal energy or "body and mind" completely in a certain activity [7], direct psychological experience and emotional feelings of immersion. Flow of mind, although it is only a temporary and subjective experience, is the psychological reason for people to continue to invest in this kind of activity [8]. When you enter the flow state, you will have a high sense of excitement and fulfillment, you will lose your sense of self-existence and time, and you will have a sense of participation from the heart and a desire to learn the experience.

User skills and challenges are two important indicators in the flow model. Csikszentmihalyi [9] believes that the flow state will appear only when skills and challenges are balanced; when challenges are greater than skills, users often experience anxiety and skills are greater than When challenged, the bored mental state will be highlighted. In-depth analysis of the flow model to form eight user experience states, but there will be two unified presentations on the matching degree of their skills and task difficulty, that is, the combination of low skill level and low task difficulty, high skill level and high With the unification of skill difficulty, the former brings more indifference and the latter will produce flow experience. Only when the user has a subjective high skill level and perceives a difficult challenge will a flow state be formed, as shown in Figure 2.

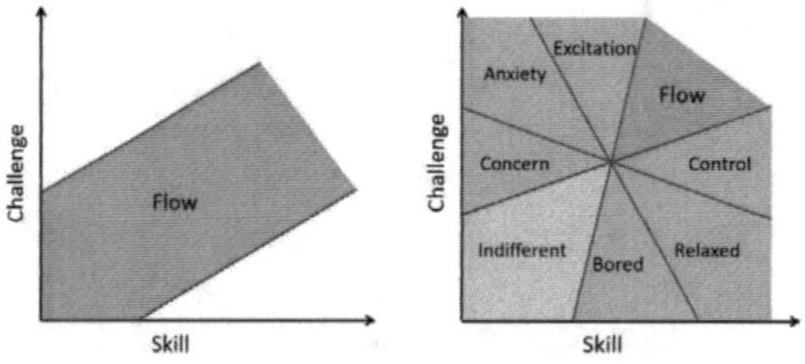

Figure 2. The model of flow and depth refinement.

The formation of the flow experience will present 9 features [10]. Applying feature transformation to the design of somatosensory interactive science exhibits, it is divided into the control layer, guide layer and effect layer. The control layer is the basic part and controllable part of the design of somatosensory interactive science exhibits, which provides the possibility for the application of flow theory to somatosensory interactive science exhibits, and plays a decisive role in the final presentation of the flow experience; the guide layer belongs to the intermediate layer, which controls The degree is low, while rationally using the control layer to push the user's flow state to a high possibility, the role of the guide layer is to advance the opening point of the flow state and maintain a stable state; the effect layer is the performance result of the user's flow state , When the three are satisfied at the same time, the best science experience will be presented,as shown in Table 1.

Table 1. Feature transformation of flow experience.

Csikszentmihalyi	Somatosensory interactive popular science exhibits flow characteristics	
Clear and clear goals	Control layer	Clear science goals and sense of substitution
Even valuable feedback		Instant and effective feedback
Match skills with mission challenges		Unity of user skills and task difficulty
The fusion of action and consciousness	Guide layer	Integration of interaction and consciousness
High concentration of attention		High concentration of attention
Potential sense of control		Potential sense of control
Loss of self-consciousness	Effect layer	Loss of self-consciousness
Change in time		Change in time
A sense of participation from the heart		Highly spontaneous learning

4. Design method

The design structure of somatosensory interactive science exhibits based on flow theory is shown in Figure 3. Reasonable use of the control layer and the guidance layer can help users generate a flow experience.

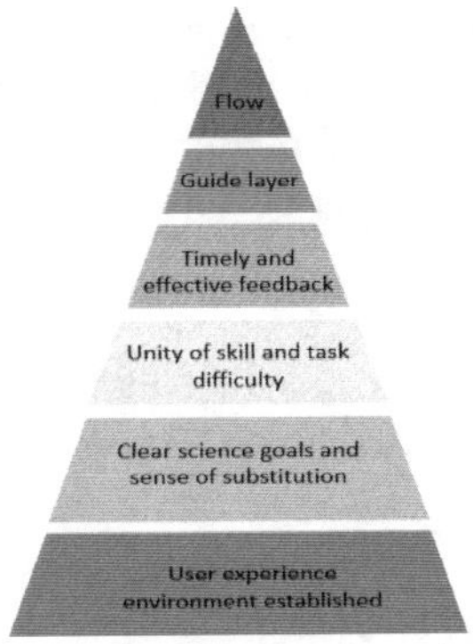

Figure 3. Structure of somatosensory interaction of the science and technology exhibition.

4.1. Establishment of user experience environment

The establishment of the user experience environment is a prerequisite for the construction of interactive scenes and one of the main reasons for the selection of somatosensory interactive hardware.The user's actual environment determines the choice of the carrier of the somatosensory interactive product.Carrier and the actual environment also determine the proportion of reconstruction between the virtual scene and the real scene in the interactive scene. If there are many users in the experience environment and situation is complicated, the product carrier should take the no-carrying device as the main direction, and the interactive scene should be built with the screen imaging as the main technology to reduce the user experience pressure and reduce the equipment maintenance cost .

Jack software can provide human biomechanics models, quickly obtain and use human body data from various countries. Using Jack as the experience environment establishment layer, the optimal interaction area and interaction distance can be determined.

4.2. Clear science goals and sense of substitution

Science education is an informal education with a high degree of user spontaneity. As the main target user group of science popularization sites, teenagers are mature and not yet perfect. In the experience, they should clearly indicate the tasks, clues, information and interaction methods to the users to ensure that the goals are easy to understand and that the information given by the users is their own needs or task requirements. Completed.

The establishment of the sense of substitution is the opening point and instantaneous peak of the flow state of the somatosensory interactive science exhibits users, but the flow state obtained by the pure sense of substitution is very easy to lose, and it tends to be stable after repetitions. Analyze and design in the manner, manifestation and existence time of the establishment of the sense of substitution, which can make this flow state peak and remain relatively stable. When there are few somatosensory interactions in the exhibits and the interaction frequency is low, the user's current substitution manifestation should be increased. For example, the user's real-time skeleton model should be established and displayed for a long time; otherwise, the user substitution manifestation should be properly avoided to avoid high frequency and multiple actions.

4.3. Unification of user skills and task difficulty

Maintaining the unity of dynamic skills and difficulty is a design method that maintains high skill level and a high task difficulty. It is also the focus of the design method of somatosensory interactive science exhibits based on flow theory. Due to the large age range of exhibit users, there is a big difference between physiological and psychological maturity, and the design of the task difficulty needs to be divided. Combined with the characteristics of somatosensory interaction and human body model for information input, the current data terminal AI algorithm and pre-set combination are used to obtain the current user judgment, evaluate the user's psychological and physiological skills, and give the corresponding interactive task and interactive action difficulty level At the same time, the information input feedback in the interactive process is used to judge and adjust the reasonableness of the current difficulty and balance the challenges. Suppose the accuracy of the user's action is A, the standard interaction action is B, the optimal completion time of the interaction is S1 to S2, and the actual completion time is T, then the dynamic balance process structure can be introduced, as shown in Figure 4.

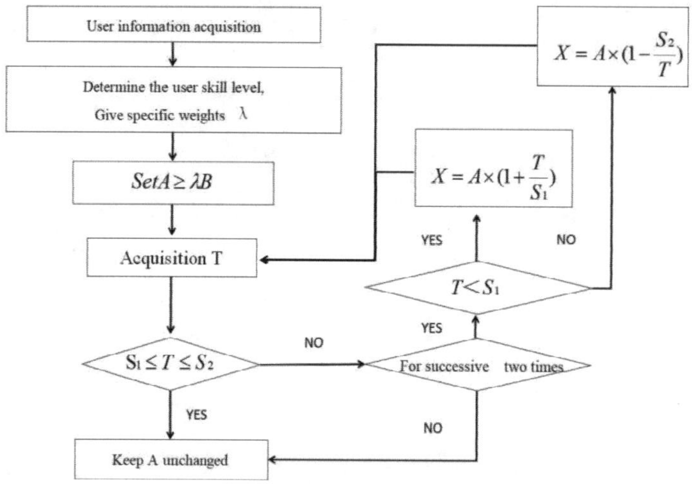

Figure 4. Structure Structure of dynamic balancing process.

4.4. Instant and effective feedback

When the interactive action is completed, the user is quickly given relevant feedback, which can help the user to deal with the relationship between the interaction and the result, strengthen the learning and adjustment mechanism, reduce the somatosensory interaction lag, anxiety, and avoid interruption of flow.

4.5. Advancement and promotion of the guidance layer

The advancement and promotion of the guide layer can maintain and enhance the flow state that has been generated. The integration of interactive actions and consciousness, high concentration of attention, and a potential sense of control are the three indicators of the guidance layer.

An interactive action setting follows natural logical thinking, which is the difference and advantage of somatosensory interaction compared to touch interaction. Combining proprioception such as the movement of muscles and joint positions, body postures with vision and human brain, it evokes the mental model established daily, which is more helpful for the understanding and learning of popular science content. Due to the high dimension of the interaction scene, the non-linear process structure will increase the user's memory pressure and judgment ability. The experience process of somatosensory interactive science exhibits should be based on a linear structure to reduce unnecessary user anxiety and avoid task failures. Of users are frustrated.

At the same time, reasonable design needs to be made in terms of action frequency to ensure the necessity of interactive actions, maintain a high concentration of user attention, and avoid meaningless actions. In terms of control, it is very important to broaden the user's control limitations and develop skills. If the user cannot complete the task because of the low degree of freedom of control and the narrow field of vision, it will generate pressure and reduce the execution power, resulting in a state of flow. The loss of science and the destruction of the experience of popular science, and therefore unwilling to continue to try.

5. Design case

This article takes the case of the somatosensory interactive popular science exhibit "Life of Seeds" of the Guizhou University Product Innovation and Digital Interactive Technology Research Studio.

First, set up a user experience environment for the Guizhou Science and Technology Museum. Due to its large user base and high liquidity, combined with the interactive actions extracted from the exhibits, projection imaging technology and Kinect equipment were selected as interactive hardware. Use Jack simulation to analyze user accessibility, visibility and posture [11]. Establish a human model of the age group of the target user group in Jack, rely on imaging technology to get the best interaction distance, and mark on the ground to delineate the best interaction range.After determining the content, environment and carrier, design the control layer. Connect Unity3D with Kinect, at the same time the background AI judges the input information, based on the height and gender to determine the difficulty, give the accuracy of the interactive activities, and then open the dynamic balance process structure; then the science goal is presented, that is, short-term tasks and long-term tasks, Show users short-term interactive action tasks such as seed selection, sowing, watering, fertilization, etc., and inform them that science knowledge can be obtained after successfully completing a series of long-term planting operations. After the user confirms, the image of the skeleton model captured by Kinect is optimized and projected into the interactive interface to establish a sense of substitution, which is the highest peak of the flow state at this moment.

The design of the propulsion layer is used to maintain the flow state. After collecting data on standard actions such as seeding and watering, a large number of target users of the product should be captured for the second time, combined with comparison, balance and optimization, to ensure that the action and consciousness are consistent and integrated, avoid deviation, and avoid frustration; Multi-frequency scene conversion will cause environmental distortion, so "Seed of Life" action feedback is established in the only scene, and the restrictions on the selection of objects and functions in the scene are appropriately relaxed, and timely and timely feedback is established to establish a sense of potential control and maintain user attention. Is highly concentrated, see Figure 5.

Figure 5. Scene graph of The Life of The Seed.

6. Design evaluation

To evaluate the users of the somatosensory interactive science popularization exhibits of the "Life of Seeds" Likert flow measurement table, see Table 2. There are 15 evaluation questions in the table, and the result adjective scale is divided into 5 corresponding levels and their scores can be provided, such as against very much (1 point), against (2 points), difficult to judge (3 points), agree (4 points)), agree very much (5 points), and collect statistics.

Table 2. Likert scale of flow.

Serial number	The question
1	I know exactly what I want to do with this popular science exhibit
2	My goal is clear
3	I can concentrate on what I do
4	I didn't pay attention to what other people think of me during the experience
5	During the experience, I feel I can control what I am doing

6	Some things seem to happen naturally (subconsciously enter certain pages; subconsciously make feedback actions)
7	I will naturally perform and complete the task
8	My ability is enough to cope with difficult and complicated operation situations)
9	The time seems to have changed during the experience (slow/speed up)
10	During the experience, I didn't realize that I was in the real world
11	I did not feel the pressure from the outside during the experience
12	In the process I can clearly focus on what is happening
13	I really enjoy the experience brought by this popular science exhibit
14	I know I'm doing well
15	I like to use popular science exhibits to cope with the challenge, I want to experience it again

A total of 120 questionnaires were distributed, and 114 valid questionnaires were collected. According to the score, users are divided into Ⅰ (less than 45 points, no flow state), Ⅱ(45-54 points, not sure whether there is a flow experience), Ⅲ(55-64 points, have a flow experience but interrupted), Ⅳ(65-75 points, strong flow experience) Four states, see Table 3. The results show that this design method can effectively improve the immersion and pleasure of users.

Table 3. Statistics of user' flow status.

Sample number	Recovered number	Effective number	Ⅰ	Ⅱ	Ⅲ	Ⅳ	average value	Standard deviation
1-40	39	38	0	6	1	31	67.92	7.02
41-80	19	19	0	1	5	32	69.57	4.69
81-120	18	18	0	2	2	34	69.46	5.95
total	117	114	0	9	8	97	69.00	5.89

7. Conclusion

Somatosensory interactive science exhibits based on flow theory have advantages in improving user immersion and pleasure, making it easier for users to enter the flow state, achieving efficient and pleasant science knowledge learning purposes, improving user science experience, and establishing a new somatosensory interactive science exhibit design model.

Acknowledgments

This work is supported by Guizhou Science and Technology Fund Project "Science and Technology Museum Exhibition Engineering and Science Exhibition Teaching Innovation Design Theory and Methodology" (Qiankehe J word [2014] No. 2057).

References

[1] Cabero Julio and Barroso Julio J 2016 *Journal of New Approaches in Educational Research* **5(1)** p 44-50
[2] Cai S, Chiang F K and Sun Y J 2016 *Interactive Learning Environments* **25(6)** p 1-14
[3] Ma Huan J 2018 *Packaging Engineering* **39(18)** p 229-233.
[4] Li Hui and Wang Wei J 2014 *Modern education technologyv* **24(07)** p 120-126.
[5] Liu Hong-jian and Yang Yu-bao J 2013 *Remote education magazine* **31(01)** p 54-59.
[6] Chen Jin-liang,Zhao Feng and Zhang Qian J 2018 *Packaging Engineering*, **9(18)** p 146-153.
[7] Csikszentmihalyi M 2008 *New York:Harper Perennial Modern Classics*
[8] Ou Xi-fan and Tan Hao J 2016 *Packaging Engineering* **37(04)** p 70-74.
[9] Deng Peng J 2006 *Education* **03** p 74-78.
[10] CHANG Chiao-chen J 2013 *Telematics and Informatics* **30(4)** p 311-321.
[11] Yin Hang J 2017 *Journal of Machine Design* **34(08)** p 120-123.

IOP Conf. Series: Materials Science and Engineering **926** (2020) 012034 doi:10.1088/1757-899X/926/1/012034

Research on Modeling Design of Ear Thermometer Based on User Image

HuLei[1] and ZouXin[1,*]

[1]School of Mechanical Engineering, Guizhou University, Huaxi District, Guiyang City, Guizhou Province,550025, China

[*] E-mail: 1162221909@qq.com

Abstract. The purpose is to enable users to obtain a better perception experience and enhance the user's perceptual needs recognition of the appearance of the ear thermometer. The method is guided by the kansei engineering theory, through quantitative analysis of perceptual needs evaluation and eye tracking test. Obtain the mapping relationship between the vocabulary representing the sexy image and the modeling features of the ear thermometer, and conduct the example design of the product form of the ear thermometer based on user needs. Conclusion The design practice verifies that the ear thermometer design method based on user image preference has high feasibility. Provide product development directions for related companies and designers.

1. Introduction

A product perceptual design is a product development technology that prioritizes consumer imagery, a product design technology that transforms consumer perceptions or images of products into new product design elements, and continues to develop and optimize[1]. Su Jianning[2][3] and others put forward an optimal design method based on the perceptual image through the future development direction of the product. Djatna[4] Data carried out the construction of perceptual space based on fuzzy quantitative theory and carried out product development. Feng qing[5] ting uses the semantic differentiation method and semantic variables to obtain the perceptually perceptual relevant data of the target product, summarizes the user needs with the data statistics, and performs perceptual design. Based on the perceptual characteristics of consumers, Wang Xinting[6] integrates the theory of object knowledge classification, thus establishing modeling knowledge base and designing the scheme. The use of Kansei engineering theory to integrate multi-disciplinary domain knowledge to obtain design elements has gradually attracted widespread attention. Zhang Yue[7] conducted a perceptual demand survey on consumers. In order to obtain the perceptual needs of consumers for product packaging, key design elements that affect consumer emotions were extracted through eye movement test indicators and consumer perceptual needs; and And through quantitative class theory and The the relationship between design elements establishes a mathematical model for packaging design. Feng Yixiong[8] and others conducted in-depth research on performance-oriented customized products, and combined the sensory image evaluation method in product research, using EEG to collect signal from the user's sensory image, and used this as the main basis to verify the product's performance. Based on the sensitivity of user, Qian Hao[9] combined the product image bionics method with fuzzy evaluation method. By screening the perceptual vocabulary pairs, the mathematical matrix was established and analyzed to obtain the mapping relationship between the characteristics of the biomimetic and the product modeling elements, and the user's comprehensive evaluation value of the scheme is used to improve and optimize the design scheme. Yuan Shuzhi[10] and others used the Kansei engineering method to establish a perceptual index evaluation system. According to the fuzziness of perceptual engineering evaluation, a comprehensive evaluation model combining intuitive fuzzy set theory and TOPSIS method was established, and a multi-image evaluation method for human-machine interface based on Kansei engineering was proposed. To quantify people's sensibility to things or products through statistical methods, the key point is how to accurately grasp consumers' sensibility to products and establish the relationship between sensibility and product elements.

At present, due to the continuous development of technology, people's awareness of medical care is constantly improving, the Minamata Convention restricts mercury-containing products, and the spread of new coronavirus pneumonia. A variety of new electronic thermometers are rapidly entering user 'life, and attention should be paid to the improvement of the user' perceived experience brought about by related products. This article uses the theory of Kansei engineering to take the ear thermometer as the research object, through qualitative and quantitative analysis, establishes the modeling feature mapping relationship, obtains the user's perceptual image modeling features, guides the innovative design of the ear thermometer, and provides a new design idea.

2. Product samples and emotional vocabulary selection

2.1. Image sample screening

Ear thermometer samples are widely collected on websites at home and abroad. During the collection process, 180 samples were collected based on research principles, diversified product shapes and suitable image angles, and interference factors such as trademarks were removed by using image processing software. According to the first impression of the product, the approximate shape classification was carried out, finally, 50 sample pictures were retained. In order to further select samples, a research group (industrial design students, 10 males and females) was organized to classify 50 sample into 10 categories. According to the classification results, the similarity was described by formula(1), where S formula (1), which represents the similarity of the two pictures, F pictures. f represents the number of times two samples appear in the same category, T it represents the number of times the same sample appears in the same category. According to similar situations, 10 samples were finally selected as shown in Figure 1.

$$S\frac{F}{T} \tag{1}$$

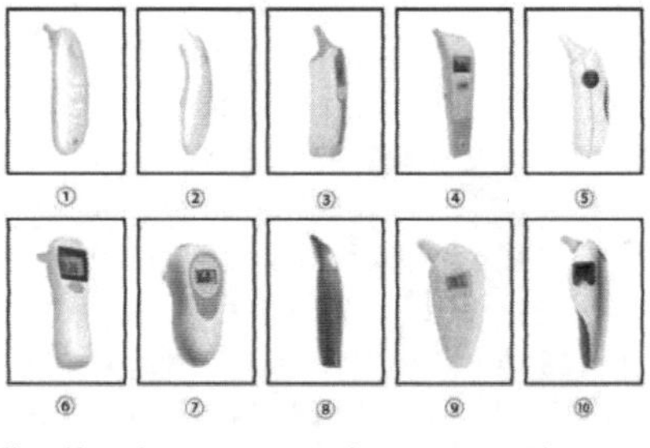

Figure 1. The finalized representative 10 ear thermometer samples.

2.2. Perceptual vocabulary screening

A large amount of perceptual vocabulary was collected from relevant research literature and research materials, and 160 suitable vocabularies were collected. After the classification and summary of the research team, 50 perceptual words were obtained and 25 pairs of perceptual words were formed. In order to further refine the 25 perceptual vocabulary pairs, the KJ method was used to extract the perceptual vocabulary. After extraction, 6 groups of perceptual vocabulary pairs were finally obtained, as shown in Table 1.

Table 1. 6 groups of vocabulary after screening.

Safe-dangerous	Soft-hard	Lightweight-heavy
Technological-backward	Soothing-nervous	Exquisite-rough

3. The mapping relationship between perceptual image vocabulary and key modeling features

3.1. Evaluation of styling image

Taking the 10 representative image samples and 6 sets of perceptual image vocabulary selected above as the evaluation experimental materials, a total of 50 testers (undergraduate and graduate students majoring in industrial design) were selected. The testers evaluated through their their subjective feelings. The experiment was conducted in the form of a questionnaire. The Semantic Differential (SD) 7-level scale was used to grade the emotional vocabulary. There are 7 levels in total, -3~3,0 is neutral, which means no special feelings, -3 and 3 corresponds to the two ends of the scale, representing the weakest and strongest feelings. After sorting out the results of the questionnaire, the average value of image evaluation is obtained, as shown in Table 2.

Table 2. The average value of each image evaluation of representative samples.

Target sample	Safe-dangerous	Soft-hard	Lightweight-heavy	Technological-backward	Soothing-nervous	Exquisite-rough
Sample 1	0.77	1.22	0.11	0.36	1.22	1.76
Sample 2	1.99	2.24	0.24	1.51	1.09	1.21
Sample 3	-1.82	-1.84	-1.2	-1.5	-1.44	-1.6
Sample 4	-1.88	-2.62	-2.04	-2.42	-2.48	-2.46
Sample 5	1.03	0.51	2.44	0.22	0.26	1.61
Sample 6	-1.22	-1.04	-1.26	-1.4	-0.82	-1.74
Sample 7	0.8	1.32	2.36	-1.22	0.8	0.4
Sample 8	2.22	2.06	1.1	2.52	2.02	1.92
Sample 9	-1.38	-0.62	1.14	-1.8	-0.94	-1.98
Sample 10	-0.28	-0.8	-0.58	0.76	0.88	0.42

3.2. Evaluation results of styling imagery

The perceptual value is the preference of each adjective. The positive value indicates that the sample is biased toward the positive adjective; the negative value indicates that the sample is biased toward the negative adjective in the perceptual direction; according to the data obtained by the above-mentioned perceptual image evaluation experiment, the sample 8 is outstanding in terms of safety, technology, soothing and refinement, sample 2 is outstanding in softness, sample 5 is more representative in terms of portability, and sample 1 is balanced in all aspects. The overall performance is sample 8> sample 2> sample 5> other samples.

3.3. Eye tracking experiment

Psychological research shows that vision can help humans obtain 80% of the outside world's information, and can make preference judgments on gazing things, and can effectively obtain users' mental activities through eye-tracking technology[11]. This eye-tracking experiment equipment uses experiment-related data recording and analysis software such as the German SMI RED desktop eye tracker and experiment center. After screening, 10 sample pictures were processed by removing color, uniform background, uniform size and other influencing factors, numbered from 1 to 10, as shown in Figure 1. In this experiment, a total of 20 subjects with normal eyesight were recruited, and relevant operation guidance was given to them to ensure the smooth progress of the experiment. This experiment was divided into two phases. In the first stage, ten sample pictures were arranged on the same plane. After eye movement calibration, the subjects were free to observe the picture of the sample for 30 seconds. In the second stage, the positions of the 10 sample pictures arranged on the plane were reordered and used as experimental materials. After eye movement calibration, observe the sample pictures on the plane for 30 seconds again. The eye-tracking experiment process is shown in Figure 2.

Figure 2. Subjects are conducting eye tracking experiments.

3.4. Eye tracking experiment results

According to the above eye-tracking experiment, many eye movement indicators have been obtained. Including eye-tracking area of interest (AOI), heat map (Heat Map), gaze time (Dwell Time [ms]) and other related gaze indicators, the experimental heat map is shown in Figure 3. According to the purpose of the experiment, the eye movement indicators are screened. Among the many eye movement indicators, gaze time, number of fixations, and saccades are the best indicators to reflect the attention distribution of the subjects. The Event Statistics module is used to export the relevant eye movement index data. According to the data, in terms of the average number of fixations, sample 8>sample 2>sample 5>other samples; in terms of average gaze time, sample 8>sample 2>sample 5>other samples; the results of the two eye movement indicators are consistent, It is consistent with the experimental results of the modeling imagery evaluation conducted previously. According to eye-tracking technology and cognitive psychology, when a subject observes freely, the greater the interest in a certain area, the longer the observation time for that area[12]. Through the above experimental results, it can be seen that the mapping relationship between the representative image vocabulary of the ear thermometer and the key modeling features is effectively established. The subjects' interest in Sample 8, Sample 2 and Sample 5 are significantly greater than that of other samples, Sample 8, Sample 2 and The characteristics of the ear thermometer model of sample 5 need to be studied and designed. The data of eye movement experiment are shown in Table 3.

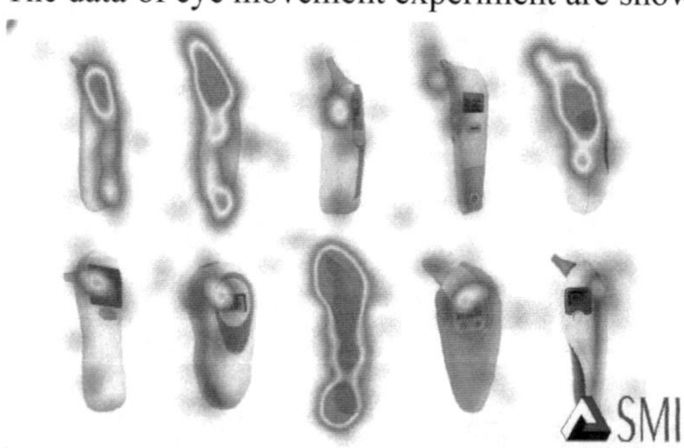

Figure 3. Eye tracking experiment heat map.

IOP Conf. Series: Materials Science and Engineering **926** (2020) 012034 doi:10.1088/1757-899X/926/1/012034

Table 3. Likert scale of flow.

Eye tracking indicators	Sample 1	Sample 2	Sample 3	Sample 4	Sample 5	Sample 6	Sample 7	Sample 8	Sample 9	Sample 10
Average number of fixations	4.6	5.8	2.2	2.1	5.5	2.4	4.1	7.1	2.7	3.4
Average gaze time/s	0.28	0.36	0.18	0.16	0.33	0.21	0.25	0.42	0.19	0.24

4. Examples of ear thermometer design

Through the above research, the user's preference of the ear thermometer shape is excavated, and the sample 8 shape is selected to carry out innovative product design. According to the difference between the perceived value of sample 8 and the modeling characteristics of sample 8 and other samples, the softness and form of the whole line were summarized. Integrity, the curved surface of the hand-held part hand-held and the fusion and coordination of the shape of the probe and the hand-held part. Attention should also be paid to the outstanding performance of other samples in perception mode. For example, the overall shape of Sample 2 gives a strong sense of softness. As an in-ear temperature detector, softness can bring good perception and user experience to users. Similarly, sample 7 is a light and representative form, and can also be used as a reference in the innovative design of odometer design. Based on the above, the innovative design of the ear thermometer design was carried out, and CINEMA 4D was used to build a three-dimensional model of the product, see Figure 4.

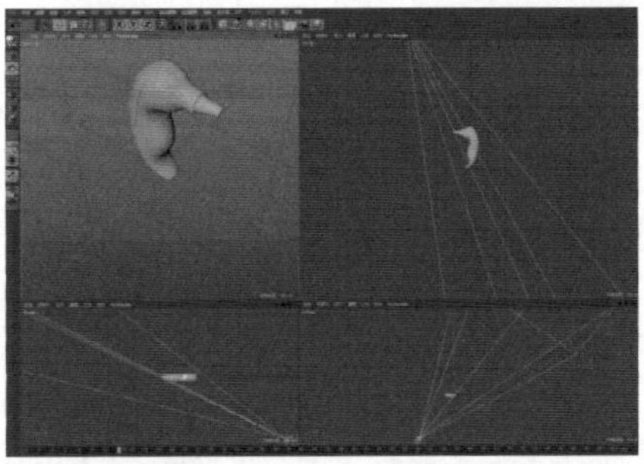

Figure 4. 3D model design drawing.

In order to verify whether the product design scheme meets the user's needs, the final effect diagram of the perceptual image evaluation sample was drawn, 30 questionnaires were distributed, 27 valid questionnaires were collected, and data statistics were carried out, and the average image evaluation value is obtained, as shown in Table 4. Through the image average value, we can see that the design scheme of the ear temperature gun product has a high score in the safety, technology, ease and refinement of the six groups of perceptual vocabulary, and the soft and light ones have also been improved to meet the design requirements summarized in the early stage. The rendering effect diagram is shown in Figure 5, and the actual user experience display of the design results is shown in Figure 6.

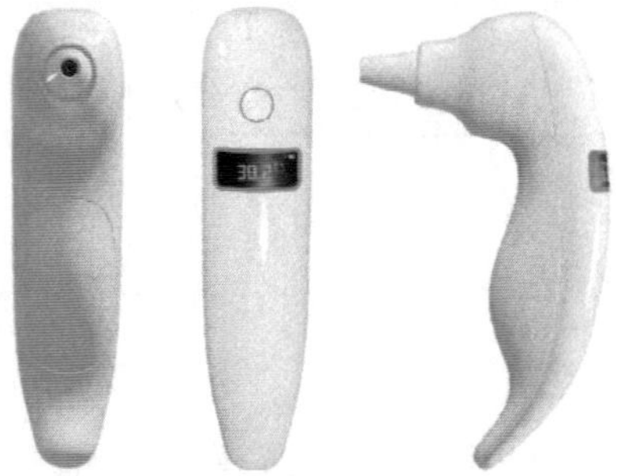

Figure 5. Rendering of the shape design of the ear thermometer.

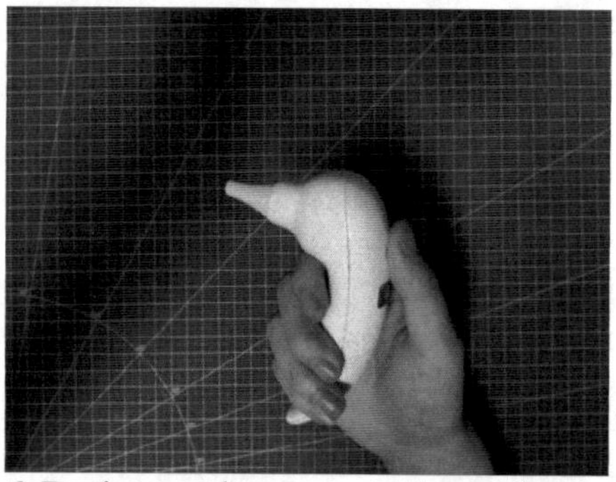

Figure 6. Design results user actual experience display.

Table 4. Final product image evaluation average.

Perceptual vocabulary	Image evaluation average
Safe-dangerous	2.28
Soft-hard	2.35
Lightweight-heavy	2.11
Technological-backward	2.47
Soothing-nervous	2.09
Exquisite-rough	1.82

5. Conclusion

According to the user's perceived demand for the ear thermometer, combined with the eye movement test method, the abstract perceived demand is quantified as the design basis, and the mapping relationship between perceived image vocabulary and the key modeling features of the ear thermometer is mined, the perceived features of the product are extracted, the user's perceived image is obtained, and the innovative design of the ear thermometer is carried out based on the perceived image. The effectiveness of the modeling design method of ear thermometer based on the perceptual image is verified, which can provide a certain reference for the modeling design of similar products.

Acknowledgments

This work is supported by Guizhou Science and Technology Fund Project "Science and Technology Museum Exhibition Engineering and Science Exhibition Teaching Innovation Design Theory and Methodology" (Qiankehe J word [2014] No. 2057).

References

[1] LUO Shi-jian J *Journal of Mechanical Engineering* **2007(03)**: p8-13.
[2] SU Jian-ning and LiHe-qi J *Journal of Mechanical Engineering* **2004(4)** p 164-167.

[3] SU Jian-ning and LIU Zhi-jun J *Journal of Machine Design*, **2017, 34(11)**: p112-116.
[4] Djatna T C 2015 *International Conference on Industrial Engineering and Service Science 2015* **2015** p 176–183.
[5] FENG Qing and WU Zhi-ling J *Food & Machinery* **2019, 35(03)** p120-123.
[6] Wang Xinting and WANG Kang J *Packaging Engineering*,**2019,40(04)**: p140-145.
[7] ZHANG Yue and HOU Ya-dan J *Food & Machinery* **2019, 35(04)**: p113-119.
[8] FENG Yi-xiong J Journal of *Mechanical Engineering* ,**2020, 56(09)**: p181-190.
[9] QIAN Hao and GAO Yang J *Packaging Engineering* **2019, 40(14)**: p144-149.
[10] YUAN Shu-zhi and GAO Hong-ni J *Chinese Journal of Engineering Design* **2017, 24(05)**: p523-529.
[11] TANG Bang-bei and GUO Gang J *Computer Integrated Manufacturing Systems* **2015, 21(6)**: p1449-1458.
[12] ZHANG Ling-yu and WU Jian J *Packaging Engineering* **2019, 40(18)**: p40-47.

IOP Conf. Series: Materials Science and Engineering **926** (2020) 012035 doi:10.1088/1757-899X/926/1/012035

RecyGlass, A Low-cost System for Recycling Glass Bottles in a Distinctive Way

Andrée Montoya[a], Leonardo Bravo[b], Mario Velasquez[c], Hector Oscanoa[d] and Jhon Portella[e]

Engineering Department, Pontifical Catholic University of Peru, Universitaria 1801 Avenue, San Miguel 32, Lima, Peru

[a] andree.montoya@pucp.pe; [b] leonardo.bravot@pucp.edu.pe; [c]velasquez.mario@pucp.edu.pe; [d] h.oscanoa@pucp.edu.pe; [e] jportella@pucp.pe.

Abstract. Traditional glass recycling processes exhibit difficulties when it comes to segregating carefully the glass material by type and color. Having this in mind, new recycling alternatives that avoid this stage in the entire process have arisen. Nevertheless, they are still dangerous for operators or novel users. To surpass both problems, this paper presents in detail the mechanical, electronic and control design and implementation of RecyGlass, a low-cost autonomous glass recycling machine with 6 different sub-systems: clamping, rotational, translational, cutting, polishing, and serigraphy. Real and 3D model images are used to present in a better way the design and result of the RecyGlass project in its different design levels. Finally, this work also analyses, in the conclusion section, the energy consumption of our alternative compared to traditional glass recycling processes.

1. Introduction

Environmental pollution is a matter of latent concern because of the consequences related to climate change, the depletion of natural resources, and the extinction of living beings. Therefore, waste recycling has appeared as a possible solution to this problem.

Despite the fact that many countries have encouraged waste treatment, the recycling rate of generated solid waste (GSW) is still low. This can be illustrated by the low recycling rates of solid waste that Europe, the U.S.A and Peru exhibited in 2016: 37.8 % [1], 35.2 % [2] and 1.9 % [3], respectively. One of the reasons that explains why these numbers remain low is that consumers do not have proper resources on hand to recycle in an easy way. Moreover, the authorities of different countries do not operate in an articulate manner to generate incentive recycling policies valid for different territories [4]. To ease the recycling operation for the average user, many efforts have been made to conceive alternative processes to traditional recycling ones, especially for glass bottles (5 % of GSW) [4].

It is important to note that in the traditional glass recycling process, the glass needs to be segregated carefully by both material and color. On the one hand, material segregation is required because certain kinds of glass such as ovenware, Pyrex, or crystal can cause production problems and defective containers [5]. On the other hand, color segregation is necessary to ensure a final product with a specific shade of color [5], [6]. An alternative to avoid material and color segregation is the transformation of glass bottles into different glass products without the melting process. This is especially important because the melting stage consumes the highest amount of energy (approximately 4.72 kJ/g) out of the 6.66 kJ/g that the entire process consumes [7].

In the literature review, there are some examples of machines designed to transform bottles without a melting stage, which implement machine learning algorithms [8] and the use of FPGA's [9]. However, these machines only transform plastic bottles or metal cans because of profitable reasons [10]. In the case of glass bottles, there are not complete existing machines designed to transform them. Thus, companies like Green Glass [11] perform this process using a bench saw which could eventually cause severe damage to the operator. Additionally, DIY tools such as [12] are available to cut glass bottles; nonetheless, these are still dangerous in the sense that users manipulate these tools during the entire transformation process. In this light, this paper presents an innovative low-cost solution for transforming a glass bottle into a glass.

RecyGlass is an autonomous system that allows obtaining glasses with serigraph designs from used glass bottles. This system can receive as an input round glass bottles regardless of the type or color of the glass material. RecyGlass performs a cutting process, followed by a polishing stage to soften the edges of the glass, and finally performs a serigraphy process to print different designs on the surface. All this transformation process is executed autonomously, the user only needs to feed the machine with glass bottles, press the start button, and wait for the glass to be ready. Even more important, Recyglass represents a safe alternative for operators, since they will not be exposed to cuts or glass dust.

This document describes the design and implementation details of the first functional prototype of RecyGlass in the following organizational structure: Section 2. analyses the design and manufacturing of the system based on 2.1. the logic of the system, 0. the description of the main subsystems, and 2.3. the electronic design. Section 3. presents preliminary tests and results. By last, Section 4. lists the conclusions of the present work.

2. Design and Development of the system

The design of RecyGlass considers the following requirements:

- Incoming glass bottles must have a diameter between 65 and 90 mm without labelling and embossing.
- The system must have a maximum operating time of 12 minutes per bottle.
- The dimensions of the system should not exceed 1.5 (L) x 1 (W) x 1 (H) m.
- The electronic components must be inside an enclosure with a minimum rating of IP55, according to IEC 6052 standard.
- The system must have an enclosure to prevent glass dust from coming out of the working space of the prototype and for general safety.
- The electronic components should work properly between 0° and 30°.

Figure 1 shows the final design of the system, and Table 1. presents its main specifications.

Table 1. General specifications.

Specification	Value
Total length	0.82 m
Total width	0.72 m
Total height	0.72 m
Material	AISI 6061
Weight	60 kg
Input energy	220 VAC – 60 Hz

2.1. The logic of the system

As soon as the operator turns RecyGlass on, all variables and GPIO's restart to the initial setup. After this, all the actuators move to a pre-defined home position (Position A in figure 2). Then, the user should insert the bottle and press the start button, the system clamps the bottle and the cut process starts. When

IOP Conf. Series: Materials Science and Engineering **926** (2020) 012035 doi:10.1088/1757-899X/926/1/012035

the cut process finishes, the system removes the top half of the bottle, and the remaining half is moved to the polishing and serigraphy area in point B (figure 2). After those processes, the polished glass with a serigraph design moves back to the initial position (Position A in figure 2). The user should remove the final product and press the end-process button. Then, if required, the user will be able to insert another bottle.

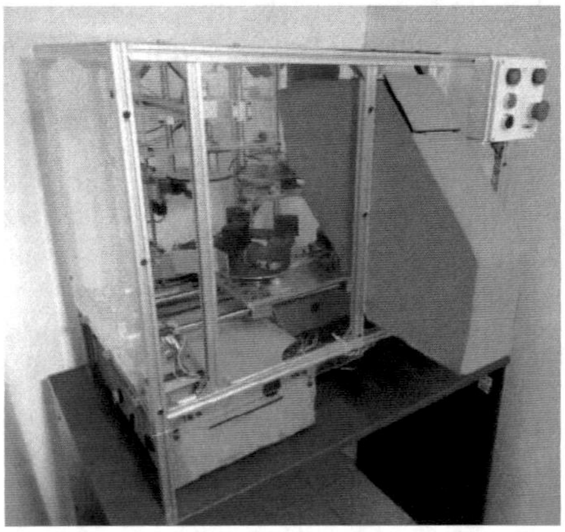

Figure 1. Implemented system.

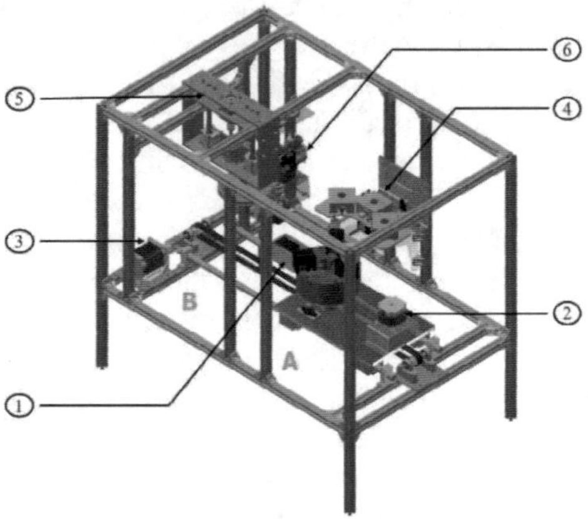

Figure 2. CAD model of RecyGlass subsystems: 1) Clamping, 2) Rotational, 3) Translational, 4) Cutting, 5) Polishing and 6) Serigraphy Subsystems.

2.2. Description of the subsystems

RecyGlass consists of different subsystems with specialized tasks. Figure 2 shows the location of these subsystems.

2.2.1. Clampling subsystem. This subsystem is responsible for holding the bottle after it enters. The bottle remains clamped during all the processes: cutting, polishing, and serigraphy. After the entire transformation process finishes, RecyGlass releases the final product and the user can remove it. This subsystem is able to determine the bottle diameter during the clamping process. The clamping mechanism consists of a 3-jaw chuck [13] with HDPE extensions. The subsystem uses a stepper motor with a 5:1 gearbox reduction to drive the jaws. RecyGlass uses a hall sensor and a magnet to detect a complete turn of the base. It also uses them to determine the bottle diameter. Apart from that, the subsystem uses another hall sensor and magnet to detect when the chuck is completely open. In this way, the machine can determine both the bottle diameter and the moment when the bottle is completely holded by the clamping mechanism. Figure 3 shows the subsystem configuration.

2.2.2. Rotational subsystem. This subsystem controls the bottle's rotational direction and speed during the cutting and serigraphy processes. This subsystem uses an axial bearing, a belt, two pulleys, a chuck, and a stepper motor to transmit power. The Arduino UNO controls the subsystem using the TB6560 driver.

2.2.3. Translational subsystem. This subsystem controls the translational motion of the clamped bottle from stage to stage. The subsystem has a synchronous belt transmission with a stepper motor, bearings, and a couple of pulleys to transmit power.

2.2.4. *Cutting subsystem.* This subsystem uses three glow plugs to cut the glass. The structure of the cutting subsystem has three parts: a glow plug holder, a mobile base, and a vertical backrest. The structural material of the subsystem is acrylic PMMA, according to NBR-ISO 7823-2. Furthermore, bakelite rods are used as thermal insulation elements. Figure 4 shows the 3D model of the cutting subsystem.

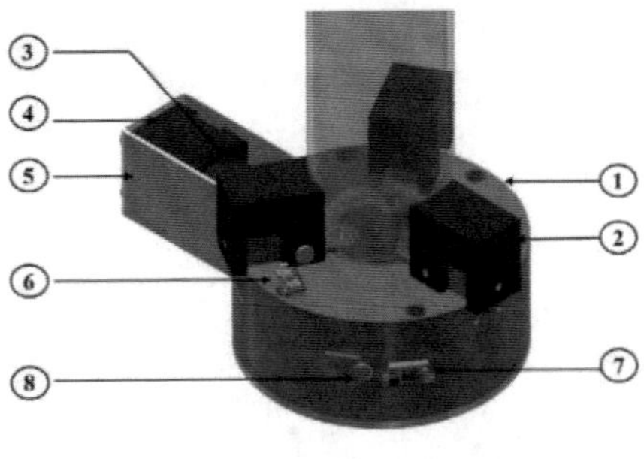

Figure 3. CAD model of clamping subsystem: 1) Chuck, 2) HDPE extensions, 3) 5:1 Gearbox, 4) Stepper motor, 5) Motor holder, 6-7) Hall sensors, 8) Magnet.

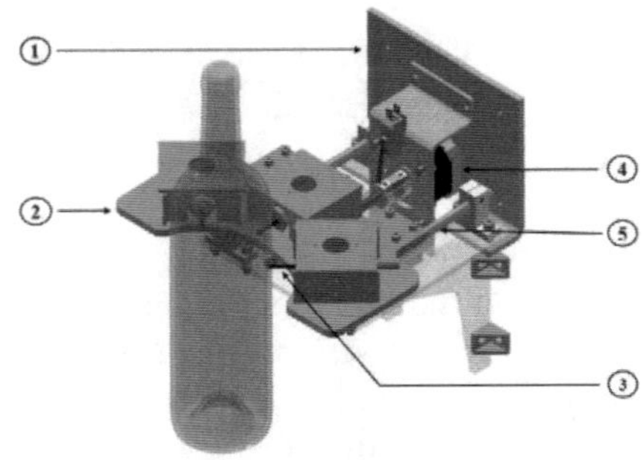

Figure 4. CAD model of the cutting subsystem: 1) Vertical backrest, 2) Mobile base, 3) Glow plug, 4) Stepper motor, 5) Shaft.

When the cutting process starts, the rotational subsystem turns the bottle gradually both clockwise and anticlockwise. After two minutes, a horizontal crack appears on the surface of the bottle, which indicates that a thermal shock has occurred. This subsystem was designed taking into account important properties of the different glass materials to obtain the most symmetrical and horizontal possible cut (according to the UNE-EN ISO 7459 standard). Some of these properties are wall thickness, impact resistance, resistance to vertical compression, and resistance to thermal shocks.

2.2.5. Polishing subsystem. This subsystem uses a rotational mechanism, which consists of a DC motor, a mechanical chuck, and a polisher. The polishing subsystem performs vertical movement thanks to the combination of a stepper motor and a T8-trapezoidal screw. A limit switch detects the maximum available displacement of the structure. Figure 5 shows the 3D model of the polishing subsystem.

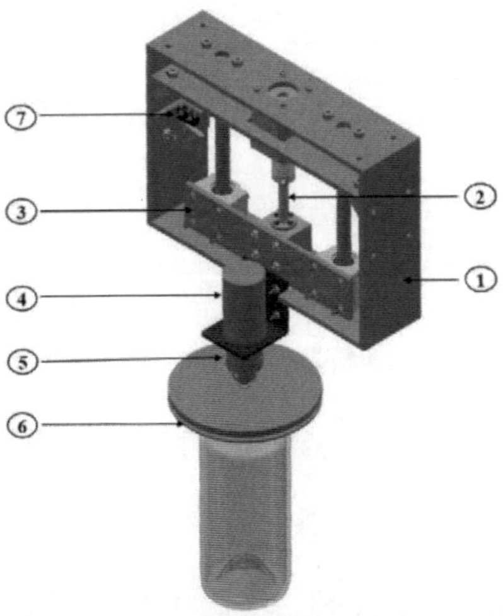

Figure 5. CAD model of the polishing subsystem: 1) Structure, 23) Trapezoidal screw mechanism, 4) DC motor, 5) Mechanical chuck, 6) Polisher disc, 7) Limit switch.

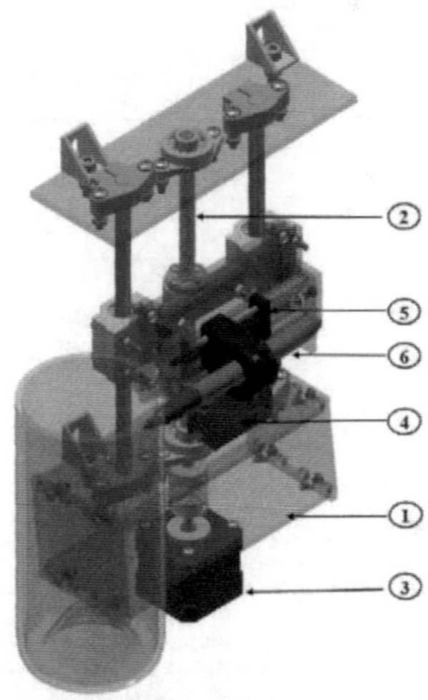

Figure 6. CAD model of serigraphy subsystem: 1) Structure, 2) Trapezoidal screw, 3) Stepper motor, 4) Servomotor, 5) Marker holder, 6) Marker.

2.2.6. Serigraphy subsystem. This module prints a personalized design with a permanent marker on the glass. Additionally, it prints the capacity of the glass in mm^3. This subsystem allows a 2-axis movement by using a stepper motor and a T8-trapezoidal screw. It also consists of a servomotor and a 3D-printed holder made of PLA to perform the serigraph printing. Figure 6 shows the model of the serigraphy subsystem.

The serigraphy subsystem uses an adaptation of the open-source G-Code parser and CNC milling controller, GRBL [14]. This permits drawing any figure stored on a micro SD card. However, it is necessary to generate a specific G-Code for each new image before turning on the machine.

2.3. Electronic Design

Figure 7 shows the electronic architecture of the system. RecyGlass has an Arduino Mega as a master microcontroller that uses a state-machine logic. The states represent the six subsystems described above. The Arduino Mega controls the translational subsystem and decides which one should work. Depending on the state, it communicates with two slave Arduino UNO microcontrollers using UART communication. One of them controls the clamping, cutting, and polishing subsystems; and the other one, the rotational and serigraphy subsystems.

Each microcontroller generates PWM and digital signals to drive and control the speed and direction of the rotation of the motors. Besides, they can handle analog and digital signals coming from the different sensors allocated in each subsystem.

The user interface takes into account allowing the user to supervise the machine at each stage. The master microcontroller receives a digital activation signal from the start button. The finish button sends also a digital signal, and the operator must press it when the transformation process of a bottle finishes. There is also an emergency button that disconnects the energy source from the outlet. Additionally, two LEDs indicate the condition of the machine; when RecyGlass is waiting for a new bottle, the green LED is on, and when it is working, the red LED is on.

The outlet energy provides 220 VAC at 60 Hz. Switching and ATX power supplies transform the alternating voltage into continuous voltage. The 5 V-switching power supply is connected to the processing unit, the user interface, the sensors from the subsystem, the stepper motor drivers, and the relays. On the other hand, the stepper motors, their drivers, the DC motor, and the relay receive energy from the 12 V-switching power supply. Finally, each of the glow plugs receives energy from the ATX power supply because of the high current that they require.

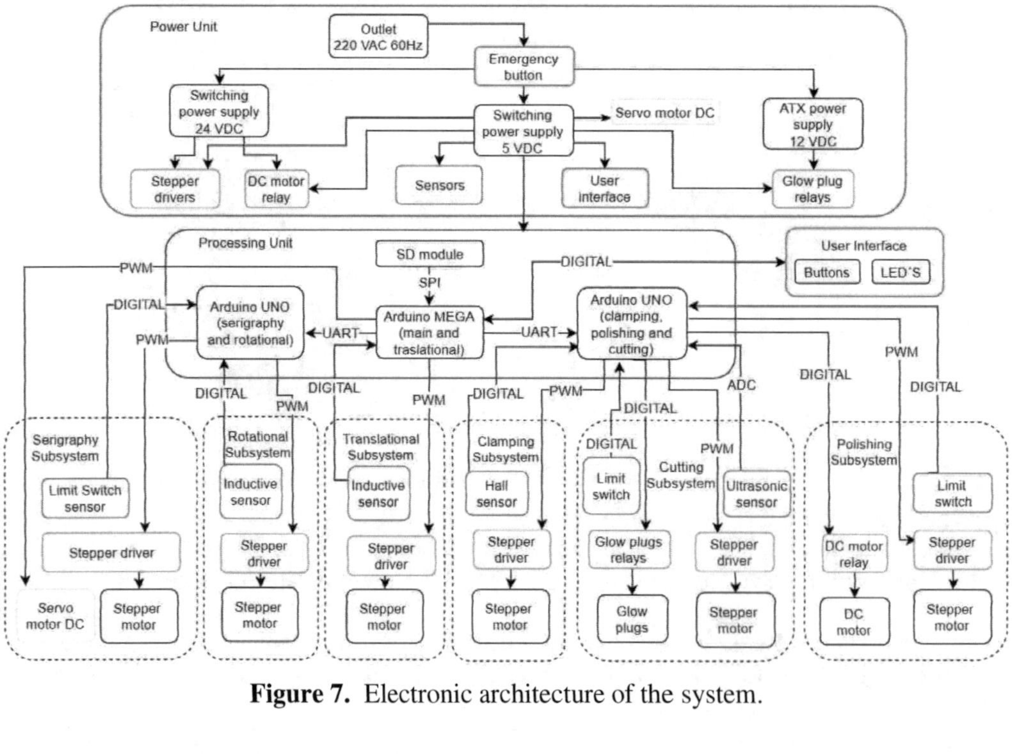

Figure 7. Electronic architecture of the system.

3. Implementation and results

In this section, preliminary tests, the experimental setup and results are presented. It is important to mention that RecyGlass is a low-cost system built with a budget of 1000 USD.

3.1. Preliminary tests

These tests allowed us to carry out some modifications. For instance, during the cutting process, it was identified that the glow plugs needed to be perfectly aligned in order to obtain a uniform cut. If the glass rim obtained from the cutting process is not uniform, the polishing will not perform correctly and it may fail during the process.

Another aspect to take into account was the rotational movement of the bottle during the cutting process. While it is convenient to have a continuous rotational movement until the bottle is completely cut, it was not possible to implement this because the wires would have gotten tangled. Thus, to avoid this problem, the clamping chuck was programmed to rotate2 revolutions clockwise, and 2 anti-clockwise, until the cut of the bottle was finished.

Finally, the linear speed of the translational subsystem was determined through several tests in order to allow a fast movement from point A to B and vice versa, keeping the clamped bottle stilled. Likewise, the rotational speed of the rotational subsystem was determined to allow a proper cut and an acceptable serigraph drawing. These parameters are presented in Table 2.

Table 2. System parameters.

Parameter	Value
Translational speed	0.1 m/s
Cutting rotational speed	0.5 rad/s
Serigraphy rotational speed	0.4 rad/s

3.2. Experimental setup

A sample of 10 different 85-mm-diameter glass bottles with an average wall thickness of 2.6 mm were tested. The setup included a video camera to get the amount of time of every process, and a DC-current clamp meter to measure the current consumed by the components of each subsystem. The total energy consumption of RecyGlass was determined using the supply voltage and current of all the electrical and electronic components, and the operating time provided by the camera.

Figure 8. Cutting process.

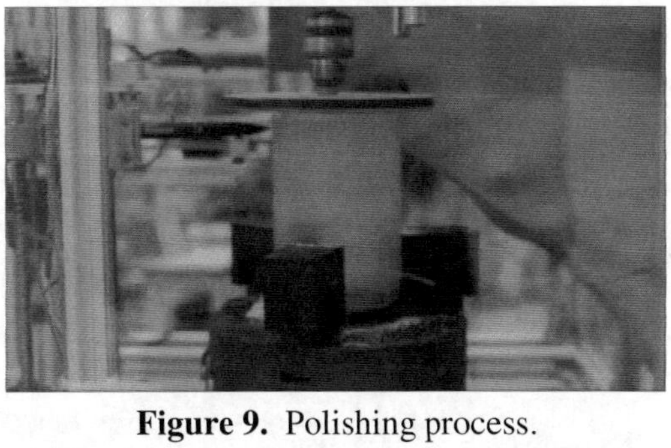

Figure 9. Polishing process.

3.3. Results

The tests performed in RecyGlass with the 10 different bottles lasted in average 10 minutes, which can be approximately extrapolated to a 6 bottles per hour processing capacity. The bottleneck of the entire process was the cutting operation, which lasted approximately 4 minutes (almost the 40 % of the full process). Energy wise, it was found that the system consumed in average a total amount of energy of 144.33 kJ/bottle (0.328 kJ/g), and the subsystem that required more energy was also the cutting subsystem, 47.89 kJ. Most of this energy was consumed by the three glow plugs located in this subsystem, which each require 4 amperes to reach around 1000 °C. The average operating time and energy consumption for the ten tests are presented in Table 3.

To illustrate the overall process, a sample of photos of the most representative operations are presented as follows: Figure 8 shows the exact moment when the bottle is cut. The spout of the bottle is pulled by the glow plugs to the ramp, which at the end has a recipient to store these remains. Figure 9 presents the stage of the polishing process where the polisher is in contact with the in-process glass. Figure 10 shows the stage of the serigraph process where the pen is pressed against the bottle. The input bottle and the final product of one of the tests can be seen in Figure 11.

Table 3. Average measurements of tests.

Subsystem	Operating time (s)	Energy Consumption (kJ)
Clamping	22.2	14.77
Cutting	242.1	47.89
Polishing	155.4	18.22
Serigraphy	175.2	15.97
Translational	17.8	23.91
Rotational	-	22.99
Electronics	-	0.58
Total	612.7	144.33

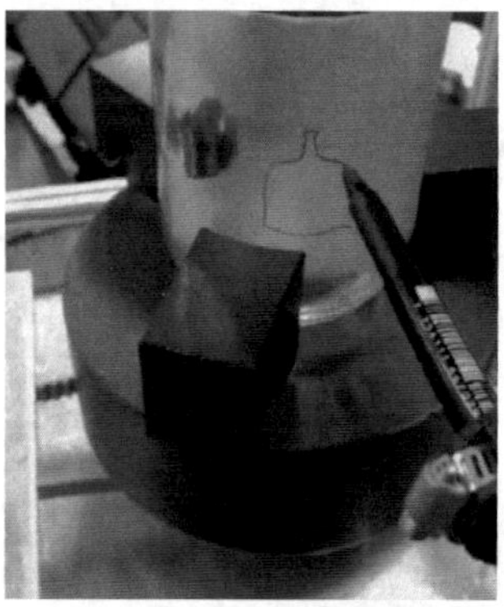

Figure 10. Serigraphy process.

(a) (b)

Figure 11. (a) Test bottle, (b) Final product.

4. Conclusion and future work

We have presented the design and implementation of the first functional prototype of RecyGlass. The autonomous design of the system avoids the dangers of manipulating the in-process glass, and of inhaling glass dust coming out of the cutting and polishing stages. We have also noticed that the use of this prototype represents a less consuming alternative to the classic recycling options: 0.328kJ/g 6.659kJ/g .100 = 4.9 % of the total energy consumption of typical glass recycling processes (for more details see the introduction and results sections). Furthermore, color and material segregation is not required to obtain a final product of acceptable quality level. The design of RecyGlass comes with no scalability or replicability restrictions. Future work will elaborate on reducing the operating time and increasing the quality level of the final product. To do this, the configuration of the cutting subsystem needs to be redesigned to focus even further on the heat applied on the bottle. In the same way, a laser will be implemented to obtain a better resolution on the serigraph designs.

5. Acknowledgment
The authors acknowledge the support from the Mechatronics Section of the Engineering Department of the Pontifical Catholic University of Peru.

References

[1] Euroestat, "Waste statistics," 2020. [Online]. Available: https://ec.europa.eu/eurostat/statistics-explained/index.php/Waste_statistics. [Accessed: 11-Jul-2020].
[2] United States Environmental Protection Agency - EPA, "National Overview: Facts and Figures on Materials, Wastes and Recycling," 2020. [Online]. Available: https://www.epa.gov/facts-and-figures-about-materials-waste-and-recycling/national-overview-facts-and-figures-materials#Recycling/Composting. [Accessed: 11-Jul-2020].
[3] Ministerio del Ambiente - MINAM, "En el Perú solo se recicla el 1.9% del total de residuos sólidos reaprovechables | SINIA | Sistema Nacional de Información Ambiental." [Online]. Available: https://sinia.minam.gob.pe/novedades/peru-solo-se-recicla-19-total-residuos-solidos-reaprovechables. [Accessed: 11-Jul-2020].
[4] World Bank Group, "Trends in Solid Waste Management," 2020. [Online]. Available: http://datatopics.worldbank.org/what-a-waste/trends_in_solid_waste_management.html. [Accessed: 11-Jul-2020].
[5] Glass Packing Institute - GPI, "Glass Recycling Facts," 2020. [Online]. Available: https://www.gpi.org/glass-recycling-facts. [Accessed: 11-Jul-2020].
[6] American Chemical Society, "Why glass recycling in the US is broken," 2020. [Online]. Available: https://cen.acs.org/materials/inorganic-chemistry/glass-recycling-US-broken/97/i6. [Accessed: 11-Jul-2020].
[7] L. E. Carrión Domenech and R. Royo Pastor, "Optimización energética de una empresa de producción de envases de vidrio para alimentación Portada normalizada descargada 1," Universidad Politécnica de Valencia, 2016.
[8] P. Dhulekar, S. T. Gandhe, and U. P. Mahajan, "Development of Bottle Recycling Machine Using Machine Learning Algorithm," *2018 Int. Conf. Adv. Commun. Comput. Technol. ICACCT 2018*, pp. 515–519, 2018.
[9] M. F. Karin, K. S. Noor, and H. U. Zaman, "Hardware based design and implementation of a bottle recycling machine using FPGA," *Proc. - 2016 IEEE Conf. Syst. Process Control. ICSPC 2016*, no. December, pp. 43–46, 2017.
[10] E. R. Vieitez, P. Eder, A. Villanueva, and H. Saveyn, *End-of-waste criteria for glass cullet: Technical proposals*, no. 1. 2011.
[11] Green Glass, "Productos," 2020. [Online]. Available: https://www.greenglass.cl/collections/all. [Accessed: 11-Jul-2020].
[12] Bottle Cutting Inc., "The Kinkajou Bottle Cutter is easy and fun to use," 2020. [Online]. Available: https://www.bottlecutting.com/. [Accessed: 11-Jul-2020].
[13] J. C. Harvey, "US2598165A-Universal Chuck," 1952 Available: https://patents.google.com/patent/US2598165A/en.
[14] Simen Svale Skogsrud and S. K. Jeon, "GitHub - grbl/grbl: An open source, embedded, high performance g-code-parser and CNC milling controller written in optimized C that will run on a straight Arduino," 2020. [Online]. Available: https://github.com/grbl/grbl. [Accessed: 11-Jul-2020].

IOP Conf. Series: Materials Science and Engineering **926** (2020) 012036 doi:10.1088/1757-899X/926/1/012036

Innovative Design of Household Body Composition Analyzer Based on INPD and QFD

Huan Jia-mei [1], Chen Yang [1,*] and Luo Fei [2]

[1]School of Mechanical Engineering, Guizhou University, Huaxi District, Guiyang City, Guizhou Province,550025, China
[2]AECC Guizhou Liyang Aero-engine, Co, Ltd., Baiyun District, Guiyang City, Guizhou Province, 550014, China

[*] E-mail: 892168711@qq.com

Abstract. In order to solve the problems that the household body composition analyzer has not yet established a good experience relationship with users and the diversity of user needs, a new design method is proposed to get a more complete solution. Taking INPD as a research framework, re-examining the changes of SET factors can create new product opportunities and find product opportunity gaps; Through market surveys and user surveys, in-depth digging into the needs, preferences and expectations of target user groups, understanding product opportunity; Combining QFD models quantitatively analyze user needs and refine design requirements, establish a mapping relationship between user needs and design requirements, and quantify key design requirements. Through the combination of INPD and QFD, a design scheme driven by customer needs is given, which solves user needs to a certain extent, and its feasibility provides a reference for the innovative design of such products.

1. Introduction

With the improvement of people's awareness about health and the popularization of smart health care products, household body composition analyzer has become a research hotspot. However, the current trend of homogenization of household body composition analyzers on the market is obvious, lack of pertinence, and non-professionals are not easy to recognize, resulting in low user satisfaction. Therefore, developing products driven by customer needs is an urgent problem to be solved. According to the data, there are few studies on this product, mainly focusing on function and technology. Li Zhao [1] and others designed a body composition analyzer based on the data analysis technology of the Android platform, which improved the human-computer interaction of the product; Xiao Xiao-ming [2] and others studied the software and hardware system of the body composition analyzer, which improved the accuracy and reliability of the measurement results; Sung-Hoon Bae [3] and others introduced a chip design of the body composition analyzer based on the BIA method technology. It can be seen that these studies tend to be guided by the realization of functional technology, lacking in-depth analysis of the needs of consumer groups, and user-centric product development ideas. This article proposes a design method based on the combination of INPD development method and QFD model. Through systematic qualitative and quantitative analysis, users' needs are sorted out accurately, the importance of needs and key design requirements are quantified, and the design scheme that meets users' needs are put forward according to the quantitative results.

IOP Conf. Series: Materials Science and Engineering **926** (2020) 012036 doi:10.1088/1757-899X/926/1/012036

2. Design methods and procedures

2.1. INPD

INPD (Integrated New Product Development) is a development method proposed by Professor Cagan J and Vogel C M of Carnegie University, which is suitable for the early stage of product development ambiguity. It emphasizes the multi-professional combination of the development team based on the needs, requirements and wishes of users and the interests of other stakeholders. User- centered INPD includes four stages: identifying product opportunity, understanding product opportunity, product concepts formed by product opportunity, and Realization of opportunity [4].

2.2 QFD

QFD (Quality Function Deployment) is a quantitative analysis method that transforms user needs into design requirements. Its core content is to create products based on customer needs, and truly implement user needs into design practice [5]. As a scientific method for analyzing user needs, its key role is to correlate customer needs with product engineering requirements, construct a relationship matrix between the customer needs and product engineering requirements, and effectively convert the user's demand language into the technical language of development personnel [6]. To reduce the repeated changes and development of cycles in the design process, improve product quality.

2.3 Integration of INPD and QFD

Both INPD and QFD are user demand-oriented design methods. The advantage of the former is that it can accurately reflect the needs of users, but it can only achieve the effect of qualitative analysis, which has a certain subjective color. The latter uses matrix calculation to quantify the importance of user needs and calculate the key design requirements. Therefore, the combination of INPD and QFD can combine the common advantages of the two methods and give full play to their respective advantages to improve the design efficiency and product quality. Taking the design of the household body composition analyzer as an example, discussed the application of INPD and QFD in household body composition analyzer, so as to solve the shortcomings of the existing household body composition analyzer.

The four stages of INPD are used as the theoretical framework for research, which analyzes the early stage of product design fuzzy, identifies and understands product opportunities, comprehensively obtains user needs; In the conceptual opportunity stage, use QFD to determine the mapping between user needs and design requirements, quantify key design requirements and prioritization, and accordingly transform the product concept formed by the opportunity into a concrete design plan, and then guide the innovative design of the body composition analyzer. The application process of the design methods of INPD and QFD as shown in Figure 1.

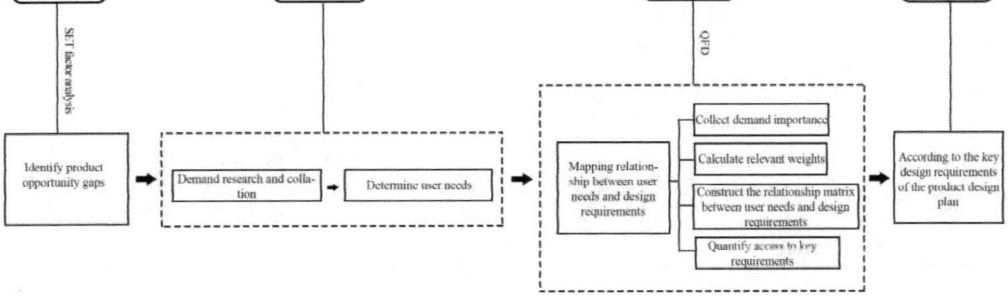

Figure 1. INPD and QFD design method application process.

3. Identify opportunity gaps

When the existing products are difficult to meet the needs of users under the new trend, the re-examination of changes in social factors, economic factors and technological factors can create new product opportunities [7]. By reviewing books, news and online literature, the product opportunity gaps of body composition analyzer were identified by using the SET factor analysis method, as shown in Figure 2.

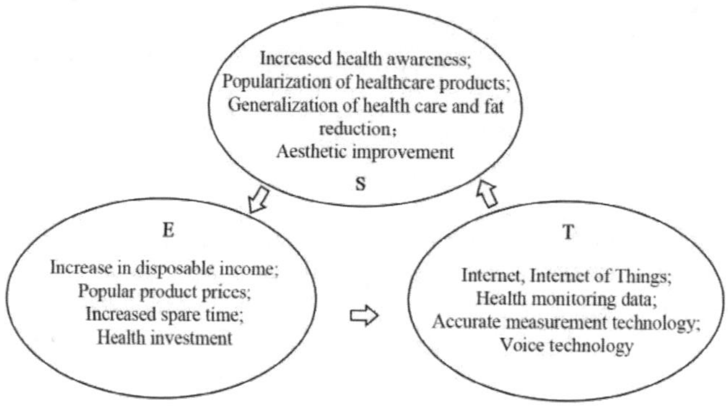

Figure 2. SET factor analysis.

Through the analysis of SET factors, the potential needs of users under the new trend are found: In the aspect of social factors, according to the concerns and physiological characteristics of different groups, different display modes are designed and switchable data display modes are designed; In terms of economy, the household body composition analyzer can be used as a health investment in spare time. The middle-aged and elderly people can pay attention to the health indicators and warning signs, while young people can pay attention to body shape, image and health. As far as technology is concerned, there are problems in data reading of body composition analyzer, which leads to poor user experience and insufficient intelligence information. Internet, Internet of Things and other intelligent technologies can be applied to the products of body composition analyzer to enhance the user's Experience and convenience.

4. Understand product opportunity

At this stage, through market and user research, in-depth understanding of the needs of the target user, to understand the use and experience of the existing body composition analyzer, as well as potential users' lifestyle and preferences needs. The previous market research will be combined with questionnaires and user interviews to collect users' needs for body composition analyzers. This questionnaire designed 10 questions around the use of the body composition analyzer, and 76 valid questionnaires were collected. According to the questionnaire survey, interviews were set up. At the same time, we interviewed 12 consumers in the school and the community. The age range is mainly 18-65 years old, with a certain proportion of each. Collect and collate the user's demand for household body composition analyzer, as shown in Table 1.

Table 1. Identify the needs of users.

Number	User needs	Number	User needs
R1	Easy to store	R6	Easy operation
R2	Easy interaction	R7	Beautiful shape
R3	Accurately measure	R8	Affordable
R4	Security	R9	Small size
R5	Feedback of data	R10	Durable

5. Establish a mapping relationship between user needs and design requirements

For design and development, determining the importance of needs in the minds of users is the key to guiding innovative design in the later period [8]. To meet the needs of the user group for household body composition analyzers to the maximum extent, 20 users are invited to participate in the important assessment of household body composition analyzers according to 1～5 points. The records of demands that have no impact on the function realization are 1, those that have no impact on the main function realization are 2, those that have a great impact on the function realization are 3, those that have an important impact on the function realization are 4, and those that are basic, safety-related and particularly important are 5 [9]. By collecting the user's evaluation and analysis, the importance of the user's demand for household body composition analyzer is obtained, as shown in Table 2.

Table 2. Importance and demand of household body composition analyzer.

Demand No.	R1	R2	R3	R4	R5	R6	R7	R8	R9	R10
Importance I	3	4	5	5	5	4	2	3	4	2

Based on the important results given in Table 2, calculate the relative weight of user needs according to formula (1) [10].

$$w_i = I_i / \sum_{i=1}^{n} I_i \tag{1}$$

After calculation, the relevant weight matrix of demand is:

$$W = [0.0811\ \ 0.1081\ \ 0.1351\ \ 0.1351\ \ 0.1351\ \ 0.1081\ \ 0.0541\ \ 0.0811\ \ 0.1081\ \ 0.0541]$$

In order to implement the needs of household body composition analyzers and integrate them into every design detail, these needs should be translated into design requirements. Using the KJ method to sort out the design requirements according to the relevant design requirements that the product discussion may have a certain impact on customer needs. The KJ method can effectively organize the survey data and the qualitative ideas of the participants according to the similarity between user needs and design requirements, and unify the opinions of the participants [11]. Through analysis and sorting out the design requirements, as shown in Table 3.

Table 3. Design requirements.

Number	Design requirements	Number	Design requirements
D1	Can be stored	D7	Product structure
D2	Provides multiple display modes	D8	Measurement method
D3	Provides data analysis and recording	D9	Comfortable grip
D4	Voice broadcast data	D10	Product form
D5	Key data are clear	D11	Man-machine size
D6	Stable to prevent rollover	D12	Material selection

Construct the relationship between the user needs and the design requirements of the body composition analyzer. According to the correlation between them, the relationship values are given in the range of [-2, 2], which corresponds to strong positive correlation, positive correlation, irrelevance, negative correlation, strong negative correlation. And adding the relevant weight matrix of the above user needs to the relationship matrix table, as shown in Table 4.

Table 4. Matrix table of relationship between user needs and design requirements.

Number	W_i	D1	D2	D3	D4	D5	D6	D7	D8	D9	D10	D11	D12
R1	0.0811	2	0	0	0	0	0	1	0	0	1	0	1
R2	0.1081	1	2	1	2	2	0	0	1	0	2	0	0
R3	0.1351	0	2	0	0	0	0	0	2	0	0	0	0
R4	0.1351	0	0	0	0	0	2	1	0	0	1	0	1
R5	0.1351	0	2	2	1	1	0	0	0	0	0	0	0
R6	0.1081	0	1	1	1	1	0	1	1	1	1	0	0
R7	0.0541	1	0	0	0	0	0	1	0	0	2	0	0
R8	0.0811	-1	-1	-2	-1	0	0	-1	-1	0	-2	0	-2
R9	0.1081	2	0	0	0	0	0	1	0	0	2	0	0
R10	0.0541	1	0	0	0	0	0	1	0	0	1	1	2

According to the relationship matrix in Table 4, the user demand weight matrix and the design requirement matrix are quantified by the matrix calculation method to quantify the key design requirements, and the overall evaluation degree and preferred ranking results of the household body composition analyzer are obtained, as shown in Table 5.

Table 5. Evaluation degree and optimal ranking of household body composition analyzer.

Design requirements(D)	D1	D2	D3	D4	D5	D6	D7	D8	D9	D10	D11	D12
Rating	0.5136	0.7836	0.3242	0.3783	0.4594	0.2702	0.4595	0.4053	0.1081	0.7568	0.0541	0.1622
Prioritization	3	1	8	7	5	9	4	6	11	2	12	10

6. Scheme design

According to the ranking results in Table 5, the design requirement D2 " provides multiple display modes" is the most important thing for the user community. Because different people have different data and information, satisfaction of the product is not high at present. Therefore, product design should aim at user groups and meet the needs of people with different attributes to the maximum extent. The target population is mainly divided into three categories: the first category, body fat measurement is irregular, focusing on simple weight. For this group of people, a simple reading mode should be set. When it is necessary to measure the whole body fat, you can set it yourself, as shown in Figure 3 (a); The second and third categories are middle-aged and elderly, respectively, while focusing on the health index, each has its focus. The middle-aged and elderly people pay more attention to disease prevention, and their body functions decline, and their eyesight is poor. It is necessary to consider the display font size; Young people pay more attention to the body shape and body fat rate direction, but the product display function area is limited, and it does not cover all the data well. According to the data which they are most concerned about, different display modes are presented to reduce the complexity of the interface and make it easier to recognize. as shown in figure 3 (b) and (c).

The design requirement D10 "product form" is second only to D2. The use of body composition analyzer is irregular, and users will use it more in their spare time. Good product shapes can give users a beautiful visual sense. Current product shapes are mostly rectangular and round, which lack interesting shapes. Extracting the relevant elements of such products for the morphological design of the body composition analyzer can increase the popularity of the product. This design is mainly extracted as a human body curve, using it as a modeling element, as shown in figure 4.

The design requirement D1 "can be stored" is the third most important. The frequency of use of the body composition analyzer is relatively low. When people use it, they hope it can be stored reasonably and save space. Therefore, the storage method of figure 5 can not only satisfy the different properties measured by different people, but also store the line reasonably.

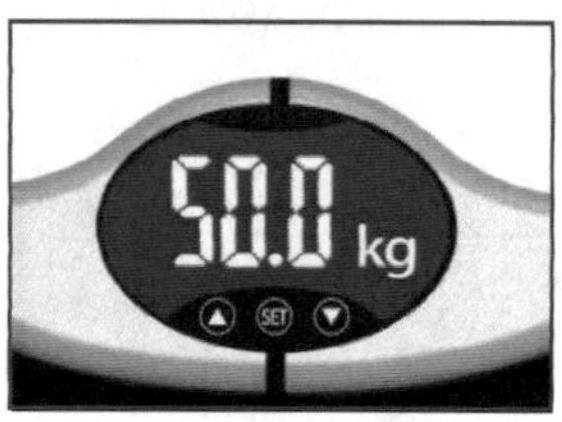

(a) Simple weighing reading mode.

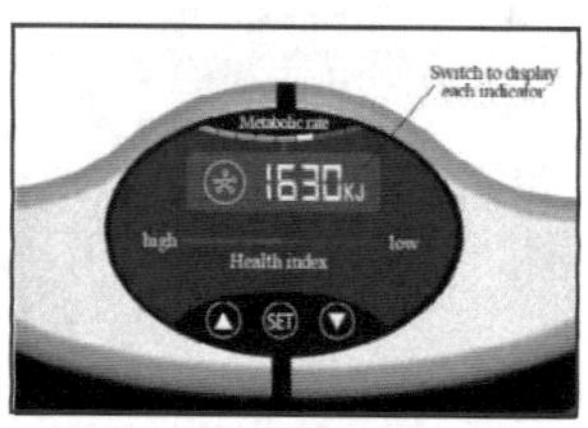

(b) Reading patterns of the middle-aged and elderly people.

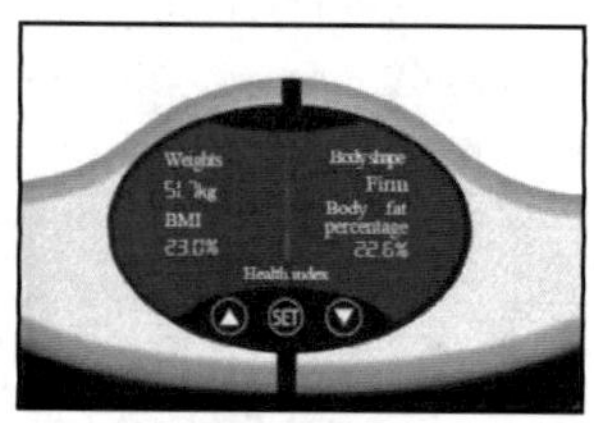

(c) Reading patterns of young people.

Figure 3. (a) and (b) (c) three display modes.

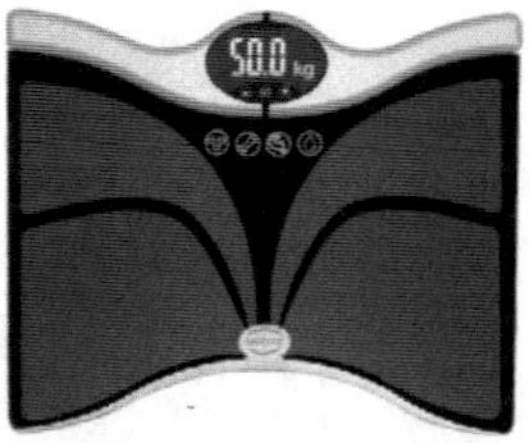

Figure 4. Product rendering.

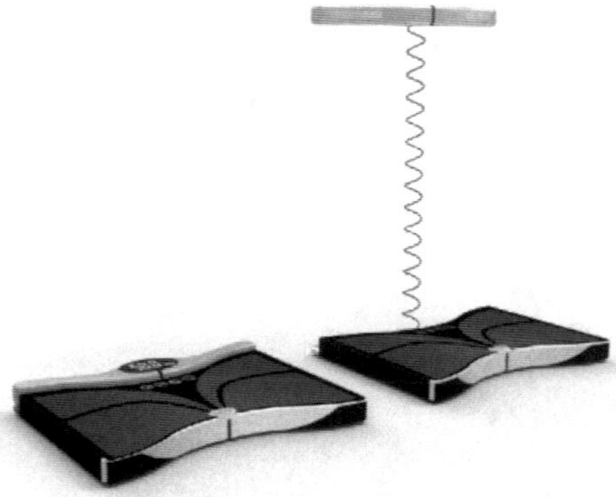

Figure 5. Storage method.

Based on the ranking results and in-depth analysis, the design requirements are implemented into the product design, the key design requirements of the household body composition analyzer are optimized, and the design expectations that are closer to the user's needs are proposed.

7. Conclusion

The INPD and QFD models are combined to fully analyze the user needs, which can not only ensure effective investment in the analysis of user needs in the early stage of development, but also quantify the importance and key design requirements through the QFD model, It avoids the subjective deviation of the analysis of user needs causing by the INPD carding requirements. Through qualitative and quantitative analysis in different stages, the key design requirements of the body composition analyzer are obtained, and the key development direction is pointed out for the designers and developers. According to the key design requirements, a brand-new design scheme was put forward, and the design of the household body composition analyzer for users' needs was finally completed, which provides new ideas for the innovation of related products.

Acknowledgments

This work is supported by Guizhou Science and Technology Fund Project "Science and Technology Museum Exhibition Engineering and Science Exhibition Teaching Innovation Design Theory and Methodology" (Qiankehe J word [2014] No. 2057).

References

[1] Li Zhao, Shan Ke, Deng Shao-long J 2015 *Shandong Science* **28(06)** p 127-132
[2] Xiao Xiao-ming, He Wei, He Yu-cheng J 2015 *China Medical Devices* **30(8)** p 9-13
[3] Bae S H, Moon B S, Lee W J J 2006 *IEEE 2006 Biomedical Circuits and Systems Conference Healthcare Technology* p218-221
[4] Cagan J and Vogel C M M 2003 *Creating breakthrough products* (Beijing: Mechinery Industry Press)
[5] Tang Lin, Chang Yu, Wang Zi-rui J 2019 *Packaging Engineering* **40(24)** p199-204
[6] Ren Gong-chang, Zhang Bo-wen, Zhang Shao-zhi J 2020 *Packaging Engineering* **41(07)** p 157-163
[7] Chen Liang, Feng Zhi-wen C 2016 *International Conference on Information Technology for Manufacturing Systems (ITMS 2016)* p91–94
[8] Liu Zong-ming, Ge Bi-hui J 2018 *Journal of Graphics* **39(04)** p 695-699
[9] Zhao Xiang-tao, Xing Jun-jie J 2014 *Journal of Graphics* **22(01)** p 126-129
[10] Liang Hao, Liu Zhuo, Zhang Fang-yan J 2015 *Journal of Machine Design* **32(08)** p126-128
[11] Ren Guang-pei, Wang Zi-qing, Yu Dong-jiu J 2018 *Packaging Engineering* **39(08)** p161-165

AUTHOR INDEX